ENDING DIRTY ENERGY POLICY

Climate change presents the United States, and the world, with regulatory problems of a magnitude, complexity, and scope unseen before. The United States, however, particularly after the midterm elections of 2010, lacks the political will necessary to aggressively address climate change. *Ending Dirty Energy Policy* argues that the country will not adequately address climate change until we transform our fossil fuel energy policy. Yet there are signs that the country will support the transformation of our country's century-old energy policy from one that is dependent on fossil fuels to a low-carbon energy portfolio. A transformative energy policy that favors energy efficiency and renewable resources can occur only after we have abandoned the traditional fossil fuel energy policy, have redesigned regulatory systems to open new markets and promote competition among new energy providers, and have stimulated private-sector commercial and venture capital investment in energy innovations that can be brought to commercial scale and marketability.

Joseph P. Tomain is Dean Emeritus and the Wilbert & Helen Ziegler Professor of Law at the University of Cincinnati Law School, which he joined in 1987 and where he held the deanship for fifteen years. He practiced general litigation in New Jersey before beginning his teaching career at Drake University School of Law. Dean Tomain has also held positions as visiting professor at the University of Texas Law School; distinguished visiting energy professor at Vermont Law School; visiting scholar in the program of liberal studies at the University of Notre Dame; visiting Fellow at Harris Manchester College, Oxford University; Fulbright senior specialist in law in Cambodia; and visiting environmental scholar at Lewis & Clark Law School. His most recent book is *Creon's Ghost: Law, Justice, and the Humanities* (2009).

Dean Tomain is chair of the board of KnowledgeWorks Education Foundation, founder and principal of the Justice Institute for the Legal Profession, and a board member of the Greater Cincinnati Foundation. He is also a Fellow of the American Bar Association, is actively involved with the ABA Section on Legal Education and Admissions to the Bar, and is a member of the American Law Institute. He has published widely in the field of energy law and policy.

Ending Dirty Energy Policy

Prelude to Climate Change

JOSEPH P. TOMAIN

University of Cincinnati College of Law

CAMBRIDGE
UNIVERSITY PRESS

CAMBRIDGE UNIVERSITY PRESS
Cambridge, New York, Melbourne, Madrid, Cape Town,
Singapore, São Paulo, Delhi, Tokyo, Mexico City

Cambridge University Press
32 Avenue of the Americas, New York, NY 10013-2473, USA

www.cambridge.org
Information on this title: www.cambridge.org/9780521127851

First published 2011

Printed in the United States of America

A catalog record for this publication is available from the British Library.

Library of Congress Cataloging in Publication data
Tomain, Joseph P., 1948–
Ending dirty energy policy : prelude to climate
change / Joseph P. Tomain.
p.   cm.
Includes bibliographical references and index.
ISBN 978-0-521-11109-6 (hardback)
1. Energy policy – United States.   2. Fossil fuels – United States.   3. Climatic
changes.   4. Renewable energy sources.   I. Title.
HD9502.U52T66   2011
333.790973–dc22   2011001819

ISBN 978-0-521-11109-6 Hardback
ISBN 978-0-521-12785-1 Paperback

To the students and teachers of
Christian Brothers Academy,
Lincroft, New Jersey,
from a member of the Class of 1966.

Contents

Preface

In both the *Apology* and the *Republic*, Plato has Socrates issue the Delphic command: "Know Thyself." Throughout the dialogues, Plato, for himself as well as for his teacher Socrates, issues another command: "Question everything." I had the great good fortune of attending Christian Brothers Academy, CBA to its devotees, in Lincroft, New Jersey, where both commands were embedded in our education. CBA was, and is today, an all-boys high school that offered a college prep curriculum and so much more. As I remember our second day of classes, our homeroom teacher, Brother Brian, rolled up the sleeves of his cassock, which always meant business, stared at us for a moment, then asked: "How many of you were taught by the nuns?" From where I sat, all hands were raised. He paused and then said: "Well forget everything they taught you."

Forget? What did Brother Brian mean? How could we forget? After all, we just graduated from primary school. And, isn't the purpose of education to remember all that we had learned? Was he simply taking a not-so-sly dig at the nuns? Was he otherwise preparing us for a different regimen of thought, a regimen taught by the Christian Brothers? Or did he have a deeper purpose? Was he challenging us to unlock the psychological mysteries of education, which, as revealed by Milan Kundera and Jorge Luis Borges, require memory *and* forgetting? Perhaps all of the above.

After his shot at the good sisters, there was a bit of nervous laughter in the classroom; these guys were pretty tough, and we were awfully young, so we were wary. Still, I felt that the message Brother Brian was delivering was crystal clear: "Question everything." Indeed, as we attempt to know ourselves, we should be prepared to question everything our new teachers were about to teach us as well; otherwise, they would have failed their central mission as teachers.

As a 14-year-old, I was unaware that Brother Brian's comment constituted the two Socratic commands. CBA was a laboratory of learning. It was, as the Persian scholar Avicenna said of libraries, a school of many rooms, and our core curriculum opened many doors to those many rooms. I dedicate this book to CBA because of the Socratic injunctions and because it was a high school that instilled in us all a great passion for learning through various guises. Brother Bernadine chastised us in French class for not knowing who Jean Paul Sartre was the day Sartre won the Nobel Prize in Literature. Brother John taught us advanced calculus and physics and Brother Andrew taught us advanced biology for no credits because these after-school classes were offered before Advanced Placement courses were invented. Brother John took us to Manhattan for a Federico Fellini film festival. Who knew that the reforms of Vatican II included *La Dolce Vita*? My education at CBA was decidedly not another brick in the wall. Hopefully, *Ending Dirty Energy Policy* is written in the spirit of Socratic inquiry and is, then, true to the spirit of CBA.

The book was written both at the University of Cincinnati and at Lewis and Clark School of Law. At Cincinnati, I bothered our librarians endlessly and thank them for their unstinting help. I thank Jan Smith, Lisa Britt Wernke, Alan Wheeler, Bill Kembelton, and Ron Jones especially for all of their help finding books and reports and keeping the technology working. I also owe special thanks to two classes of Cincinnati law students in my course *Energy Policy and Climate Change* for letting me test out the ideas of the book. I owe a special thanks to James Sproat and Christine Flanagan for their superb research assistance.

I also acknowledge with great appreciation the kindnesses shown to me by the faculty, students, and staff at Lewis and Clark School of Law where I was appointed the Visiting Distinguished Scholar of Environmental Law for the spring 2010 semester. Linda D'Agostino was immeasurably helpful in getting me oriented and guiding me through the university. Faculty members Melissa Powers and Chris Wold were particularly supportive as was Visiting Professor Francine Rochford, who had the unfortunate experience of having the office next door to me and hearing me either curse at the computer or talk about "the book" way too often. Thank you all.

Introduction

Ending Dirty Energy Policy was completed while the stories and the federal investigations of the Upper Big Branch Mine disaster and the Deepwater Horizon offshore explosion, which appeared intent on killing the Gulf of Mexico, were unfolding. The corporations responsible for these tragedies, Massey Energy and BP, through their CEOs Don Blankenship and Tony Hayward, sadly exemplify the dominant energy policy of the United States: Fossil fuel profits are to be made at the expense of the safety and lives of workers and at the risk of catastrophic, sometimes irreversible, environmental degradation. This callous attitude cannot be blamed on corporations alone. The United States government served as partner to constructing a fossil fuel policy intent on bringing to market cheap and dirty energy. As consumers of cheap fossil fuels, we are complicit as well. The century-old, fossil-fuel-based U.S. energy policy must be transformed for a clean and economically healthy energy future.

The thesis of this book is straightforward. Regardless of one's position on climate change, traditional energy policy and its regulation must be dramatically reformed. In this way, energy policy transformation is a prelude to an effective climate change response. *Ending Dirty Energy Policy* proposes two dramatic changes. First, traditional energy policy with its fossil fuel favoritism must be rejected. Instead, new energy markets and new entrants that generate energy more cleanly through energy efficiency, and with renewable resources, must be promoted and supported. Second, the twentieth-century model of government regulation must also be rejected.[1] It has served its purpose and is now outdated because it is incompatible with many of today's problems, including energy policy and climate change. New market structures, new products, and new technologies require new, and in many instances dynamic, regulatory responses. Energy regulation must move away from the narrow market correction model of the administrative state that has been in place since the last third of the nineteenth century.

An energy transition away from fossil fuels can be made today, just as a transition to fossil fuels from whale oil and wood occurred in the mid-nineteenth century. The story proceeds by first recognizing that economic markets, in this instance fossil fuel markets, do not exist independently of the political support structure that has built them and do not exist independently of the bureaucracy that sustains them. Therefore, in order to understand how a transition can proceed, it is necessary to understand how the traditional dominant energy policy came to be in the first instance. In Chapter 1, the regulatory history of fossil fuel energy policy will be described, and then the protectionist political and economic assumptions underlying it will be explained in Chapter 2. Understanding traditional policy assumptions helps reveal the necessary assumptions for a low-carbon energy future.

Chapter 3 goes on to describe the thinking behind a transformative energy policy. The elements underlying that policy are widely accepted today and form the basis of a consensus energy policy as explained in Chapter 4. After examining, in Chapters 5 and 6, how the transportation and electricity sectors of our energy economy can be transformed, the book concludes, in Chapters 7 through 9, with a discussion about new forms of energy regulation, a new politics of energy, and a set of strategies for accomplishing the transformation.

For the first two-thirds of the twentieth century, U.S. energy policy, and the assumptions on which it was based, largely served the country well. A national energy infrastructure has been built, the economy has expanded together with expanded energy production, and technological innovations have made all of our lives more comfortable. There has been a downside, though. As we enter the twenty-first century, we find that the infrastructure is aging; there is good reason to question the continuing correlation between an expanding economy and energy production; and pollution externalities threaten our health and comfort. There is a deep disconnection between the positive contributions of U.S. energy policy for most of the twentieth century and the problems and challenges of the twenty-first century. The disconnection is made all the more daunting by the political and regulatory support systems that have been constructed to further old energy policies and that now constitute major impediments to meaningful and effective policy change, as demonstrated by the congressional failure to enact serious energy policy/climate change legislation.

The legal rules, government institutions, and political alliances that sustain and support traditional energy policy present substantial barriers to a better energy future. For more than four decades, though, there have been voices warning about the folly of continuing down the old path and advising that change can be accomplished efficiently and consistently with the country's commitment to market values. "Said differently, adoption of a robust energy policy is the fastest and cheapest way to improve the economy, environment, health, and equity and increase security."[2] Fortunately, these voices are now being heard. Unfortunately, established institutions greatly weaken the impact of these voices. Consequently, the voices for a new energy policy must become louder, and they must be more clearly heard as we design an energy future built on new assumptions and supported by new institutions and by new forms of government regulation.

Our sharpened awareness of the need for a new energy policy is taking shape in no small part as a result of the challenges presented by climate change. The assumptions on which a better energy future can be based are relatively well known, and we have a fairly clear picture of the contours of a responsible energy policy. We can identify the essential variables for a new policy, and we can articulate a new set of policy objectives. The contours of the new institutions and the political paths to creating them, however, are less clear; yet they are beginning to emerge.

The Model of Government Regulation

Ending Dirty Energy Policy is an extended case study of the relationship between law, policy, and politics. The interaction of these three variables constitutes a model for the study of government regulation.[3] Each of these elements is essential. If one element is missing, then there can be no legitimate government regulation. By way of example, I argue that there is a consensus in society that we should change energy policy. The legal rules and institutions are in place that enable government to do so because we have been regulating energy since the latter part of the nineteenth century. I further argue that there is ample policy support for the proposition that an energy transformation to a low-carbon economy is necessary and affordable. Thus, the law and policy elements of the model are easily satisfied. However, it is equally clear, at the moment, that the political support to aggressively take on that transformation is lacking – fossil fuel catastrophes notwithstanding. Until political support for an energy transition is galvanized, government regulation will fall short. This book argues that bottom-up political support for a transition is building.

The legal element in this model of government regulation imposes two requirements. First, any legislative proposal must pass constitutional muster. Second, once a statute authorizes agency action, then that action must conform to the statutory authorization. Over the course of the last century, regulations and laws that have set prices, allocated resources, created an interstate infrastructure for the delivery and transportation of energy resources, and allowed for the correction of market failures such as monopoly and information asymmetries have all been held to be constitutional.[4] Inevitably, there will be legal challenges as legislation is passed to accomplish an energy transition. It can be expected, for example, that states, at times, will seek to preserve their rights against federal legislation. There have already been legal challenges to how the federal government allocates costs for infrastructure investment.[5] Nevertheless, the basic legal principles are well-established, and most challenges will address the application of those principles rather than their underlying constitutional legitimacy. In short, there is more than ample legal authority for a new energy policy.

The policy element is directed to the factual bases for legislation or regulation. Quite simply, the legal test is whether or not the reasons offered for either legislation or regulation are reasonably related to the ends that are pursued. In less stilted language, in support of a legislative or regulatory proposal, the proponent must assert that a reasonable set of scientific, technical, economic, or other data-based reasons exist for its enactment. If the government, for example, wishes to limit carbon dioxide emissions, then the reasons for the limitations and limits it suggests must be reasonable. Although policy is often contestable, this book will demonstrate that the policy arguments behind an energy transition are not only sound, they are well supported by data and are generally accepted as reasonable.

Finally, the political element of the model is clearly the most contentious. The politics of energy have a long history behind them. Additionally, the politics of energy have trillions of investment dollars behind them. Further, the partnership between the fossil fuel industry and government regulation is deeply entrenched in the political and economic marketplaces. A full energy transition will only be possible once that partnership is renegotiated so that low-carbon producers become the dominant energy providers and consumers realize greater choice and value.

Overcoming Barriers to an Energy Transition

One consequence of having a century-old energy policy is that transition is difficult. Moving from a fossil fuel economy to a low-carbon energy economy is made particularly difficult by three substantial barriers to change. The first barrier is that government regulation has been based on a fatally defective theory of markets. Second, industrial and bureaucratic incumbency favors old ways of doing business and favors old actors. Third, energy policy transformation, as well as climate change, present regulatory problems that are different in kind than most of the regulatory problems the government confronted in the past. A successful transition, then, requires us to rethink our theory of markets, break down the advantages of incumbency, and bring new thinking to future forms of regulation.

Theory of Markets and Their Regulation
Whether we refer to the governmental structure of the United States as a form of democratic capitalism or as a liberal democracy, the relationship between state and market is the same. Markets are seen as the most desirable form of social ordering because markets have their virtues. Markets can create wealth, stimulate innovation, and efficiently distribute resources – in theory. In practice, markets have been subject to control by special interests who have successfully garnered government favors, thus distorting markets by reducing competition. There is nothing particularly shocking about this idea; it is as old as the Founding. After all, politics is about who gets what.

What is troublesome about a strong commitment to markets is that such a commitment is too narrow. Markets do achieve their virtues when they are competitive. Markets, however, are often imperfect, and those imperfections can often be fixed through government regulation. A theory of government regulation, though, that limits itself to fixing markets is too narrow and ignores other matters of social importance such as the fair distribution of wealth and power and the long-term construction of a better society.

Government regulation throughout the twentieth century was based on the *ex post* fixing of markets when they were broken. Market fixes, in turn, were justified with arguments from efficiency. To be sure, we continue to experience market failures, but we must respond to them differently for two reasons. First, for the last half-century the very idea of "the market" has been corrupted and politicized in ways that have distorted the role of government in general, and energy and environmental policy in particular.[6] Second, and more importantly, the energy and environmental problems we face (climate change is the paradigmatic example) are structurally different from the types of market imperfections we have historically addressed. Twenty-first-century regulation must look *ex ante*, must anticipate problems, and must act pragmatically and responsively to construct a better society justified on principles of equity and fairness as well as efficiency.

Thus, the market failure model of the past is too narrowly focused, favors a limited number of incumbent actors, misdirects resources, and cannot effectively respond to today's economic and environmental challenges.

The Power of Incumbency

Traditional energy policy has erected roadblocks for new energy policy and has impeded the entry of new energy actors. Traditional energy policy not only favors fossil fuel firms, it also has created a regulatory mindset that privileges short-term economic growth over long-term environmental protection and rewards business leaders, investors, and government regulators for sticking to the old path.[7] In no small part, this narrow focus is reinforced by the bounded rationality of decision makers; the path dependency of their previous investments; and generally in a regulatory psychology of bureaucratic inertia, at least, and regulatory capture, at worst.[8] In short, the pull of tradition is not irrational. It is also not optimal.

The demand of shareholders and credit markets for quarterly profit and loss statements necessitates that business leaders not only look to their bottom lines but also look for immediate gains. This financial focus is understandable, but it is not at all realistic to continue to focus on old dirty industries while ignoring emerging technologies and new energy providers and strategies. Incumbents may appear to be safe investment bets, but incumbency alone is not safe enough – witness the U.S. automobile industry over the past thirty years and the necessity of the 2009 bailout.

The institutional constraints on markets that demand quarterly returns on investments affect regulators as well. Government regulators, particularly political appointees, may not have quarterly demands placed on them, but their time horizons do not exceed two- or four-year election cycles. Again, this short-term focus is understandable because, like all organizations, government agencies operate under constraints. Information problems, for example, abound. Given the magnitude of designing an energy policy for any one sector demands accurate information, and in the dynamic and contentious field of energy, information is difficult to come by. Regulators work closely with their regulatees and must, often by necessity, look to industry for its data. Government agencies are constrained by limited staffs and budgets, by significant caseloads, and by the changing winds of politics. Consequently, regulators focus on the here and now, not on some hypothetical future. The current structure of administrative agencies reinforces a narrow industry focus additionally because both the regulators and the regulatees have a certain sympathetic understanding of one another's job assignments.

Institutional inertia is unacceptable. Narrowly focused, short-term thinking, and the reluctance to change, must be resisted if the country is to enjoy a healthier environment and a more vibrant economy. Incumbency, then, constitutes a significant barrier to change. Incumbency in the economic marketplace places a demand on investors for a return on those investments. Incumbency in the political marketplace places a demand on bureaucrats to conform to the political temper of the times, which too often looks backward for financial support rather than forward to challenging old ways. Yet a deeper problem exists. Transforming energy policy or responding to climate change are problems that are categorically different from the types of regulatory problems we have addressed in the past.

Twenty-First-Century Regulatory Challenges

Both energy policy and climate change are extraordinarily difficult and categorically different from standard regulatory issues.[9] Standard regulatory problems are discrete; their costs and benefits are reasonably determinable; and they are often susceptible to technical fixes. Traditional regulatory responses will not solve the problems of energy policy transition or climate change, because both are: (a) complex and not discrete; (b) not open to short-term, simple solutions; and (c) the costs and benefits are highly contested and, not infrequently, incommensurable. If we examine these problems more closely, we find that they contain a series of problems of cascading uncertainties. These uncertainties involve spatial, temporal, interdisciplinary, technical, economic, scientific, geographical, and jurisdictional dimensions. Is the earth warming? Is there a human contribution to warming? Do we risk catastrophe by doing nothing? Can we have an energy economy without coal or oil? What are the costs of transition? What are the costs of doing nothing? What are the benefits? What steps should we take? Will those steps be effective? Is it possible for government to change energy policy comprehensively? Can government sustain the necessary political will to address a problem that will outlive us all? And so many, many more.

The need for decision making in the face of uncertainty is not new, although energy policy and climate change are orders of magnitude greater than the run-of-the-mill regulatory issue. The proper path to finding solutions amid a host of uncertainties is unclear and is susceptible of being obscured by a tendentious politics that attempts to sabotage hard science and sound policy.

The multi-faceted dimensions of such large and complex problems can lead to stasis. Why act in the face of so many unknowables? As individuals, is the problem too big to comprehend and too invisible to appreciate? As a society, how can any institution, public or private, come to grips with problems as long-term and as multi-dimensional as an energy transition and climate change? Can we respond to "Giddens' Paradox?" It states that the dangers of global warming may not appear tangible, immediate, or visible in our day-to-day lives, therefore, we can choose to do nothing concrete; yet, if we wait until these problems are visible and acute, then action may be too costly and too late.[10] Inaction is always an option, but it is not a wise one and must be resisted.

Energy policy and climate change are paradigmatic of other contemporary problems such as health care, public education, and financial markets for which there are no quick fixes or one-shot solutions. Instead, these issues require a re-envisioning of government regulation in which we no longer retreat to the Reaganesque slogan that government is not the solution, it is the problem. Government must be seen as an active participant in generating broad-scale, creative responses to such vexing social and economic problems. If we are to have any hope for even a partially responsive solution, we need a new environmentally sensitive energy policy and we need a new regulatory regime.[11]

The United States will neither serve its domestic agenda nor serve as a world leader for climate change initiatives until our traditional fossil fuel energy policy is transformed into one that acknowledges environmental costs and capitalizes on the economic opportunities available in moving away from a traditional policy to a smart one. *Ending Dirty Energy Policy* argues, not only that an energy policy transition is necessary for any effective climate change response, but that an energy policy transition is good in and of itself. Whereas climate change and smart energy policy are complementary regulatory initiatives, this book concentrates on transforming energy policy. A transformed energy policy will take us away from old habits of fossil fuel favoritism and will lead us toward a new competitive and innovative energy portfolio.

Energy and Climate Change Observations

One can accept the science behind the Earth's warming and even accept, as scientifically sound, the anthropogenic contribution to warming.[12] Still, one might be forgiven any skepticism about questioning whether any response is economically feasible.[13] There is not a high degree of comfort, let alone a guarantee, that a response will necessarily be economically efficient or even energy efficient. And yet, what alternatives exist? Can we wait and see? Or as Judge Richard Posner writes, we can ignore the threat of climate change to our peril.[14]

Ending Dirty Energy Policy can be read as an agnostic text insofar as the book does not concentrate on the fears associated with melting ice caps, the spread of arid farmland with its attendant displacement of the world's poor, or the possible extinction of hundreds of species. Nor does the book extol the benefits of a warmer climate for some parts of the globe or the human inevitability of mitigation through migration. Issues, pro and con, surrounding climate change are addressed in other precincts. Still, without a new energy policy we cannot adequately address climate change. The complementary assertion is that a new energy policy is a valuable public good.

Throughout this book, significant amounts of data are presented, but they should not be overwhelming. Instead, the data should be read as signaling the direction and magnitude of change. In brief, there are only a handful of data points that set the context for a discussion of energy policy. The first data point is the current situation of the electricity and transportation sectors of our energy economy. Electricity accounts for 60 percent of the energy produced and consumed in this country and oil accounts for the other 40 percent. At the moment, these sectors operate almost exclusively independently of one another. Very little oil is used to generate electricity, and very little electricity is used to move cars. That situation will have to change.

The second data point concerns the fuel mix for electricity generation. Roughly 50 percent of all electricity is generated by coal, followed by 20 percent generated by natural gas, then 20 percent generated by nuclear power. Renewable resources, including hydropower, constitute about 7 percent of the electricity generated with less than 1 percent generated by solar or wind power. That situation too must change.

The third data point necessary to understand the current situation is that 85 percent of our energy is fossil fuel based and 10 percent is based on nuclear power. The traditional energy path of large-scale, highly centralized, capital-intensive, and largely fossil fuel firms control the energy market. The consequence of such a large market share, then, should be obvious. A transition away from the traditional path necessarily means that economic and political incumbents must face more competition and the loss of market share. Change has its costs.

The Deepwater Horizon disaster captivated the country in 2010. It was impossible to watch the online video of tens of thousands, if not hundreds of thousands, of barrels of oil a day and associated methane gas being released from a mile below the water's surface without a queasy, helpless feeling. The dangers of offshore drilling are not new, even though the dangers of deep water drilling have apparently been underestimated. It seems shocking, then, that even though we have been talking about the need to "break our oil addiction" for more than fifty years, we have done precious little to do so. It would not be difficult to be pessimistic about the future of our energy economy and the environment. Indeed, it would be easy to blame Washington for its failure to act and its apparent unwillingness to act aggressively.

Ending Dirty Energy Policy, however, is not written from a point either of frustration or despair. Instead, the driving theme of the book is that there is a significant policy consensus about the need for, and path to, a low-carbon energy economy. Further, there are a variety of public and private actors in both the political and economic marketplaces that are pushing forward a clean energy agenda. The book argues that Washington is a necessary player for a transformed energy future and Washington will be brought there from bottom-up activities, not from top-down edicts. There are positive political and economic signs about our energy future; but to get there, we must study, then dismantle, the past.

1

A Regulatory History of Dirty Energy
Law and Policy

Plus ça change, plus c'est la même chose

Thomas Edison would clearly recognize today's electric grid. John D. Rockefeller would also clearly recognize the desire to drill for oil in remote places. Both would recognize our passion for fossil fuels. Neither would recognize the extent of the government regulation of energy or the development of the field of energy law. Energy law, as a legal discipline, grew out of the energy crises and legislation in the 1970s. The first legal casebooks were published at that time, energy law treatises and journals were begun, and the organized bar and law schools began to treat energy law as a recognized field.

At the time, energy issues, particularly high oil prices and the costs of nuclear power, were regular headline news. President Carter staked his presidency, and lost it, on his aggressive, and largely failed, approach to redefining U.S. energy policy. Now, a generation after Carter, we find ourselves confronting another energy crisis in the form of global warming and environmental catastrophes. To date, the energy legislation that has passed Congress has not dramatically reformed U.S. policy. Nevertheless, energy policy makers and energy politics have taken on new configurations as new think tanks, new university-based research centers and institutes, and new non-governmental energy organizations (NGEOs) have entered the energy policy arena. Energy issues are again headline news.

Energy law has its predecessors, and energy industries have been well regulated for over 100 years. Much of what today constitutes the field of energy law is based on the rationale and forms of public utility regulation first taken up by municipalities and states and, then later, by the federal government in the first third of the twentieth century. As a direct result of a long history of regulation, energy industries have developed an identifiable structure, and the country has developed an identifiable, but loosely coordinated, traditional energy policy. The regulated energy industries, the firms within those industries, and their regulators have formed a working and mutually supportive relationship over that period.[1] Today, that relationship inhibits the development of an energy policy for our times.

This chapter describes the regulatory history of our traditional and dominant energy policy, which developed from the end of the nineteenth through the beginning of the twenty-first centuries, and argues that the old policy has outlived its usefulness and must be changed. Change, though, is occurring as a new energy thinking takes hold, as new coalitions are formed, as new energy financing enters the market, and as a new energy politics develops. Going forward, our new energy policy must support those new ways or, as economist Jeffrey Sachs points out, a business-as-usual economic and environmental strategy will result in an "ecological crises with calamitous results."[2] We, the world, he writes, have a narrow window of opportunity to redesign the way we do our energy business. Redesigning energy policy is a central task that must be addressed.

What is Energy?

Most simply defined, energy is the ability to do work. Heating, cooling, lighting, and transportation are all examples of the types of work that energy does. In order to harness energy, natural resources must be transformed from their original state into their useful forms. It is during this transformation process that the key laws of physics take over. The First Law of Thermodynamics, the conservation of energy, holds that we do not "lose" energy, rather it changes from one state to the next – ice melts, water boils into steam vapor, and plants decay. The Second Law of Thermodynamics, entropy, means that energy moves from a more concentrated state to a less concentrated one; consequently, some of the useful energy in a natural resource is lost in the transformation process. Both laws are important for the use, and regulation, of the resources used to produce energy. By way of example, if the *potential* energy in one ton of coal can provide electricity to 100 homes for one day, by the time that coal is burned, to heat water, to turn the rotors on an electric generator to generate electricity, and then to have that electricity distributed to a home, there will be energy loss throughout the cycle. After the transformation and the operation of both laws of thermodynamics, less than one-third of those potential 100 homes will be served by that one ton of coal.

The example of coal is instructive. When we speak of energy, we are also speaking about the natural resources used in its production – energy and natural resources are inextricably linked throughout the fuel cycle from exploration and extraction to end use and disposal. Mining coal dramatically changes the landscape, destroys topsoil, and spoils mountain streams, to say nothing of endangering the health and lives of miners.[3] Burning coal pollutes the air and atmosphere, contributes to global warming, and those pollutants affect the health and lives of people around the world. Energy laws, policies, and regulations, then, should not ignore the environmental effects that occur throughout the fuel cycle. However, for nearly four decades, energy and environmental laws have been treated as separate disciplines and have been largely uncoordinated. Global warming significantly challenges this old way of doing business. The country cannot continue to ignore the direct connection between energy and the environment if we are to address either the need for a transformed energy policy or climate change in meaningful and effective ways.

There are several reasons to regulate energy resources and industries. First, we should attempt to maximize the potential energy in each resource, to have it distributed most efficiently, and to have it consumed most sensibly. Second, regulation can address problems of scarcity, and regulators must pay attention to price and the availability and reliability of energy. During the first few months in 2008, oil was trading at over $147 a barrel (and about $4 per gallon of gasoline at the pump). Just the year before, oil was priced at below $40 a barrel. This price volatility is illustrative of the problem of scarcity. Whether global oil production has peaked or not, as demand increases, so will prices. The third reason for regulation is to minimize the social, health, and environmental costs of energy production, distribution, and consumption.

The United States consumes roughly 100 quadrillion BTUs (quads) per year, mostly in the form of oil, coal, and natural gas. Fossil fuels dominate our energy economy, and with 5 percent of the world's population, the United States consumes 25 percent of the world's fossil fuels and contributes 25 percent of the world's carbon emissions. As the world's largest producer and consumer of energy, the United States bears responsibility for its energy policies.

Energy Policy History

Since 1950, our energy consumption has more than tripled from 30 quads to 100 quads per annum, but production has grown only from 30 quads to slightly over 60 quads. Imports, particularly oil imports, make up the difference. Calls for independence from foreign oil are not new. They began in the Eisenhower administration when oil consumption exceeded oil production for the first time. Our concern with oil independence, however, has been exacerbated since 1970. In 1970, domestic oil production peaked and has since been declining continuously as the gap between domestic oil production and consumption has grown dramatically. Today, for example, we consume roughly 20 million barrels of oil per day (mbd). From that total, we consume approximately 15 mbd of crude oil, producing slightly over 5 mbd, thus importing the remainder.[4] Also since 1950, the use of renewable resources for energy production has remained essentially flat. In short, U.S. energy policy has not changed to fit our consumption patterns, and this failure has made the United States vulnerable to economic dislocations as well as national security threats. The story of this predicament began in the mid-1800s.

Regulation in the Nineteenth Century

In the middle of the nineteenth century, our energy economy experienced a transition from wood and whale oil to coal and oil.[5] Imagine America at mid-century. Agricultural life was thriving, the country was expanding west, manufacturing was literally gaining steam, and immigration into the cities was beginning in earnest. The combined forces of population growth, urban development, and industrial expansion meant that whale oil and wood were insufficient energy resources. New energy resources were needed, and they were found.

In 1859, Colonel Edwin Drake drilled the first commercially successful oil well in Titusville, Pennsylvania, and in 1882, the first electric generating station was switched on by Thomas Edison on Pearl Street in New York City. Drake's discovery recovered 4,450 barrels of oil in the first year, followed by over 220,000 barrels the next. Oil, in large and easily recoverable quantities, quickly replaced whale oil. Equally important, the discovery of oil necessitated the development of ancillary industries such as refining and transportation. At the time of the discovery, the oil industry was unregulated and would stay unregulated until the end of the century. The turning point for oil regulation came in 1911 with the antitrust case brought against the Standard Oil Company, which resulted in the breakup of John D. Rockefeller's oil monopoly.[6] Perhaps not so curiously, the breakup of Standard Oil doubled Rockefeller's wealth because he had stock in each of the subsidiary companies and all of his stock increased in value.[7]

The Standard Oil antitrust case revealed the market imperfection behind big energy. Rockefeller's Oil Trust was busted because it was a monopoly, and monopoly is the classic example of a market failure. Industry concentration and market power negatively affect the public because a monopolist can simultaneously raise prices, reduce output, and reduce consumer choices in the market. Further, the Oil Trust adversely affected other industries as it manipulated rail rates and oil prices to the detriment of the consuming public. The electricity industry also exhibited monopoly characteristics. However, instead of breaking up the electric industry, utilities were given government protected monopolies. In both instances, big energy was here to stay as firms within those industries inexorably consolidated.

Edison's Pearl Street generating station served only fifty-nine customers in a fairly small area because Edison's direct current distribution system could not travel far without risk of burning the transmission lines. In the span of a few decades, though, the local electric utility evolved into a regional monopoly. In 1896, George Westinghouse, a former Edison employee, harnessed the hydropower of Niagara Falls to serve the city of Buffalo twenty-nine miles away. Shortly thereafter, another former Edison employee, Samuel Insull, created a complex electric utility holding company capable of manipulating electricity prices and stock holdings to his benefit with costs to the public.[8] Just as the government stepped in to break up the oil trust, regulators stepped in to address the market abuses in the electric industry.

Government regulation of energy was largely non-existent throughout the nineteenth century. Firms were small and competitive, markets were local, and common law rules were adequate. However, in the latter decades, modern government regulation began to take shape. Government had long been aware of its oversight role in markets. In the early years of the nineteenth century, government played an activist and mercantilist role by constructing roads, bridges, and canals.[9] This infrastructure enabled businesses to grow as markets expanded. Later in the century, as firms grew in size and obtained strategic advantages in the economy, inefficiencies became apparent.

In the 1876 case of *Munn v. Illinois*,[10] the Supreme Court, for the first time, articulated a rationale for direct government involvement in markets through price setting. At issue in *Munn* was the operation of grain storage facilities in Chicago and other large Illinois cities. At the time of the case, there were fourteen grain warehouses in Chicago that were controlled by nine firms. The Court found that prices were collusively set among the nine firms and that these firms exerted monopoly power over Midwest grain markets. Farmers were forced to sell their grain at a price set by the monopolists. In order to avoid the wages of monopoly, the Illinois legislature passed a statute that empowered it to set rates at the grain silos in the name of the public interest. The United States Supreme Court upheld this law on two bases: (1) government price setting regulation could be exerted in the face of monopoly power, as long as (2) the product or service was affected with a public interest.[11]

The regulation of monopolies had taken on two forms at the beginning of the twentieth century. First, antitrust laws were applied to check the abuse of market power by disaggregating monopoly firms. Second, regulatory agencies, at different levels of government, began to regulate what are known as natural monopolies. Loosely defined, a natural monopoly is a firm that experiences falling costs over a long range of output. In other words, a single large firm can produce a product more cheaply than multiple firms.[12] The classic example of a natural monopoly is a network industry, such as oil and natural gas pipelines, rail lines, electric power lines, and cable TV lines.[13] It is more efficient to have one firm engage in distributing these goods than multiple firms because multiple transmission, distribution, and rail lines would be duplicative and, therefore, wasteful.

The second significant event giving birth to the modern administrative state occurred with the passage of the Interstate Commerce Act in 1887. The statute created the first modern, and now defunct, regulatory agency – the Interstate Commerce Commission (ICC). The ICC was the expert regulatory body designed to oversee the rates charged by railroads and later by trucking companies. The primary purpose of the ICC was to regulate these common carriers, specifically with the authority to set fair rates for transportation and to eliminate rate discrimination.

These two developments, the constitutional permissibility of price setting and an expert regulatory body to set rates, formed the basis for the state's authority to intervene in private energy markets. Clearly, energy resources, particularly electricity and natural gas, were resources affected with the public interest. Equally clearly, some segments of these industries, such as oil and natural gas pipelines and electric transmission lines, had monopoly characteristics. Together, the public interest in energy resources and the potential for private energy industries to exercise monopoly power satisfied the Supreme Court's test for government regulation and, consequently, justified state intervention into those markets.

In the early decades of the twentieth century, local and state regulators began to set electricity rates to avoid the monopoly abuses in the electric industry. The various monopoly sins of increased prices and reduced quantity could be corrected by setting prices at reasonable levels, and state regulators engaged in rate making exactly for that purpose.

Around this time, state regulators also stepped in with laws to prop up the developing oil market. Oil and natural gas are fugacious, that is, moveable, resources. The general common law rule was that whoever captured a fugacious resource, like someone capturing a hare running across a field, had legal title to that resource. The problem, known as the tragedy of the commons, should be obvious – the race goes to the swiftest. If the first captor wins the prize, then he who hesitates is lost; thus, there will be more oil exploration companies than the market can reasonably suffer as multiple would-be captors attempt to be the first in the ground to extract the oil and natural gas resources lying underneath.

The tragedy in the tragedy of the commons is that it pays to rapidly consume and exhaust a resource before someone else does. This is exactly what happened in the flush fields of Texas, Louisiana, and Oklahoma in the 1920s and 1930s. Oil exploration yielded vast quantities of oil to such an extent that oil could be purchased for less than 10 cents a barrel. The common law rule of capture had to be replaced by state regulation in order to sustain reasonable and competitive markets. Market stabilization was accomplished through state unitization and proration laws that limited the amount of oil that could be put on the market at any one time. Regulations also rewrote the rules of property ownership so that multiple owners could share the proceeds from oil and natural gas discoveries from a specific oil field that ran across the property lines of multiple owners.

At the end of the nineteenth century, energy was produced and distributed on local or regional bases and regulations were developed at the local and state levels, mirroring the structure of the energy industries themselves. The history of energy regulation in the early years witnessed a transition from local common law regulation to state administrative regulation. It is also worth noting that there was no policy coordinating the development and use of natural resources. Instead, individual resources were, and continue to be, regulated independently of one another. Nevertheless, we can see the development of traditional energy policy taking shape. Traditional policy viewed the regulation of energy resources as a series of rules and regulations that supported private ownership as well as the expansion of the energy economy. Regulations and market forces worked in tandem to promote the production and distribution of energy. As energy industries moved from local to regional to national and, finally, to international markets, energy firms changed accordingly and so did the regulations intended to monitor them.[14] However, in all instances, the regulations were intentionally designed to bring cheap and abundant fossil fuels to market.

Early Federal Regulation

During the first two decades of the twentieth century, modern energy industries and federal energy regulation began to take shape. Corporate and regulatory structures ran parallel to one another largely, but not exclusively, along technological lines. Technological advances meant that the country was moving from a low-energy society to the beginning of a high-energy one dependent on large-scale, capital-intensive, centralized, interstate energy production and distribution. The move to scale was most notable in the oil, natural gas, and electricity industries, and regulators behaved accordingly. The general intent of federal energy regulation was easily discernible – regulation was intended to promote energy production and the industrial stability of these large energy firms.

Coal remained dominant in the first quarter of the twentieth century, but by 1925, oil constituted almost one-fifth of the market. Still, the government never abandoned coal during the transition from solid to liquid and gaseous fossil fuels. Instead, regulations supported the coal industry, as well as the emerging oil and natural gas industries, particularly during World War I. The war expanded the use of oil in the transportation sector and in international markets. During this time, the United States established its first energy agency, the United States Fuel Administration (USFA). The USFA was authorized to regulate oil and coal prices, transportation, and distribution. The agency concentrated on the coal industry, especially in the face of coal shortages for the war. The USFA supported coal for war mobilization by lightly regulating the industry, relying primarily on voluntarism and on public appeals to keep prices stable and to conserve coal use, and by leaving regulation in local, firm-friendly hands.

As the war ended, oil and natural gas began their dominance as energy resources. Even though Standard Oil had been broken up by antitrust enforcement, large firms still controlled the oil industry. In 1920, for example, thirty firms controlled 72 percent of the country's oil refining capacity. Later that decade, twenty-one major oil companies controlled 60 percent of oil production, ten firms controlled 60 percent of the refining, and fourteen firms controlled 70 percent of the oil pipelines. In the natural gas industry, eight holding companies controlled 85 percent of production. Twenty-two electric holding companies controlled 61 percent of the country's electricity generation.[15]

Federal regulation concentrated on the transportation bottleneck segments of these network industries. The ICC regulated railroad rates, the Hepburn Act regulated oil pipelines, the Federal Power Act regulated interstate electricity transmission, and the Natural Gas Act regulated natural gas transportation. At first glance, it would seem that federal price setting was a dramatic and heavy intervention into markets. However, as it developed, rate regulation had the effect of virtually guaranteeing profits for the private firms owning these transportation segments because rates were based on a reasonable cost of service plus a reasonable rate of return, which, in effect, is a pass-through of costs to customers and is a form of cost-plus pricing. As long as a firm operated prudently, a fairly low business standard as it turned out, then it would reap its profits.

In the early decades of the twentieth century, energy markets developed a structure that persisted throughout the century. These markets had the following characteristics: (1) seemingly inexhaustible supplies of fossil fuels; (2) a shift from local to interstate production and distribution; (3) continuing growth in the size of markets; (4) continuing realization of energy efficiency; (5) increasing industrial concentration, integration, and large-scale production; and (6) transportation bottlenecks in each industry.[16] This pattern, set in the beginning of the twentieth century, continues today. Federal energy regulations were designed to react to market conditions by supporting the specific industries being regulated. Federal regulators also ignored the social costs of pollution.

After World War I, energy markets began to experience a transformation as oil replaced coal as the nation's dominant fuel. Coal markets experienced a decline until after World War II when they rose with the expansion of the electric industry. The rise of the oil industry came with its own set of problems that needed government support. The oil-producing states suffered from overproduction when flush fields destroyed oil prices and produced economic waste. At the state level, oil conservation boards helped regulate the market. At the federal level, the Federal Oil Conservation Board (FOCB) performed a similar function and was charged with the responsibility of stabilizing the oil market. It did so by favoring major oil companies through financial incentives, such as the oil depletion allowance, and by opening up public lands under the Mineral Leasing Act of 1920. The FOCB also allowed large oil firms to control production and reduce the amount of oil on the market, which allowed these firms to capture economic rents.

Modern Federal Regulation
As energy industries continued to expand and grow across state lines, federal regulation followed. The first major federal regulation was the Federal Power Act of 1920 under which the federal government exerted authority over hydroelectric projects. With the coming of the New Deal, federal regulation expanded further. For many, the New Deal was a constitutional moment

in our political history that literally changed the way the government did business.[17] New Deal legislation shored up a broken market economy and enabled the development of a middle class consumer society that thrived for most of the rest of the century.[18] Energy, in no small part, was integral to the economic progress that the country enjoyed.

The challenge for FDR and his New Deal hotdogs[19] was to stabilize the economy, provide a safety net for citizens generally, and construct a consuming middle class. Oil markets continued to be a problem during the New Deal, especially after the U.S. Supreme Court invalidated one of FDR's economy-wide programs, the National Industrial Recovery Act (NIRA), a portion of which attempted to prohibit interstate and foreign trade of petroleum products in excess of state quotas.[20] After the Court ruled that the legislation was unconstitutional, Congress passed the Connally Hot Oil Act of 1935, which reenacted the oil provisions of the NIRA and allowed government to regulate the flow of oil between the states, thus increasing oil prices in a fairly concentrated industry. Similarly, the administration created two National Bituminous Coal Commissions that were directed to promulgate minimum prices and to enforce fair trade practices, which greatly aided mining companies. Further, during the war, two new agencies, the Petroleum Administration for War and the Solid Fuels Administration for War, were established in order to support the oil and coal industries, respectively, as part of war mobilization efforts.

Oil and coal regulations continued to support both industries, but the larger impact on energy markets occurred in the electricity and natural gas industries. In the electricity industry, the New Deal expanded public power projects and enacted two major statutes regulating private electric utilities. The Public Utility Holding Company Act of 1935[21] was enacted specifically to address shareholder abuses by holding companies such as Samuel Insull's.[22] In the same year, Part II of the Federal Power Act[23] was enacted in order to give the federal government rate-making authority over interstate wholesale sales of electricity.

The electricity industry is divided into three segments. Electricity is: (1) *generated* from a power plant at high voltage; (2) *transmitted* to local utilities; then (3) *distributed* to end users. Private investor-owned utilities (IOUs), these are the local utilities with which we are most familiar, found it most profitable to vertically integrate and to own all three segments – generation, transmission, and distribution. As the vertically integrated electric IOUs grew in scale, they began to sell electricity across state lines, and the transmission system began to exhibit monopoly characteristics with the usual perverse economic consequences.

As energy industries moved from local to regional to national, transportation played an increasingly important role. Simply, private oil, natural gas, and electricity firms could most profitably operate at scale through interstate transportation of those resources before they were distributed to end users by local utilities. Each industry can be pictured as having multiple producers at one end of the transportation network and millions of consumers at the other end. However, it makes no economic sense to have multiple and competing oil and natural gas pipelines or multiple and competing high voltage electric power lines constructed across the country. Neither producers nor consumers benefit from such wasteful expenditures. Because of economic necessity, then, the transportation system was limited to a handful of private owners, creating a transportation bottleneck that gave private firms monopoly power that, in turn, demanded regulation by federal and state governments. The federal government was left regulating interstate wholesale electricity and natural gas rates, and retail rate regulation was left to state regulators. This bifurcation of regulatory authority led to severe problems at the end of the century for both industries.

The transportation segment of the natural gas industry also exhibited monopoly characteristics. In 1935, the Federal Trade Commission found that natural gas pipeline ownership was highly concentrated; that finding led to the passage of the Natural Gas Act of 1938[24] to regulate the interstate sales of natural gas. The Natural Gas Act mimicked Part II of the Federal Power Act by regulating interstate sales through a rate-making process intended to set rates that were just, reasonable, and non-discriminatory.[25] The rates were also set at levels that returned costs, plus a return on investments, to the utilities that constructed and laid the natural gas pipelines and the electricity transmission and distribution lines.

From Post-World War II to the Energy Crises of the 1970s

After World War II, there were four notable developments in energy markets until the energy crises in the 1970s. First, the coal industry regained its place in the country's energy portfolio as a result of the growing demand for electricity. Second, the natural gas industry experienced a period of destabilization. Next, domestic oil production was insufficient to satisfy growing demand, so the country turned to foreign imports. Finally, commercial nuclear power promised an endless supply of clean, cheap energy for the foreseeable future; it did not deliver on its promise.

When the Natural Gas Act of 1938 was passed, regulators focused on the transportation segment of the industry and ignored wellhead prices. As a consequence, wellhead prices were of little concern to natural gas sellers because these prices would be passed through to the consumers who had to bear the costs regardless of their reasonableness. This pass-through was challenged, and ultimately the Supreme Court ruled that the Federal Power Commission had full authority over wellhead rates of gas to be sold in interstate commerce and, therefore, could set natural gas prices.[26] The dramatic effect of this ruling was to require the Federal Power Commission to hold individual rate hearings to set the wellhead prices of natural gas for all gas producers, a task that it could not complete.[27]

The ensuing problem was not merely a matter of case management or docket control. Instead, the effect was to create a dual market in natural gas prices. Under federal regulations, interstate sales were based on the historic average cost of producing gas. Intrastate natural gas sales, however, were set at the world market price. The direct consequence was a significant price disparity between intrastate and interstate natural gas markets. With rising world prices, natural gas producers preferred to sell their gas in the higher-priced intrastate market. Smart gas producers moved their gas from the interstate market into the intrastate market, thus causing a national gas shortage.[28] The story of the distortion of natural gas markets is a darling of regulatory critics because it is a clear example of government price regulation causing a shortage rather than repairing a problem.[29]

In the early years of the Eisenhower administration, crude oil imports accounted for about 10 percent of domestic consumption but, more significantly, represented a doubling of imported oil in less than a decade and began our failed quest for oil independence.[30] Increasing oil imports revealed two points of vulnerability. Because domestic producers were unable to supply all of the country's demand, the United States did not have full energy security and was economically vulnerable. More troubling, the Middle East was becoming more destabilized as oil producers began to nationalize oil companies and attempted to operate production and refining facilities on their own. Thus, political destabilization in the Middle East revealed our national security vulnerabilities.

In order to reduce oil imports and shore up our economic and political vulnerabilities, Eisenhower initiated the Mandatory Oil Import Quota Program (MOIP), which set a limit on the amount of oil that could be imported. The consequence was not economically favorable, as consumer prices rose and as domestic resources were depleted. The MOIP was conducted under the auspices of national security, but the move backfired because it became the catalyst for creating the Organization of Petroleum Exporting Countries (OPEC), an oil pricing cartel with which we have been engaged ever since.[31]

After World War II, the national economy expanded in ways that increased the demand for fossil fuels. As the interstate highway system was built and as those following the American dream constructed single-family homes in suburbs further and further from central cities, automobile use increased dramatically, putting increasing demands on oil. As air transportation increased, so did demand for aviation fuels. Moreover, as the suburbs added housing development after housing development, the demand for coal-fired electricity grew. Fortunately, both electricity and oil were in abundance, and even with increasing demand, relative prices were flat or declining. Expansion and flat prices in the electric industry continued until 1965. Expansion and flat prices in the oil industry continued until 1970. Still, during this period, troubling signs in energy markets began to appear.

Through the post-World War II period, the fundamental economic assumption behind energy policy was the belief that there was a direct and positive relationship between energy production and consumption, and the economy. From the end of World War II, looking into the future it would have seemed that: (1) our demand for electricity would continue unabated at an annual rate of 7 percent; (2) coal, although a dirty-burning fuel, was abundant and cheap; and (3) nuclear power promised so much – it was clean, abundant, and expected to be cheaper than coal.

Eisenhower took the lead in moving nuclear power out of military hands and into civilian hands under the banner of Atoms for Peace, which led to the development of commercial nuclear power, once touted as being "too cheap to meter." Commercialization also had the effect of transforming atomic energy from the destructive force witnessed at Hiroshima and Nagasaki to the peaceful "Our Friend the Atom." In 1954, the Atomic Energy Act[32] was passed with the intent of promoting nuclear power for electricity generation. Testifying in Congress in favor of nuclear generation, utility executives, however, were well aware of the potential danger of a catastrophic event, and Congress responded with the Price-Anderson Act of 1957,[33] which limited the liability of utilities in the case of a nuclear disaster. Commercial nuclear power was to produce vast quantities of cheap electricity. However, it was also accompanied by vast economic and safety risks that regulators had to address.

The post-World War II era was a time of economic prosperity across the country, and energy consumers, as well as policy makers in the early 1960s, might well be forgiven for failing to miss signs of impending difficulties in energy markets. Fuel oil prices had been declining since 1953, domestic oil reserves appeared relatively healthy, electricity rates were flat or declining, nuclear power was energy's rising star, prices were not expected to rise, and energy resources appeared limitless – costs to the environment were simply not part of the national energy equation.

The beginning of our troubling energy times began in the mid-1960s. In 1965, electric power production peaked with the result that marginal costs began to outstrip average costs for the first time. In other words, when economies of scale were no longer being realized and when power production reached a technological plateau, then electricity prices began to escalate. Additionally, in 1970, domestic oil production peaked, and we began to rely on imported oil for most of our oil needs. Even before then, the Johnson administration issued a report on foreign oil, which signaled serious concerns about Middle East control. The report stressed that there was only so much oil the country could import before raising national security concerns specifically due to supply disruptions.[34] The report was prescient in this regard. Also around this time, we became aware of the environmental harms of unchecked resource use and began enacting environmental legislation, such as the National Environmental Policy Act of 1970 and other associated legislation to protect air, water, and land. Nevertheless, environmental legislation and regulations remained unconnected to energy policy despite the fact that the energy fuel cycle is fraught with environmental impacts. To the public in 1970, the energy economy seemed fairly healthy; it would not be healthy for long. It also seemed unaffected by environmental concerns; it could not ignore the environment for long either.

Energy Crises

President Nixon inherited a shaky economy as a result of Vietnam War expenditures. Inflation was rising to unacceptable levels, and in response he instituted a series of wage and price controls, which were not only overly complicated, they were applied unevenly and awkwardly. The oil industry experienced perhaps the most complex set of price regulations. Nixon rescinded Eisenhower's foreign tariffs, which thus rescinded the mandatory oil import quotas. The elimination of these tariffs had a positive effect on the energy economy as prices floated closer to market levels and supplies increased.[35] The net gains achieved by this move, however, were eliminated by the continuation of oil price controls, which had the opposite effect and restricted supply. Oil price controls outlasted Nixon's economic stabilization program and required an elaborate bureaucratic machinery for their administration. Not surprisingly, like the natural gas regulations before them, oil price controls distorted oil markets rather than stimulated them and were ultimately dismantled, but not after billions of dollars of abuse by oil companies.[36]

Nixon was aware of the delicate situation of our energy vulnerabilities and made several proposals to Congress to increase research and development (R&D) funding and to establish administrative agencies with an energy mission. Congress ignored Nixon's requests, so he then created a White House Energy Policy Office, the first in the country's history, in June 1973.[37] Little did he know that the United States was about to be embroiled in an energy crisis.

From 1973 through 1980, a series of events throughout different energy sectors occurred to which we are still trying to respond. In 1973, in reaction to the United States' decision to supply Israel with weapons, the Organization of Arab Petroleum Exporting Countries stopped oil shipments to the United States, reduced its oil production levels, and began to close the Suez Canal, which was essential for oil transportation. As a direct consequence, world oil prices quadrupled and the Arab Oil Embargo incontrovertibly demonstrated our reliance on, and our vulnerability to, the Middle East oil states. It was from these events that Project Independence, ostensibly our first effort to wean ourselves from foreign oil, was initiated.

As oil prices increased, double-digit inflation ensued, and energy became headline news as people waited in gas lines, on alternate days, to fill their tanks. This first Arab Oil Embargo was Nixon's baptism into the energy crisis. Congress passed emergency energy legislation, but Nixon vetoed it because it would have had the negative effect of increasing imports. He transformed the White House Energy Office into the Federal Energy Administration (FEA), the predecessor of the current Department of Energy. The FEA was charged with allocating fuels and administering the oil price controls. These regulations had multiple and conflicting purposes. The regulations were intended to distribute oil throughout the economy, reduce foreign dependence, stimulate conservation, manage inflation, and encourage domestic production. They did not succeed.

After Nixon left office, President Ford inherited Project Independence as well as significant inflation. The challenge for Ford was to allow some oil price controls without creating inflationary pressures. President Ford attempted to balance these conflicting goals by promoting oil self-sufficiency through a schedule of reduced oil imports together with increasing domestic production on the Outer Continental Shelf. Additionally, Ford had an ambitious program that involved energy efficiency standards including tax credits for households, the strategic oil program, increased R&D for synthetic fuels, the construction of 200 nuclear power plants, the opening of 250 coal mines, and the construction of 150 coal-fired power plants together with the construction of oil refineries and synthetic fuel plants. Congress

ignored Ford's grand requests. However, Congress did pass notable legislation including the Energy Supply and Environmental Coordination
Act,[38] which authorized the FEA to order power plants to substitute coal
for oil as their boiler fuel. In 1975, Congress passed the Energy Policy and
Conservation Act,[39] later supplemented by the Energy Conservation and
Production Act,[40] which gave the president the authority to remove oil price
and allocation controls according to a phased schedule and to adopt certain
conservation measures.[41]

The energy programs of Presidents Nixon and Ford were ambitious in
design but were never fully implemented. Amidst political problems for
President Nixon, and given the inflationary concerns that plagued both
presidents, neither Project Independence nor Ford's Whip Inflation Now
Program were successful, and neither led to any semblance of energy independence let alone any coordinated energy policy. More notably, however,
is the fact that traditional energy policy, concentrating on the fossil fuels of
oil and coal while paying scant attention to conservation and environmental concerns, continued. The decade of the 1970s was roughly at midpoint
before the second great energy crisis hit; in response, our next two presidents attempted to change energy policy along broad fronts.

Energy in the Carter and Reagan Administrations

The history of energy law and policy until the mid-1970s demonstrated the
strength of the traditional model. Traditional energy policy was tested significantly during that decade as world energy markets, especially in oil, went
through cataclysmic shifts. In response to those changes, President Carter
attempted to coordinate energy policy and shift it dramatically away from
fossil fuels and toward conservation. President Reagan moved in a decidedly
different direction. Instead of centralizing and coordinating energy policy
at the federal level as Carter wanted to do, he attempted to dismantle it. One
plank of his campaign platform was to abolish the Department of Energy;
and as one of his first acts in office, on January 28, 1981, President Reagan
decontrolled oil prices, which had already been scheduled for decontrol
on September 30, 1981. In short, the country rejected Carter's attempted
centralization, and it rejected Reagan's attempted dismantling.

President Carter's Energy Agenda
The first significant event of the Carter administration was to create a cabi-
net-level Department of Energy (DOE). Ideally, DOE would create a coordinated and comprehensive national energy plan and was given the authority

to do so. Although DOE biannually issues an energy plan, a comprehensive plan eludes us for several reasons. First, energy decision-making and policy-making responsibilities are scattered over several branches of government and among the several states. Second, even within the DOE, authority is fragmented.[42] Additionally, energy policy making has been divorced from environmental policy making up until, and including, now.

President Carter's defining energy moment was his "moral equivalent of war" speech delivered on April 18, 1977.[43] The speech outlined the substantive principles of Carter's energy policy and led to the passage of the National Energy Act (NEA)[44] in October 1978. The NEA consisted of five major pieces of legislation. The act mainly addressed conventional fossil fuels as it attempted to move the country toward energy independence, promote the use of coal,[45] increase energy efficiency, modernize utility rate making,[46] stimulate conservation,[47] and fix the distorted natural gas market. For the most part, the NEA addressed conventional energy resources as independent industries and not in any overarching or coordinated way. The NEA simply continued the traditional energy path on which the country had been traveling since the turn of the twentieth century.

The NEA had one great surprise and a significant impact.[48] The great surprise came in the Public Utilities Regulatory Policies Act (PURPA). The main focus of PURPA was to redesign state electricity rate making in order to get away from promotional rates that increased the demand for electricity and to set electricity prices closer to the true cost of production.[49] PURPA's surprise, however, was that its provisions provided for a new form of electricity generation and opened up new and unanticipated markets.

Going back to the second law of thermodynamics, to the extent that heat is produced in any process including manufacturing, that heat, if recaptured, can be used to produce energy. Thus, the idea of co-generation entered energy policy discussions. If a company can recapture heat and use it to generate electricity, then that company would not require as much electricity from the local utility. In some instances, the company would generate more electricity than it needed, and it could sell the excess electricity back to the local utility. The surprise was that there was a greater amount of co-generation and small power production than anticipated. In short, PURPA revealed that there were electricity providers, other than traditional utilities, that wanted to put cheaper electricity on the market.

PURPA's success was based on the requirement that co-generators and small power producers were able to generate electricity for their own use and then they could sell the electricity to the local utility at the local utility's cost of production. The gamble was that these new power producers could produce electricity cheaper than the local utility. The new power producers found this regulation attractive because now they had a guaranteed purchaser.[50] Simultaneously, the utility was given an incentive to its lower costs because of its buy-back obligations.[51]

The success of PURPA revealed that cheaper electricity was available; it only needed to get to market. This new electricity market challenged the existing utility order and raised a series of questions about: (1) encouraging new production; (2) avoiding saddling incumbents with stranded costs; (3) maintaining affordable electricity for consumers; (4) continuing to assure electric reliability; and (5) providing consumers with access to cheaper electricity from utilities and non-utilities.

The principle problem in satisfying these challenges involved transmission and distribution. Because traditional utilities were vertically integrated and owned the transmission lines, they were unwilling to have new providers use their lines at any rate that would threaten the price of their own product. Privately owned transmission lines could be used by other providers, but only at a price that was not competitive with the local utility owning the lines. The matter of opening access to new sources of generation over the privately owned lines of incumbent utilities has been a problem regulators have not solved since it was first revealed in the mid-1970s, and it continues to be the major obstacle to restructuring the transmission and distribution segments of the electricity industry. The access problem has become even more acute as we attempt to generate electricity from renewable resources such as wind and solar power.

The Natural Gas Policy Act (NGPA) was the intentional centerpiece of President Carter's National Energy Act. The express purpose of NGPA was to eliminate the dual natural gas market that had been created by over two decades of regulation. NGPA proceeded by decontrolling the prices of all natural gas that had not been dedicated to interstate commerce as of November 8, 1978, together with decontrolling the prices of some other categories of gas. NGPA proceeded along four lines. First, federal price controls were imposed on the intrastate market. Second, the Act created a formula for monthly increases in the wellhead price of post-1978 rates of gas. Third, the ceiling price on natural gas was pegged to the price of refined oil. Finally, the Act provided for the scheduled elimination of price controls starting January 1, 1985. In addition to beginning price decontrols, NGPA was intended to stimulate production as well as unify markets.

The NGPA regulations had the effect of increasing domestic natural gas production. The Act also unified the market and deregulated prices. Problems remained, however, because the natural gas market had developed a pattern of long-term contracts between pipelines and producers. The long-term contracts assured pipelines of supplies for them to resell. In order to assure producers of cash flow, the contracts also contained "take or pay" clauses that required pipelines to pay for gas, whether they actually took it or not. As prices rose, the pipelines wanted contractual relief from the long-term obligation to purchase high-priced natural gas that they could neither use nor resell. This problem awaited resolution in the decades to come.

President Carter experienced some success with his energy initiatives, but the problems continued. His last two years in office were particularly problematic. In 1978, the Iranian Revolution boiled over as oil production was shut down for six months, thus removing 3 million barrels of oil per day from the international oil market with nearly a 100 percent price increase. This shutdown affected the U.S. market, which was purchasing 750,000 barrels per day from Iran.

The Iranian Revolution led to President Carter's second major energy address on April 5, 1979.[52] Carter was aware that we needed to decontrol oil prices as a way of increasing domestic production. He was also aware that rising prices were politically unpopular because domestic oil producers would reap economic rents as a result of a rise in world oil prices. In response, Congress passed the Crude Oil Windfall Profits Tax,[53] which was designed to capture those windfall gains from oil companies and use that revenue for the development of alternative energy resources.

The end of the 1970s witnessed the end of the growth of the nuclear power industry. The obituary for the nuclear power industry came at 4 a.m. on March 28, 1979, with the incident at Three Mile Island. Although the accident at Three Mile Island was attributable to human error,[54] it bore an uncanny and coincidental resemblance to a then-current movie called *The China Syndrome*, which set public opinion against nuclear power in very vocal ways. The industry has gone through a very fallow period. Since 1978, no new nuclear plant has opened and all plants ordered since 1974 have been cancelled. Today, we are seeing some resurgence of interest in nuclear power because nuclear plants emit no carbon, and there are over thirty nuclear generation units in some stage of planning.[55]

President Carter delivered his third and final energy address on July 15, 1979.[56] He returned to his "moral equivalent of war" rhetoric, and Congress responded with the passage of the Energy Security Act of 1980 (ESA). The ESA, like its predecessor the NEA, consisted of several major pieces of legislation. Unlike its predecessor, however, the ESA marked a major departure from past energy policy. We can look at the ESA as the first attempt to move away from traditional energy policy even though the law was ultimately unsuccessful. The ESA was intended to promote alternative and renewable energy resources; specific provisions addressed biomass and alcohol fuels,[57] renewable energy resources,[58] solar energy and energy conservation,[59] and geothermal energy.[60] More notably, however, the ESA created the United States Synthetic Fuels Corporation,[61] which was intended to exploit geological formations containing oil shale and tar sands that could be used to create liquid petroleum as a substitute for foreign oil. The Synfuels Corp was an $88 billion failure ended by Congress in 1986 because ESA's renewable energy efforts yielded minimal results.

Neither the NEA nor the ESA resulted in a comprehensive or coordinated traditional energy policy. Nor did this legislation result in a new, smart energy policy. Superficially, it can be argued that Carter was unsuccessful in revamping energy policy because he had only one term in office. It is more likely the case that Carter's energy program did not get off the ground because it went contrary to the country's entrenched traditional energy policy favoring fossil fuels and the incumbents that provided them. His experiment in energy reform ended in 1980 with the election of President Reagan whose first two acts in office were to accelerate oil price decontrols and eliminate the Crude Oil Windfall Profit Tax Act, both to the benefit of fossil fuel interests.

The Reagan Deregulation Revolution

History often declares that the 1980s constituted the Reagan Revolution. The term is meant to denote an aggressive de-emphasis on the role of government in society, deregulation across the board, and the celebration of the ideology of the "free market." There is no question that the government was in the deregulatory mood. In fact, during the Carter years, the transportation, trucking, airline, and banking industries were all subject to deregulation initiatives. The idea behind Carter's deregulation approach was simple. The heavy-handed command-and-control regulation of the past, whether it was judged successful or not, must be market tested. Could market-based mechanisms and incentives perform better? Carter's approach to deregulation was quite simply a math problem. Can the economic case be made that less government regulation was more efficient? If so, then, deregulation would take place. Carter-like deregulation was not a knee-jerk reaction against government based on ideological sloganeering. It was a matter of applied economics. Not so for President Reagan.

President Reagan's energy policy favored the private sector, was based on supply-side thinking, and thrived on anti-government rhetoric. His intent to abolish the Department of Energy never came to pass. Nevertheless, his initiatives emphasized private markets in place of government regulation. His administration saw the end of the Windfall Profits Tax, oil price regulations, the United States Synthetic Fuels Corporation, and the effective repeal of coal conversion legislation.

During the early years of the Reagan administration, oil price decontrols had their intended effect, exactly the effect that Carter hoped for as well. Whereas higher prices increased oil exploration and production throughout the world, higher prices also served as an incentive for conservation, as well as fuel switching to lower-cost resources. By mid-decade, however, OPEC had realized that with new oil on the market, prices began falling, and they responded with production quotas. If it was not clear in the 1970s, it became clear in the 1980s that world oil markets directly threatened the U.S. economy and its national security.

Curiously, perhaps, the Carter and Reagan years tell two different versions of the same story. Carter's attempt to centralize energy policy making and coordinate energy policy failed. Yet so did Reagan's attempt to abolish the Department of Energy and leave all energy decisions to the free market. A further similarity for both presidents was the fact that their energy policies, with the exception of the Energy Security Act, concentrated on fossil fuels. The United States' traditional model of fossil-fuel-based energy policy demonstrated its resilience to change. Neither president could move away from traditional policy, except only marginally. The model resisted over reliance on the markets just as it confirmed its reliance on government support of incumbent energy industries.

Simply, the political economy of U.S. energy is based on a symbiotic relationship between government and industry. This relationship is manifest by four characteristics. First, in many sectors, energy resources are complementary so that the regulation of one resource does not necessarily adversely affect the other. Oil and electricity, for example, divide the energy economy roughly in half. Federal policy can support both segments without adversely affecting the other. Second, energy resources are susceptible to competition, but that competition has not proven disruptive to incumbents. A policy that may affect the use of coal theoretically may adversely affect nuclear power, but as long as government regulation promotes energy production and consumption, then both industries win. Third, both industry and government depend on one another for the distribution and allocation of burdens and benefits. Federal lands, for example, open up resources for private development, and private development helps fund the U.S. treasury. Finally, both business and government are stimulated by market disequilibrium. Volatile oil prices affect not only producers and refiners, they affect government responses as well. There is a vibrant interplay between government and industry, and both have relied on the sturdiness of traditional energy policy.

The End of Comprehensive Energy Policy Reform

Presidents Carter and Reagan, each in their own contrary ways, attempted to reshape U.S. energy policy across the board. Both attempts failed. Nevertheless, significant problems persisted; and in the last decade of the twentieth century and in the first decade of the twenty-first, Congress passed energy legislation that had some, but not significant, impact on the regulation of energy industries. Growing concerns about environmental impacts from energy production and distribution and worries about the growing dependency on foreign oil, together with the recognition that old-style regulation had run its course, demonstrated that new forward-thinking energy regulation was sorely needed, particularly to move energy more efficiently through the infrastructure. More particularly, the natural gas and electric industries needed to open access to producers who wished to put cheaper products on the market.

From Congressional Legislation to FERC Energy Policy

After the major energy legislation and initiatives of the Carter and Reagan years, energy policy moved out of the halls of Congress and into the hearing rooms of the Federal Energy Regulatory Commission (FERC). FERC attempted to address the deregulation and/or restructuring of the electric and natural gas industries by opening access to their distribution systems.[62] The natural monopoly regulation of both industries limited the number of actors in those markets in order to achieve a certain level of efficiency through price and profit controls. The consequent problem of this arrangement is that existing firms then become incumbents comfortable with regulation as well as interested in limiting, if not eliminating, new entrants. In both industries, the transportation segments comprised bottlenecks that inhibited increased competition from lower-cost new entrants to the economic disadvantage of consumers.

FERC Natural Gas Market Restructuring

The natural gas industry accounts for approximately 25 percent of the energy consumed in the United States, most of which is domestically produced by over 6,000 producers, then processed by over 500 plants, and distributed by 160 pipeline companies.[63] In short, although the production end of the natural gas fuel cycle is competitive, the distribution segments (i.e., the pipelines) are not and require regulatory oversight. Pipelines serve two functions. As merchants, pipelines purchase and sell electricity to consumers, including local utilities. Pipelines also serve as transmission companies charging a fee for that service.

During the 1960s and 1970s, the natural gas market became greatly distorted, first because of bad regulation and then because of onerous take-or-pay contracts between producers and pipelines. As prices rose, demand decreased and pipelines found themselves obligated to pay for gas that they could not resell. Pipelines, then, had several alternatives: (1) they could honor the contract and pay for gas that they could not use; (2) they could breach their contracts and risk litigation; (3) they could initiate litigation and attempt to alter their contractual obligations; (4) they could attempt to renegotiate their contracts with producers; or (5) they could seek regulatory protection from the federal government. Litigation was not a promising option. Litigation is not only costly, it is unpredictable. Contract renegotiation was similarly problematic because producers were under no legal obligation to do so and could simply refuse to renegotiate and hold the pipelines to their high-priced contracts. Consequently, government regulation was seen as the most promising option to correct the contractual distortions in the natural gas market.

Pipelines wanted relief from burdensome contractual obligations; producers wanted clear market signals as well as their gains from trade; and consumers wanted access to cheaper gas that was being held back from the market in the hopes of higher prices. In other words, the natural gas market was ripe for restructuring, and FERC responded with a series of rules aimed specifically at encouraging long-term contract renegotiation and opening access with the intent of promoting competition.[64]

After several years of complicated open access regulations and ensuing litigation, FERC attempted what it hoped was the final restructuring of the natural gas industry in FERC Order No. 636.[65] In brief, natural gas market restructuring has involved the deregulation of wellhead prices and the continuing monitoring and regulation of pipeline sales and rates. FERC's restructuring was based on the belief that a lighter regulatory hand can stimulate competition and stabilize prices to producers and pipelines while opening up choice for consumers. Pipeline regulation continues to be justified because of the bottleneck problem in the transportation segment.[66] Natural gas regulation continues to play an important role in energy policy as natural gas assumes a larger share of our energy portfolio and will also serve as an increasing feedstock for the generation of electricity.

FERC Electricity Market Restructuring

Whereas the natural gas market has been partially restructured,[67] it is significantly further along than the electricity market, which has been dealt crippling blows over the last two decades to the point at which restructuring efforts have not only stopped, they have been reversed. Like the natural gas market, the electricity market suffers from bottleneck problems, and the transmission segment continues to be considered a natural monopoly. PURPA revealed that the cost of producing electricity by traditional utilities was higher than competitive rates and that cheaper electricity could be produced by non-utility generators. Again, the problem was opening up the transmission segment so that consumers had access to the cheaper electricity.

Although both the natural gas and electricity industries had transportation bottlenecks, the ownership structures in the industries were different. Pipelines, for the most part, were independent of producers. The electric industry was mostly vertically integrated, and a single utility controlled and owned generation, transmission, and distribution. Because utilities owned generation and transmission facilities, they had no incentive to open their transmission lines to competitors. Nor did they have any incentive to charge competitors a reasonable transportation fee. FERC proposed to address the problem through two major initiatives.

First, FERC issued orders requiring utilities to "functionally unbundle" their transmission service from their generation and power marketing functions.[68] The unbundling was intended to reduce or eliminate anti-competitive self-dealing by utilities over the pricing of their transmission service. A utility was then required to either spin off transmission from generation or to treat them as separate corporate entities. Transmission service, then, was to be offered on a non-discriminatory basis in an attempt to create a more competitive transmission market, and it was estimated that such orders would result in approximately $6 billion of annual savings to consumers.

The new market would be facilitated by an electronic information system, which revealed prices for transmission on a uniform basis. Initially, these orders had dramatic effects on the electricity market as: (1) utilities divested themselves of either their generating or transmission units; (2) the electricity industry experienced an increase in mergers and acquisitions; (3) the number of ancillary power marketers and independent generators increased in order to facilitate transactions and sales; and (4) states began the deregulation of retail markets.

The second significant initiative involved restructuring regional utility arrangements. If the market was to behave more competitively, then regional cooperation was needed so that all the utilities in a region could efficiently participate in the transmission system that moved electricity as cheaply and efficiently as possible. FERC believed that voluntary arrangements were efficient and therefore adopted a set of rules for the voluntary establishment of regional transmission organizations (RTOs) driven by independent system operators (ISOs).[69] Basically, the RTO operates independently of generation facilities and manages the transportation system either as a for-profit or a non-profit entity. The RTO plan promised to relieve stress on the bulk power system, better assess capacity needs, engage in regional coordination, facilitate transmission improvements and expansion, and promote greater reliability.

Regional differences, particularly relative to costs of producing electricity, make it difficult for all of the utilities in a region to participate on an equal footing. FERC attempted to both further industry restructuring and smooth out market discontinuities through a rulemaking proceeding known as standard market design (SMD). The purpose of SMD was to promote reliable and reasonably priced electricity and transparent markets through fair rules of participation for producers, consumers, shareholders, and regulators.[70] Eventually, problems of federalism and regional cost disparities caused the termination of the SMD program.[71]

FERC's jurisdiction was aimed at the wholesale level. To complete industry restructuring, state regulators adopted similar divestiture regulations in an effort to promote consumer choice for electricity in the same way that consumers could choose their telephone service. Neither set of regulations have worked. Just as the federal regulations stalled, state deregulation efforts imploded. The summer of 2000 witnessed an energy crisis in California caused, in no small part, by the manipulative practices by Enron traders at their trading desks.[72] California's electricity crisis and a 2003 brownout in the Northeast rippled throughout the country. The industry experienced utility bankruptcies, and state restructuring efforts halted. In fact, several states are going back to the old form of rate regulation as deregulation efforts have resulted in increased consumer prices.[73] FERC restructuring efforts were intended to lighten regulation and support the electric and natural gas industries. They were not directed at an overhaul or restructuring of traditional energy policy. Instead, if successful, electric and natural gas industry restructuring would have promoted the use of fossil fuels because the costs to consumers would have declined, thus increasing demand.

Energy Regulation at the End of the Twentieth Century

The Energy Policy Act of 1992 (EPAct 1992) was the next major legislation intended to "gradually and steadily increase U.S. energy security in cost-effective and environmentally beneficial ways."[74] The legislation was not comprehensive and did not have that effect. Instead, EPAct 1992 smoothed out some of the rough bumps in the previous legislation by loosening controls of the Public Utility Holding Company Act; encouraging new entrants in the electricity market; opening access to the electricity grid to non-utility generators; encouraging state regulatory agencies to think about integrated resource planning; and providing subsidies and tax credits for electric vehicles as well as alternative and renewable resources. In addition, EPAct 1992 continued to subsidize traditional fossil fuels by deregulating imported natural gas and liquefied natural gas and by providing tax relief for independent oil and gas producers. In addition, the act expanded tax relief and tax credits for the production of oil from shale, tar sands, and other non-conventional fossil fuel resources.

The coal industry was also favorably addressed through R&D funding for advanced clean coal technologies; relief for coal companies under the avoided cost provisions of PURPA; and relief under the Surface Mining Control and Reclamation Act, an environmental protection statute that was more honored in the breach than in the enforcement. EPAct 1992 favored the nuclear industry by streamlining the licensing procedures of the Nuclear Regulatory Commission and creating the United States Enrichment Corporation, which was a government-owned entity created to provide uranium enrichment services in the international market.

Following President George H. W. Bush's EPAct 1992, Presidents Bill Clinton and George W. Bush also addressed alternatives in their energy policies. During the Clinton administration, the Department of Energy issued a National Energy Plan in 1995, a Strategic Energy Plan in 1997, and a Comprehensive National Energy Plan in 1998. Each of the documents intended to promote energy production and efficiency, national security, environmental sensitivity, and sustainability. Similarly, Vice President Dick Cheney chaired the National Energy Policy Development Group, which issued its National Energy Policy in May 2001.[75] The subtitle of the report captures its main theme as "Reliable, Affordable and Environmentally Sound Energy for America's Future." Whereas the rhetoric of the Clinton and Bush years clearly evinced an awareness that the language of energy policy needed to change, particularly relative to national security and the environment, in reality this legislation was tilted heavily in favor of traditional industries and fossil fuels and did little more than pay lip service to alternative and renewable energy resources.

Energy Regulation in the Twenty-First Century

The NEA, ESA, and EPAct 1992 were the precursors to the first energy legislation of the new century – the Energy Policy Act of 2005[76] and the Energy Independence and Security Act of 2007[77] – each was strong on rhetoric and moved energy policy slightly away from the traditional model but kept that model, and the industries that it supported, firmly in place.

The Energy Policy Act of 2005 (EPAct 2005)

The Bush administration's 2001 National Energy Policy was based on projections that indicated a shortfall of nearly 50 percent of our energy needs by the year 2020 and stated that those needs could be satisfied by increasing the supply of energy and improving the nation's energy infrastructure. These goals could be achieved through more oil refineries, more natural gas pipelines, and an improved electricity grid. Nuclear power should be further promoted, and federal licensing of energy facilities should be facilitated and streamlined. The policy also contained provisions for limited tax credits for alternative and renewable resources and technologies. Basically, however, the policy stayed committed to incumbent industries and to the private production and distribution of energy.

The themes and principles of the National Energy Policy formed the foundation for EPAct 2005. In fact, Energy Secretary Samuel Bodman stated that nearly 75 percent of the National Energy Policy recommendations had been implemented in that law.[78] The Policy and EPAct 2005 continued the tradition of designing energy policy based on an overriding and fundamental assumption – the more energy that is produced, the healthier our economy will be.[79] This fundamental assumption will be explored in more detail in the next chapter. What is important, however, is what follows from that belief. If there is a direct and positive correlation between energy production and consumption and economic growth, then energy policy should support production and consumption as fully as possible.

Consequently, favoring traditional and incumbent firms makes sense, and EPAct 2005 continued that policy.

The most significant portion of EPAct 2005 addressed the electric industry. Most particularly, the act directed FERC to issue a rule on electric reliability standards in order to manage congestion and avoid blackouts. The act also repealed the Public Utility Holding Company Act of 1935, which restricted utility ownership and operations and allowed for greater forms of ownership by utilities. Additionally, PURPA was amended to reduce the mandatory purchase and sale obligations that utilities owed to small power production facilities and to co-generators. Price-Anderson Act liability limitations were extended for twenty more years for nuclear power plants, and R&D funding was authorized for advanced nuclear reactor designs, as well as for the construction and operation of prototype nuclear power plants.

The coal industry also benefitted from EPAct 2005, which simplified the leasing of federal lands and authorized $1.6 billion of funding for clean coal power initiatives and other projects. In addition, the Secretary of the Interior was directed to undertake a comprehensive survey of oil and natural gas reserves on the Outer Continental Shelf and review offshore leasing practices with a view toward streamlining them. The act also provided a series of royalty incentives as well as production incentives and tax credits for traditional fossil fuel industries and for the development of oil shale and tar sands projects. The natural gas industry was addressed by giving FERC the exclusive authority to permit new liquefied natural gas (LNG) terminals with the express intent of reducing federal-state conflicts.

EPAct 2005 did address renewable and unconventional resources as well as conservation. The act provided limited production credits and tax credits for solar, wind, biomass, and geothermal resources, and it required federal buildings to use renewable technologies. Conservation was promoted through the use of clean efficiency standards for buildings and appliances. Additionally, R&D for hydrogen, fusion, and fuel cell technologies was provided.

Thus, EPAct 2005 has provisions for traditional as well as alternative energy resources, yet traditional resources are significantly favored. The funding provisions of the act are estimated at $14.5 billion, and according to a House of Representatives report over $4 billion is earmarked for the oil industry, $3 billion to the coal industry, and over $5 billion to the nuclear power industry.[80] Other studies indicate that the traditional energy industries would receive approximately $12 billion in taxpayer subsidies with approximately $3 billion directed at alternative and renewable resources.[81]

In brief, EPAct 2005 continued traditional fossil fuel policy without coordinating energy policy,[82] while continuing to favor incumbent fossil fuel energy industries, which reported record profits, and gently nodding to alternative or renewable resources. Incumbents were favored over new entrants, environmental restrictions and protections were lessened on fossil fuel companies, and global warming pollution increased. The act failed to reduce dependence on foreign oil, failed to reduce oil consumption, and failed to reduce energy costs. Indeed, early twenty-first century energy legislation again reaffirmed the resilience of the traditional model of energy regulation.[83]

Energy Independence and Security Act of 2007

The 2006 congressional elections signaled a change of party control on the Hill. The Democrats gained majorities in both houses, and President Bush's popularity continued to decline. Nevertheless, the economy was weak, and the belief continued that energy production could help revive it. President Bush addressed U.S. energy policy in his 2006 State of the Union message and in his 2007 message. In both, he promoted renewable resources and clean coal. However, going inside the numbers revealed that investments in renewable resources were not substantial. The president recommended $236 million to be used for clean coal projects, which, at the time, constituted one-fortieth of Exxon's 2007 quarterly profit of $10.8 billion. His recommendation for new battery development was an astonishing $6.7 million. The president's 2007 State of the Union message was slightly more bold, recommending that we reduce gasoline consumption by 20 percent over the next ten years and that we do so by setting a mandatory alternative fuel standard to require the use of 35 billion gallons, annually, of renewable and alternative fuels by 2017. For the remainder of his term, little happened to advance the third energy transition.[84]

The Energy Independence and Security Act of 2007 (EISA) was signed into law in December 19, 2007, with the stated purpose of greater energy independence and security through the production of renewable fuels, increased efficiency in buildings and appliances, R&D for greenhouse gas capture and storage, as well as other measures. The bill originally sought to cut subsidies to the oil industry, but these tax changes were ultimately dropped. Instead, the bill focused on increasing corporate average fuel economy (CAFE) standards, which were raised for the first time in over twenty years to 35 miles per gallon.

EISA was intended to promote renewable, alternative fuels. In addition to promoting energy independence and increasing energy efficiency, the legislation supported accelerated research and development for solar energy, geothermal energy, and energy storage, among other provisions including carbon capture and sequestration.[85] The act, however, may be more significant for what it did not do. The bill failed to adopt federal renewable energy portfolio standards and failed to significantly reduce fossil fuel tax subsidies. Overall, though, EISA continued the traditional support of fossil fuels by expanding carbon capture and storage activities and investing in electricity transmission grid improvements.

Awaiting a New Energy Policy Era

With the election of President Barack Obama in 2008, the potential exists for a meaningful change in energy policy. President Obama brought together what is sometimes referred to as the "green dream team," comprised of Nobel laureate Steven Chu as secretary of the Department of Energy; Carol Browner in the White House Office of Energy and Climate Change Policy; Lisa Jackson as administrator of the Environmental Protection Agency; and Ken Salazar as secretary of the Department of the Interior. Each of these leaders has had extensive experience with alternative energy policies. Events since the election, particularly the environmental catastrophes with which this book opens, have not been particularly helpful in effecting a transition. Still, some signs are promising.

One of the president's first acts in office was to secure the passage of the American Recovery and Reinvestment Act,[86] also known as the Stimulus Bill. Several provisions of the bill were dedicated to building a new energy policy. Several billion dollars were directed for investments in smart grid technologies; additional billions were to be made available as loan guarantees for renewable energy and transmission; renewable energy production tax credits were expanded; efficiency programs were funded; and additional research and development dollars were allocated. Many of these efforts will be discussed later in this book.

In addition to the Stimulus Bill, climate legislation has been introduced. The American Clean Energy and Security Act of 2009 (ACES), also known as the Waxman-Markey bill, is the first effort to noticeably move away from the traditional model of fossil fuel protection.[87] As reported by the Congressional Research Service, this legislation is directly intended to reduce carbon emissions through multiple tracks. The electricity industry is directly affected because a federal renewable portfolio standard is proposed in which 20 percent of the nation's electricity will be generated from a combination of energy efficiency savings and renewable energy by the year

2021. To help achieve these goals, the bill proposes that renewable portfolio standards be facilitated through a market in which renewable electricity credits will be traded and a national standard for energy efficiency and conservation will be established. Additionally, smart grid development is furthered by developing a federal policy of electricity grid planning and capability assessment.

Fossil fuel industries are also addressed as the bill requires a national strategy for carbon capture and sequestration and is to facilitate those efforts through a Carbon Storage Research Corporation. The proposed legislation also provides for the development of an electric vehicle infrastructure as well as financial assistance for the manufacture of plug-in electric vehicles. Further support is provided by the establishment of Clean Energy Innovation Centers that are directed to promote the commercial development of clean, domestic energy alternatives to fossil fuels; reduce greenhouse gas (GHG) emissions; and ensure that the United States maintains a leading position regarding state-of-the-art energy technologies. ACES continues to support the development of energy conservation and efficiency standards for buildings and appliances and establishes within the Environmental Protection Agency a Smart Way Transportation Program linking energy and environmental issues with transportation policies.

The core of the proposed legislation is a cap-and-trade program, which sets targets for carbon emissions reductions and establishes the carbon market for trading emissions permits. The carbon market will require an oversight agency, and permits will be auctioned off as revenue measures to further fund the development and promotion of renewable alternative energy technologies and resources.

Additionally, in 2010 the American Power Act, also referred to as the Kerry-Lieberman bill, was introduced.[88] The proposal centers on job creation through building a clean energy economy. The bill addresses carbon pricing through a cap-and-trade regime. Support is also provided for state programs promoting energy efficiency and renewable resources. Funding is available for clean transportation initiatives, natural gas vehicles, and nuclear power. The Clean Air Act is scheduled to increase performance standards for coal-fired electric generating plants, and the natural gas industry is supported among many other provisions.

Conclusion

Since the 2010 midterm elections, the proposed legislation is likely dead in the water. Whether the proposals were ambitious enough will continue to be the subject for debate, and passage before the next presidential election seems highly unlikely. Nevertheless, these proposals mark the fact that the energy conversation may well have changed permanently. As a nation, in the near term, we may not see a dramatic transition away from our past fossil fuel policy. Yet we cannot speak of energy policy without addressing carbon emissions and an alternative policy of clean, smart energy. A full transition appears unlikely largely because of how entrenched traditional energy policy interests have become, not only in U.S. corporate boardrooms, but also on the lending desks of major financial institutions and in the halls of Congress. The country is yet to shed its belief that fossil fuel energy production and consumption are integral to economic growth. A belief that is confirmed as new sources of oil, natural gas, and coal are discovered.[89] The assumptions of traditional energy policy run deep, and they must be revised before the third energy transition from fossil fuels to energy efficiency and renewable resources is possible. The next chapter delves more deeply into the policy assumptions behind our traditional energy path and presents an alternative set of economic and policy assumptions on which to base a transformative energy future.

2

Protectionist Assumptions

I want to remind you about the fact that this economy of ours has been through a lot. And that's why it was important to get this energy bill done, to help us continue to grow.... This economy is strong, and it's growing stronger. And what this energy bill is going to do, it's just going to help keep momentum in the right direction so people can realize their dreams.

President Bush Signs Energy Policy Act of 2005[1]

The history of energy policy reveals the tight relationship between government and industry. Over the course of the last 100 years, government regulation supported an energy industry that developed certain structural characteristics. In turn, regulatory agencies adopted a similar organizational structure. Called the "hard path" by Amory Lovins more than thirty years ago,[2] hard path energy industries were privately owned, corporate, large-scale, capital-intensive, centralized, and regionally, nationally, or internationally operated. The oil, natural gas, coal, nuclear power, and electricity industries each possess those characteristics. Additionally, federal and state regulators mirrored the corporate structure and are highly centralized, operate over large jurisdictional areas, and are departmentalized to mimic the corporate operations of their regulatees.

Energy industries are large-scale. Internationally, nearly 80 percent of the oil produced in the world is produced by fifty state-owned and private oil companies.[3] In recent years, privately owned oil companies have experienced significant mergers, such that five of the remaining super majors are all ranked in the top eleven private corporations in the world.[4] The United States is served by approximately 100 major electric utilities, and electricity is produced by 104 commercially operated nuclear power plants.[5] It is estimated that there are one trillion tons of recoverable coal in the world, and the United States controls more than 25 percent of that world estimate.

Coal remains abundant and cheap, and Big Coal's influence is pervasive in our political economy.[6]

Consistent with their size, energy industries are heavily capital-intensive. Recent estimates indicate that a new nuclear plant can cost between $6 billion and $12 billion.[7] Although less costly, coal plants themselves can cost $2 billion to construct.[8] It has also been estimated that investment for upgraded electricity transmission lines runs into the trillions of dollars.[9]

Such large-scale, capital-intensive industries also exhibit other characteristics. In order to benefit from economies of scale, energy industries are national and international in scope and are centralized. To the extent that oil, natural gas, and electricity are network industries, they rely on interstate and national transportation systems instead of operating locally. Their corporate structure requires them to operate hierarchically; and because they are privately owned, energy firms are driven by short-term returns to shareholders. Combined, 85 percent of the energy economy is comprised of fossil fuel companies.

The regulatory structure for energy markets mirrors the industrial structure of the energy firms that they regulate, thus forming a mutually supportive industrial/regulatory complex. Although regulation takes place at both the federal and state levels, energy regulation is either centralized in Washington D.C. or in state capitals. Like the energy industry itself, regulatory agencies, most often as a result of legislative directive, are not coordinated, and they compete for federal funding each budget cycle. Oil, for example, is regulated by the Department of Energy (DOE), the Department of the Interior, and the Federal Regulatory Energy Commission (FERC), as well as several other agencies and departments. Regulatory agencies are also internally divided according to task. FERC, as an example, regulates natural gas, electricity, and hydropower separately. Similarly, the DOE has divisions for oil, electricity, energy efficiency and renewable resources, energy R&D, clean coal technologies, and the like. In the private sector, such divisions promote competition and, at least theoretically, efficiency. In the public sector, they produce inefficient administration and policy fragmentation to the benefit of industry.

Additionally, energy regulation is highly politicized. As discussed in this chapter, free-market energy is something of a misnomer. Private firms cull favor with regulators; regulators look to industry for post-government jobs; and both regulators and regulatees avidly watch the stock market and election results. Whereas privately owned companies pay attention to quarterly returns to shareholders, political appointees pay attention to election returns and energy policy changes accordingly. The result of this mixed form of political economy has been an energy policy that has supported and sustained select energy industries.

The criticism here is directed less to the past than toward the future. In the past, the regulatory/industrial complex has benefitted producers as well as consumers; it has also benefitted regulators. Consumers have grown familiar with these industries and have grown fond of reliable, cheap energy. Producers enjoy the profits and the competitive advantages of incumbency. Additionally, regulators understand the businesses that they regulate so that both their government jobs and their post-government employment involve familiar work. The problem is that this coziness comes at too high a price. The assumptions and the policy goals that have designed and driven both industry and its regulators must be reassessed for a sustainable energy future.

The transition from our traditional fossil fuel-based energy policy to a smart energy path that is sensitive to environmental needs, incorporates new technologies, relies on an increasing diversity of resources, and moves away from centralized, large-scale power production will not be easy.[10] Yet, today we must answer the question: Are we now at a point where the old economic assumptions and policies no longer serve our energy and economic needs? To elaborate: Can we create an energy portfolio that is decentralized, scaled-to-task, and less environmentally harmful while contributing to a healthy and growing economy? Climate change, oil dependence, and energy insecurity each suggest that we do not have the luxury of debating the niceties of these questions. Rather, we can and must develop new approaches. Fortunately, there are signs of new energy thinking to be discussed in the next chapter. Unfortunately, traditional energy politics have a firm grip on our current policies.

Traditional energy policy is based on set economic and policy assumptions to be discussed in detail in the next section. For argument's sake, we can recognize and accept the fact that each of these assumptions was true – to a point.[11] The problem of reaching that point is simply stated: The country cannot continue to pollute at the levels that we are experiencing and not account for their negative economic effects.

The energy industry/regulatory complex has stultified policy. For nearly fifty years, we have bemoaned our reliance on foreign oil, yet we increasingly rely on imports. For decades now, we have bemoaned the dangers of carbon emissions, yet we continue to burn fossil fuels.[12] We continue to plan on nuclear power[13] despite its high cost, safety concerns, and unsolved backend waste problems.[14] Thirty years ago, Amory Lovins warned of the long-term negative consequences of relying on a hard path energy policy built on supporting energy firms of the sort just described: "The commitment to a long-term coal economy ... makes the doubling of atmospheric carbon dioxide concentration early next century virtually unavoidable, with a prospect then or soon thereafter of substantial and perhaps irreversible changes in global climate."[15]

If a generation ago energy policy analysts could envision exactly the climate change challenges we face today as a result of the human contribution to carbon emissions, then why has it been so difficult to design and implement a forward-thinking, more environmentally sensitive energy policy? In part, change is difficult, particularly after 100 years of capital investment in energy production, transmission, and distribution as well as after 100 years of government support. Such investment by the public and private sectors clearly created a path dependency that makes energy business as usual seem a wise course. However, the reasons go deeper than a fear of change or a dependency on the past. Our traditional energy program was based on a set of economic and policy assumptions that may have worked in the past but must be jettisoned if we are to have a more prosperous future.

Traditional Economic and Policy Assumptions

A low-carbon energy policy and climate change regulation require new thinking about energy and environmental policies, the application of new technologies, and the opening of new energy markets. The primary barrier to a rapid expansion of new technologies and new markets is that the economic and political playing fields between fossil fuels and renewable or sustainable alternatives, including conservation and increased energy efficiency, are not level. Fossil fuels have enjoyed and continue to enjoy the fruits of a political economy that has been built and supported by those industries throughout the twentieth century. Fossil fuel development is also based on a set of static world economic and policy assumptions that are no longer operable.

The traditional energy model holds dearly to five economic and policy assumptions, all of which can be challenged and must be rejected. The first assumption is revealed in the quotation by President Bush that opened this chapter – the more energy that we use, the stronger our economy will be. Second, we assume that Bigger is Better and its corollary Cheaper is Good. Third, we believe that private markets and private investments are key to sustained economic growth. Fourth, energy productivity and consumption is, and should be, separate from environmental protection and conservation. Finally, as stated in a report commissioned by Bush Energy Secretary

Samuel Bodman from the National Petroleum Council – we behave as if fossil fuels are indispensable to future energy growth and demand.[16] In short, our historic national energy policy depended on oil, gas, and coal. It also depended on hydropower and nuclear power. All of these energy industries follow the hard path. Thus through 2010, energy efficiency, renewable and alternative resources, and conservation have been given lip service in our national energy priorities and have received greatly disparate and unequal financial support.

The dominant model of energy policy was developed on these five assumptions with the following goals: (1) provide abundant supplies at reasonable prices; (2) limit the market power of monopoly firms; (3) promote inter-fuel competition; (4) support conventional, predominantly fossil, fuels; and (5) treat government regulators, especially federal regulators, as partners for designing and supporting the traditional path. These assumptions have led to a fossil fuel economy with its carbon emission and oil dependency dangers. More problematic, though, this century-old protectionist policy has revealed the lack of political will to change and improve not only our energy policy but also our prospects for a healthier economy.[17] These assumptions, however, are not without past merit. Together they have led to the creation of the national energy infrastructure and, for most of the twentieth century, a healthy economy, the security of the United States in the world, and its position as a world leader. None of these assumptions, nor their apparent beneficial consequences, hold in 2011. Instead, we have reached the point at which the assumptions must be replaced in order to improve our economy and regain our country's world stature.

The Energy-Economic Productivity Link
The fundamental economic assumption behind traditional energy policy is that there is a direct and positive correlation between energy consumption and economic growth. The assumed energy-economy linkage means that the more energy we consume, the healthier our economy would be. Strong claims have been made for the relationship. For example, authors Peter Huber and Mark Mills write, "[I]n America, [as opposed to Western Europe] we still pursue energy. And because we use the most, we are the most productive and the most powerful." And, "The paramount objective of U.S. energy policy should be to promote abundant supplies of cheap energy and to facilitate their distribution and consumption."[18] Julian Simon, in an influential book written to debunk *Global 2000 Report to the President*[19] written for President Jimmy Carter, writes: "The prospect of running out of energy is purely a bogeyman. The availability of energy has been

increasing, and the meaningful cost has been decreasing, over the entire span of humankind's history."[20] Additionally, Fred Singer, writing in favor of free markets says: "A free market is well suited to supply energy, because energy resources are owned, either by individuals or by corporations.... Under competitive conditions, management for individual profit also benefits the general population."[21] All of these authors are referencing our fossil fuel economy.

To some extent, these quotations can be interpreted as caricatures of extreme pro-market, anti-government rhetoric. Any sophisticated economic analysis, for example, can maintain a pro-market position; yet to be true to the neoclassical market model, the analysis must also recognize the costs of pollution in order to arrive at accurate price signals. In other words, sound economic analysis acknowledges the existence of market failures and also acknowledges the role of government in correcting them. In fact, the argument made throughout this book is that, ideally, energy markets should operate as freely and as unconstrained by government regulation or government support as possible.

However, that ideal goes unrealized for two reasons. First, to date, traditional energy markets have not operated and do not operate freely. Instead, government regulation has significantly distorted them. Second, energy markets contain significant economic imperfections including negative externalities and monopoly power. Consequently, policy and regulatory corrections, including shifting government subsidies away from the hard path and toward energy efficiency and renewable resources, must be made to move our energy economy in a more reliable, competitive, and sustainable market direction.

We cannot be too glib, however, and simply write off free-market rhetoric as sloganeering or as a caricature because of two problems. The first problem lies with public discourse. Most public discussion, and too much of the academic conversation, treats markets and government as separate spheres and tends to favor either market or government over the other without recognizing their interconnections. This narrow view of our political economy is detrimental to consumers as well as producers because competition is constrained and choices are limited. Instead, we must be clear about what markets can do and be clear about their limits. Under the present regulatory scheme, energy markets have been created in which consumers have fewer energy alternatives and new, as well as potential, energy producers encounter economic barriers to entry. Further, unconstrained fossil fuel markets cause environmental and health harms that have not been incorporated into the price of energy. Markets do work, and they have economic benefits. However, those benefits can only be realized as long as a proper and efficient balance between government and markets is struck.

The second problem with the free-market rhetoric lies in the history of regulation. There has rarely been a time in our nation's history in which government was absent from energy markets. In capsule form, we can argue that fossil fuel industries have been treated well by government and that consumers have fared well until the last third of the twentieth century. Looking into the twenty-first century, however, future energy markets will function neither smoothly nor efficiently unless traditional energy policies are dramatically reformulated by incorporating environmental costs to more accurately account for economic gains. In other words, past and current regulations send distorted market signals that must be corrected.

In general, all economic activity has its costs. In particular, energy productivity produces environmental harms that must be accounted for in the price of energy products. Certain consequences follow from the recognition of negative externalities. First, there is an optimal level of pollution because energy industries are not pollution free. In other words, energy industry pollution must not be overproduced. Instead, it must be limited by including all social costs in its production. Although we might posit that the optimal level of pollution might be achieved in free competitive markets, history indicates that such is not the case primarily because private actors do what they can to avoid incorporating pollution costs into the price of energy in order to maintain their own competitive advantages. Consequently, government regulation must be relied on to draw the line as close to the optimum point as possible.

The second lesson is that government regulators, largely because they are political actors, will draw the line imperfectly and either allow too much pollution or impose too many costs on business. Recognizing the inherent imprecision of regulatory solutions to economic problems, however, does not lead to the abdication of government responsibility for improving the general welfare. If private actors insufficiently protect health and the environment, then government must play its proper role.[22] Thus far, government has failed in this essential mission.

The two flaws with the first assumption of a direct and positive correlation between energy and the economy, then, are: (1) the failure to acknowledge that unchecked energy growth contributes to excessive pollution and (2) the failure to acknowledge that energy markets, historically, have not operated as competitive microeconomic theory posits. Instead, energy markets have been distorted by political favoritism and, therefore, operate suboptimally.

Bigger is Better

This second assumption is based on the idea that economies of scale can reduce costs and, through increasing sales, profits will soar. It is cheaper per barrel to produce 10,000 barrels of oil rather than one barrel because exploration and drilling costs are spread more widely. Energy industries have enjoyed economies of scale for most of the twentieth century to the enjoyment of producers and consumers alike. All of the fossil fuel companies, as well as electric utilities, found it quite profitable to move from local to national production and distribution. As they did so, the size of production facilities increased and unit costs decreased. Consumers have enjoyed flat or declining energy prices for most of last century. However, neither the cost of production nor, following the cost of production, the price of energy to consumers can continue to stay flat or decline indefinitely. Oil is more costly to extract and, if we are serious about carbon emissions, then electricity will also be more costly.

Consumers have seen energy prices increase over the last few decades as producers have experienced increased production costs. In 2008, gas at the pump hit record highs in the United States.[23] The many-fold escalation of nuclear power construction costs have resulted in plant cancellations and conversions at a significant cost to consumers as well as to shareholders.[24] The promised reduction in gas and electricity prices through restructuring could not be sustained.[25] The main culprit for the problem of increasing energy costs is the continuation of the traditional model of energy policy and its regulation, which has the effect of limiting the number of actors in the market by protecting incumbents and impeding new entrants. Bigger may be better to a point, but limiting competitive markets artificially constrains the development of energy industries.

Private Markets and Private Investment

The next economic assumption is that private markets and private investments are efficient and, therefore, are the preferred method for social ordering and for determining which goods and services are bought and sold in markets. Private ordering also determines the "desirable" distribution of those goods and services. In true Adam Smith style, the preference for private market transactions is based on the notions that wealth is created, innovation is encouraged, and that individual liberty and equality are maximized because all people cast the same votes in the form of dollars.

More often than not, the rhetoric in favor of private markets and private investment tends to be code for anti-government political sentiments. The

Great Recession of 2008–2010 has given lie to the dogmatic belief in self-correcting, efficient markets.[26] More importantly, even with its distorting effects on energy markets, government regulation is a necessary element in any modern economy. Without government regulation, our air would be dirtier, our lakes and rivers would be more polluted, and we would suffer greater health risks.[27] Private markets and private capital are necessary elements; they are not, however, the only elements shaping society; responsive government regulation has its necessary role to play.

The Separation of Energy and the Environment

In one sense, energy law and policy and environmental law and policy began only a generation ago. Environmental law and policy started with the passage of the National Environmental Policy Act in 1970 and associated legislation. Energy law, as such, began as a result of the energy crises of that decade. Both disciplines had their predecessors. Energy companies were regulated by utility laws, and the environment was regulated by property and tort laws. Nevertheless, in the 1970s, a significant amount of federal legislation was passed addressing both fields.

Curiously, however, each field developed independently of the other and developed their own vocabularies, languages, and goals regardless of their obvious interconnections. Environmentalists were concerned about conservation and protection of their specific interests. Air, land, water, and endangered species, as examples, were each governed by particular legislation under the control of specified administrative agencies. The language of the environmental movement was developed by the naturalists of the nineteenth century[28] as modernized by such books as Rachel Carson's *Silent Spring* and Aldo Leopold's *A Sand County Almanac* among many others. In the early years of the environmental movement, absent from their language was any discussion of economic trade-offs between the need for energy and the protection of the environment or of cost-benefit analyses of environmental laws.

The field of energy traveled in a decidedly different direction. Energy was about the exploration, extraction, and production of natural resources to be used either as an input for manufacturing or transportation or for residential creature comforts such as heat, lighting, cooling and, now, increasingly, to drive small-scale consumer technologies. Again, the key driving force behind energy was economic growth, and environmental laws simply added an unwanted layer of costs and imposed unnecessary regulatory burdens. Absent from the language of energy, in the early years, was any awareness of the social costs of pollution.

In short, energy lawyers and policy analysts and environmental lawyers and policy analysts not only operated independently of one another, they not infrequently held one another in disdain. Further, they failed to generate a common language. At the extreme, there was no environmental regulation that an energy advocate could respect; nor, for an environmentalist, was there any economic cost too high to be paid by industry for its pollution or failure to protect a natural resource. A $100 million hydroelectric dam could not be stopped, even in the United States Supreme Court, by some unknown species called a snail darter until later legislation.[29] What is most curious about the separation between energy and the environment is that they are directly and inextricably linked along the fuel cycle from exploration to distribution, to consumption, and then to waste. A smart energy policy requires the elimination of the separation between energy and the environment.

A Fossil Fuel Economy

All of the previous four assumptions culminate in the last. Those assumptions combined to generate an energy policy based on oil for transportation, coal for electricity, and natural gas for industrial and residential uses. Together with nuclear power, these energy resources constitute approximately 95 percent of the energy consumed in the United States. More troublesome, however, is that the industrial regulatory complex has been resistant to change, and this resistance has continued through the first decade of the twenty-first century in favor of these industries. The country, of course, will continue to need energy for transportation and electricity for industrial, commercial, and residential users. The country, however, need not rely on this traditional fossil fuel mix nor need it rely on the traditional structure of energy industries for its delivery nor on traditional administrative agencies for its regulation.

Each of these assumptions is, again, sound – to a point. These economic ideas assume a static economy in which current resource use is deemed "efficient" simply because it is most immediately profitable. Further, the assumptions led to the creation of energy industries and a supporting regulatory culture that follow the hard path. Exploiting abundant fossil fuels makes economic sense, so the thinking goes, because they are available and because, for most of last century, we ignored the social costs of pollution.

What is today problematic about each of these assumptions is that they no longer operate as smoothly as they once did. The link between energy and the economy is problematic as we will see shortly. Bigger is not necessarily better as the iPod culture demonstrates. My guess is that most consumers prefer iPods to high-fidelity, stereophonic music systems. Small may be better as well as beautiful.[30] Moreover, free private markets in fossil fuels are mythical to the extent that oil, coal, and natural gas enjoy broad-based government support.

Assumptions for a New Energy Policy

The challenge of climate change is the most dramatic reason for developing a new set of assumptions for a future energy policy. However, other reasons exist. A new energy policy promises to be more competitive by increasing the number of actors in the energy market. Additionally, a new energy policy will change the nature of government regulation and, if properly designed, will reduce fossil fuel favoritism and level the economic playing field for all energy producers and consumers. The new assumptions are not purely mirror images of the past but are intended to generate the third energy transition, which will affect not only our attitudes about energy but the way we conceive of its production, distribution, and use.

Less is More – The Energy-Environment Curve

The core idea behind the dominant energy policy is that there was a direct energy-economy link so that the more energy we use, the healthier our economy will be. As we continue to pollute and consume the commons, we cannot expect continued economic growth. Jeffrey Sachs writes that "[a]nthropocentric climate change is the greatest of all environmental risks" and will "impose catastrophic hardships on many parts of the world." More to the point, he writes, "Markets alone, on a business-as-usual path, will not carry us to safety."[31]

Instead of perpetual economic growth, there comes a point at which past policies must confront new realities and must change accordingly. We cannot continue to burn fossil fuels and expect to skip out on the environmental check when it becomes due. We will, of course, need more energy as the world population grows and as countries develop advanced economies. Nevertheless, we cannot continue to use energy in the same ways; we must harness more of the potential energy in all resources while turning to resources that generate less pollution.

Continued allegiance to the traditional assumption that our economy grows with increased energy production and consumption is foolhardy, dangerous, and inefficient in the not-so-long run. The traditional assumption is that there is a direct positive correlation between the two variables of energy and the economy such that a straight-line graph can be drawn with economic growth increasing with energy production and consumption. What is painfully missing from that two-variable equation, however, is the cost of the environmental consequences of the hypothesized relationship. The assumption has no place for the negative externalities and, therefore, assumes that they have no economic consequences whatsoever. In short, the traditional economy-energy link assumption distorts any energy market because it ignores all environmental costs of production.

If environmental consequences are incorporated into the energy-economy relationship, two possibilities present themselves. Either continued economic growth will continue to create environmental harms, or economic growth will lead a society to reduce them. Policy analysts and energy thinkers have taken both positions over the last generation. The notable and much debated Club of Rome Report, *Limits to Growth*, argued that continued and unchecked economic growth will lead to collapse.[32] Critics of the report argued that their models were defective and that the best hope for a healthy environment was a healthy economy found through free markets.[33] Today, the new energy thinking notes a convergence in these polar positions. The choice is not between energy or the environment. Instead, the choice is between traditional energy policy and a smart policy of low-carbon, sustainable energy together with a healthy environment and a robust economy.[34]

The relationship between energy and the economy, when environmental costs are properly incorporated, does not resemble a straight-line graph that rises continually to the right, as the traditional assumption posits. Instead, the more realistic energy-economic relationship resembles a flattening S-curve:[35]

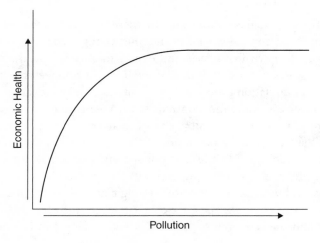

The curve flattens because no society can continue to consume its natural resources, produce harmful wastes, and expect the economy to expand indefinitely.[36] The more realistic curve will either flatten or, more likely, dip down revealing a weakening economy if environmental costs are divorced from energy prices.[37] Today's high level of carbon emissions indicates that we are at the point at which a fundamental choice must be made: We can either continue with the traditional model and expect that pollution will continue, or we can take steps to formulate a new policy to reduce pollution. Embedded in that choice is a gamble: Which choice will lead to a healthy economy?

The traditional model holds that the economy can thrive with more fossil fuel energy production. All that government needs to do is to get out of the energy regulation business and let smart, self-interested businessmen trade in markets.[38] This is particularly true for those who believe that coal, oil, and natural gas are indispensable to our country's energy/economic growth and demand.[39] This free-market approach to energy policy would be correct except for two things. First, fossil fuel firms are the quintessential political operators who have successfully won government access and favors. The second error is the failure to incorporate the social costs of pollution in the price of fossil fuels. A recent study conducted by the National Academy of Sciences, for example, estimated that total carbon pollution damages exceeded $120 billion for 2005 alone.[40]

This proposal for a better energy future is not anti-market. However, no energy policy can proceed without recognizing two dimensions of the market model that get short shrift under the traditional assumption. First, future energy policy will incorporate the costs associated with the tragedy of the commons. Energy consumers, no less than upstream factories that pitch their waste into rivers and creeks to the detriment of downstream users, cannot freely use the environmental commons of air, water, and land.[41] The related problem with the traditional market assumption involves the definition and identification of public goods. By definition, a public good is one that private markets will not supply in enough quantity or will not supply at all. Police and military protection and public education are good examples of public goods. Private markets may provide some of these services. Private education is robust and there are private police forces in select communities, but without the government contribution, these services would be under-provided and many citizens would be underserved.

Traditional energy policy treated inexpensive, abundant fossil fuel energy, in effect, as a public good. Regulations assisted the exploration and provision of these resources, and the market responded favorably for more than a century. Fossil fuel energy was inexpensive precisely because all the costs of production were not included in the price to consumers. Instead of cheap fossil fuel energy, the public good that needs government protection is clean or low-carbon energy. To the extent that we rely on markets, energy producers must account for the full costs of production across all resources. Only carbon pricing and other environmental regulations can protect this public good. Mine owners must account for the health of their workers and the land and streams surrounding their mines. Oil and gas producers must account for offshore oil spills and the carbon emissions of their products. Nuclear power must account for the full costs of waste treatment and disposal. So too must solar and wind resources account for the costs of pollution in the construction and manufacturing of their products. Consumers must likewise pay for a low-carbon energy economy, change their consumption habits, or both.

Small is Beautiful[42]

There are instances, a natural monopoly with scale economies is one example, in which the assumption *Bigger is Better (and Cheaper)* prevails – but not continuously. Electric power plants have reached a technological plateau. Not so many years ago, plants over 1,000 megawatts were being constructed. Today, the optimum size is about 600 megawatts. Further, new nuclear plants with smaller, modular units of about 100 megawatts, are becoming more economically and environmentally attractive.[43] When traditional power plants reached the peak of their capacities, then the cost of electricity began to rise. As noted in the last chapter, PURPA demonstrated that alternative power producers are able and willing to put cheaper electricity on the market as the large utility behemoths are beginning to face competition from smaller competitors including wind farms and distributed generation.

In contrast, large power plants must be replaced with smaller, decentralized electricity producers just as the large rotary dial telephones of the past have been replaced by handheld devices with greater functionality. The iPhone and iPad are not only products, they serve as a metaphor for a new way of delivering services. Music, telephony, internet access, photography, and innumerable applications are available at the touch of a finger. More importantly, such devices are customizable, are variously priced according to consumer needs, are traded in competitive markets, and use a fraction of the energy of the combined devices themselves. Big, centralized energy firms may have constructed the national infrastructure, but small, distributed energy is the future.

E. F. Schumacher's *Small is Beautiful* was written in 1973 and was notable for its critique of economics; it offered a different way of thinking about future economic growth and the natural environment. At the heart of Schumacher's critique was the notion that market values alone do not constitute a good society. In addition, beauty, health, nature, and we can also add participatory democracy, are values that cannot be monetized and, therefore, can neither be traded efficiently in markets nor can they be subjected to cost-benefit analyses.[44] In Schumacher's words, "What is ... destructive of civilisation, is the pretence that everything has a price or, in other words, that money is the highest of all values."[45]

Schumacher's critique of the application of economics to non-market values and transactions also caught the attention of policy thinkers. Social legislation, such as environmental protection, takes into account not only the economic costs of regulation, but it must also account for the "ethical, aesthetic, and cultural interpretation of the goals and purposes that underlie" it.[46] The microeconomic model, the critique argues, is limited in its analyses. Even if microeconomics can provide useful information about supply, quantity, and price, it has little or nothing to reveal about such incommensurable values as the aesthetic appreciation of nature or, even more poignantly, the value of a healthy life.[47] The small is beautiful concept supports a low-carbon energy economy by promoting innovation, energy efficiency, and renewable resources even at a higher cost because of the positive environmental benefits that may be impossible to quantify and, therefore, price.

The Tragedy of the Commons or The Myth of the Market Revealed

There are two deep problems with relying on the microeconomic model for the regulation and ordering of energy. The first problem lies with the nature of markets themselves. The conditions for competitive markets are not only difficult to achieve, market actors continually attempt to evade those conditions. In order for markets to function competitively, they must have numerous buyers and sellers, no one of which, nor small number of which, can exercise market power.[48] The products traded by those buyers and sellers must be heterogeneous, and market actors must possess perfect information. A moment's reflection exposes the underbelly of markets – all actors desire to cheat. No firm wishes to be just one among many; instead, they exist to dominate markets. Similarly, no firm wishes to sell the same product as a competitor; instead, a firm alters the price or quality or some other product characteristic so that product differentiation can be used to increase market share. Finally, no market actor willingly provides information; instead, they hoard it as a valuable resource that can be sold or used to a firm's own competitive advantage. There is nothing sinister here because it lies in the nature of markets for actors to behave in precisely these ways.

The second problem lies in the idea of relying on markets for the production and distribution of energy. There is the unstated, and incorrect, assumption that a choice exists between operating in a "free market" or operating under the guiding hand of government. There is no strict distinction between government and markets. Instead, the often referred to choice between government and markets is fundamentally a matter of political rhetoric, not sound economics. To put the point bluntly, without government there would be no market economy. There is no real choice between government or market. Instead, the choice is about the amount of government intervention into a market economy.

In order for markets to function competitively, property rules must exist. At the common law, property is: (1) defined by property rules; (2) exchanged by contract rules; and (3) protected by tort rules. Such rules of law are sometimes referred to as the common law baseline.[49] Without such rules, trades and, more importantly, planning for the future cannot reliably occur, especially in the credit economy. These common law rules, then, are a form of government regulation. There are occasions when common law rules do not allow property to be easily or effectively traded in a market or the common law rules do not effectively protect property from harm. In such cases, government regulation can be used to correct market failures.

The distinction between coal and oil or natural gas is instructive. For coal, common law rules operate quite effectively. A property owner mines a ton of coal and sells it to a buyer. Both parties can rely on the common law rules of property and contracts to effectuate the sale. However, one particular common law rule, the rule of capture, had a perverse economic effect for oil or natural gas. The intent of the rule was to reward the actor who explored for and extracted the oil or natural gas. Consequently, the incentive existed for someone to be the first to take as much oil and gas out of the ground as possible, which could have led to a significant devaluation and depletion of those natural resources. To counteract the negative consequences of this rule and to correct the market, government regulations apportioned ownership among several owners and limited the amount of resources that could be taken out of the ground at one time.

Two points are worth emphasizing. First, and most fundamentally, the choice is not between government and a so-called free market; the choice is about the proper relationship between government and markets. Second, government regulation can be effectively used to correct market imperfections and stabilize markets.[50] Further, government regulation is necessary for society to attain non-market values or difficult-to-quantify goals such as preserving clean air, clean water, endangered species, national forests, and wilderness areas to name a few.

Additionally, promoting the so-called free-market solution for our energy needs is hypocritical. Energy industries have looked to government for support and protection for over a century. Indeed, the reliance on government by fossil fuels industries is the primary and inescapable lesson of the first chapter. The coal, oil, natural gas, nuclear, and electricity industries have been the beneficiaries of decades of direct and indirect government support recounted throughout the book; they did not become as entrenched in our energy economy as they are by playing in "free" and competitive markets.

Regulation is Good

In October 2008, Joseph T. Kelliher, then chairman of the Federal Energy Regulatory Commission, stated: "[T]he notion that these legal domains are separate is deeply ingrained. In most respects, the notion that energy policy and environmental policy are separate domains is a workable fiction. But it is completely untenable when it comes to climate change."[51] Kelliher's statement in itself is unremarkable. Energy and the environment are two sides of the same natural resources coin. What is remarkable, though, is that the chairman of FERC can utter a statement about these two separate domains in 2008 as a truth with considerable staying power. The energy-environment link must now replace the traditional assumption about an energy-economy link not by abandoning a concern about a healthy economy but rather by incorporating it.

Kelliher's comment expressly identified the outworn idea that energy and the environment are separate spheres of society and, therefore, should be regulated separately. Implicit in his comment was another worn out idea that energy should be left to private market actors and not to government regulators. Private markets and private capital are good. Market competition creates wealth, stimulates innovation, and allows producers to place all sorts of goods on the market while expanding consumer choice. Resource use is maximized and efficiency is achieved, the theory goes, through market competition. However, free-market rhetoric does not operate unfettered for our energy policy and has fortified conventional energy industries. Now government must rethink its financial and regulatory policies in order to design a new energy policy.

Likewise, the third energy transition will not occur on the basis of non-market values alone. Instead, it will occur because new energy technologies can produce, distribute, and facilitate the consumption of cheaper energy in competitive markets. Similarly, to the extent that the future environmental policy is driven by climate change, that new policy must be primarily conceived of as an energy policy.

Energy Efficiency and Renewable Resources

The country's future energy portfolio will not immediately abandon fossil fuels or nuclear power even if eliminating fossil fuels is perceived as highly desirable. In part, we will continue to use fossil fuels and nuclear power at least as transitional fuels. Hopefully, our continued reliance on traditional resources will be less a matter of political intransigence than a consequence of deploying new energy resources at scale. Future energy policy, then, will depend on: (1) a diversity of energy sources; (2) an expansion of market actors, including new entrants; (3) smart energy technologies; (4) greater interaction between consumers and suppliers; and (5) more market-based regulation and less command-and-control regulation to stimulate innovation and competition instead of continuing to support energy incumbents.

Energy efficiency is at the heart of our energy future. More miles per gallon, buildings heated and cooled with less energy, more work from motors, less costly construction, and decentralized power production are all efficiency gains. Energy efficiency, together with renewable resources, particularly solar and wind power to generate electricity, can significantly further the goals of diversifying our energy mix, as well as making energy less costly and more competitive.[52]

Combined, these new economic and policy assumptions lead to a final assumption – inaction is unacceptable. We must acknowledge and accept the economic reality that the global commons, as well as clean energy, are public goods. We must also acknowledge and accept that global warming cannot be ignored; that precaution,[53] not old habits, is a valuable perspective; and that neither the competitive market nor the self-interested private actors in it can protect our commons if they operate only under so-called free-market principles. Instead, government regulation, including financial incentives, will play indispensable roles in a new energy policy.

A New Energy Policy for the Twenty-First Century

Clearly, the future calls for a change from traditional energy policy to a new policy that will be developed in more detail in the next chapter and the chapters that follow. There are two basic arguments available to support the choice in favor of a transformative energy policy. The market argument in favor of a new energy policy is that: (1) new energy technologies are available; (2) we have only begun to capitalize on our cheapest new energy resource – energy efficiency; (3) renewable resources are available and are becoming increasingly cost effective; (4) energy distribution can be localized and decentralized; and (5) we can improve national and energy security. A non-market argument is available as well. We should choose a transformative energy policy because it: (1) is more environmentally sensitive and sustainable; (2) can reduce the role of government; (3) can improve the quality of our lives; (4) can promote further technological innovation; and (5) can enable the United States to assume a position of world leadership.

Both sets of arguments are attractive, but this book recognizes that the market argument is more pragmatic and politically viable. Still, the country can move dramatically into a low-carbon energy future. Indeed, technologies exist to reduce our oil dependence by orders of magnitude; change our transportation system through the use of hybrid or fully plugged-in vehicles; and give consumers more energy choices and more control over their use of and expenditures for energy. There are two major impediments to achieving a fully transformative energy policy. First, a price must be set on carbon emissions. Second, existing technologies must be deployed at scale in order to achieve their potential gains. In no small part, both of these impediments are linked to a lack of political will that continues to be driven by the industries created and sustained by the traditional hard path energy policy. Energy regulation must be redesigned to support the smart path. There are signs that a new energy thinking and a new energy politics are appearing. Bipartisan coalitions,[54] a wide range of non-governmental energy organizations (NGEOs),[55] and a significant number of university-based research centers and institutes are emerging.[56] The new energy thinking is not limited to academic debate. New energy markets are also emerging as investors are placing increasing amounts of funds into new energy projects that are responsive to the challenges of carbon emissions. Nevertheless, these political and market efforts pale in comparison to the established institutions committed to fossil fuels that must be countered. The new energy future will be based on new variables, a new politics, and new strategies.

Energy Variables

Today, whereas we continue to believe energy is vital for the economy, we also assume that our future energy policy must be environmentally sensitive and must be sensitive to national and global security. Therefore, instead of just two variables – energy and the economy – future energy policy involves the four variables of energy, economy, environment, and security.[57] Designing a new energy policy comes with its own challenges. Aside from moving away from old policies and away from entrenched institutions and actors, it is also the case that the new energy variables are not always consonant and trade-offs are inevitable.[58] There can be, for example, a trade-off between carbon reduction and security. A full-blown reduction strategy against carbon emissions would suggest an expansion of nuclear power. However, nuclear power, in turn, has its own security issues. In short, integrating energy, economy, environment, and security requires not only a rethinking of the old assumptions, but we must also develop a new energy politics and new political and economic approaches. If we combine these variables, then we can construct a new energy policy based on the following principles. Our future energy policy will: (1) use a *diversity* of energy resources; (2) in *efficient* and effective ways; (3) while being *environmentally* responsible; and (4) while promoting domestic and global *security*.

New Energy Politics

Together with the new energy thinking, there is a new energy politics that will be developed in more detail later in the book. In addition to new research centers and new energy coalitions, private-sector actors, including commercial banks, corporations, and a variety of investors, are putting dollars into the green economy apart from government mandates to do so. Further, public and private non-partisan coalitions are working to formulate new clean energy policies and are acting to have those policies integrated into the law. These several collaborations are driven to engage business leaders, government officials, and non-governmental organizations to promote market-based economic solutions to energy and climate change issues.[59]

Future energy policy will not be tied to old ways. Instead, future energy policy acknowledges and will be responsive to climate change, will develop new fossil fuel policies and new utility policies, and, most importantly, will develop new markets and new technologies. To accomplish all of these goals, energy strategies move away from the traditional hard path structure of energy industries that are large-scale, capital-intensive, highly centralized, and significantly protected from competition. Future energy actors will face increased competition, energy production will be scaled-to-task, there will be greater energy diversity, and there will be greater decentralization.[60] Innovation is addressed more fully in Chapter 7.

What is embedded in our new energy thinking and new politics is a new economics. In the last few years, a new energy actor, if not unanticipated then surprising nonetheless, has entered the energy stage – Wall Street. In 2006, for example, more than $100 billion of sustainable energy transactions took place, and it was noted that "[t]his is more than just interesting data.... It is a powerful market signal to the arrival of an alternative future for today's fossil-fuel dominated energy markets.... This is full-scale industrial development, not just a tweaking of the energy system."[61] These new markets are directed at merging energy and the environment through such efforts as emissions trading, carbon investments, renewable resources investments, green technologies, and the like. These emerging markets involve capital investment[62] in innovation and new technology for both a healthy environment and healthy energy economy.[63]

New Energy Strategies
Predicting the future is always dicey, and predicting the energy future has been more often wrong than right.[64] However, we should draw some comfort in the fact that sometimes predictions are correct[65] and that sometimes the signals are clear at least about the general direction in which we must go. We can say, with a high degree of confidence, that we know the following about the future: (1) climate change must be addressed; (2) fossil fuels will continue to be primary fuels at least for the midterm; (3) carbon reduction strategies must occur; and (4) new technologies and new investments in energy efficiency and renewable resources are central.

What is less clear is where future challenges lie and the precise shape of our energy future. The energy future we do not know consists of uncertainty about: (1) the role of nuclear power; (2) the timing and cost effectiveness of "clean coal" technologies; (3) the proper mix of energy efficiency and renewable resources; (4) the proper carbon reduction strategies; (5) the proper level of investments in new strategies and the financial risks involved with those new investments, especially investments intended to bring new technologies to scale; and (6) the strength of the political will and the shape of the political leadership necessary for transition.

Nevertheless, we can make certain comments about the new energy politics and about future energy strategies. The federal government should continue to legislate in this area and should regain its former role as a national leader in energy policy formation and regulation. In addition to the federal government for guidance, however, we can look to cities and states and regions for responsive climate control, policies, and regulations.[66] We can also look to new bipartisan coalitions for continued energy thinking, and we can look to new players, new trade associations, and new research efforts to provide solid analyses and assessments. We can also look to new energy markets as old Wall Street meets new venture capital and new venture philanthropy. And, we can look to new energy strategies that would include a redistribution of government subsidies, an increase in energy R&D, the use of multiple public actors, proliferation of public-private partnerships, and a willingness to sustain private investment in technology and innovation.

Conclusion

As in the past, we will rely on a mix of regulation and markets, and we will not adopt a comprehensive energy policy aimed at coordinating energy industries. Unlike the past, though, we must redesign the regulatory/market mix to assure more flexible market-based regulations that promote competition, innovation, and environmental protection; we must promote a diversity of resources while downsizing producers; and we must diversify our energy portfolios with a greater investment in renewable and non-conventional energy resources. Let's assume for the moment that climate change is no threat, still a new, more efficient, more resource-protective, and more scaled-down energy system makes great sense as we move more rapidly into the twenty-first century.

The Next Generation Is Now

[I]n our murderous yet also suicidal treatment of the environment, far more than greed and stupidity[,] [m]an is possessed of some obscure fury against his own remembrance of Eden.

Claude Levi-Strauss[1]

The decade of the 1970s brought the matter of energy policy to public consciousness and made regular headline news. The immediate precipitating events were the Arab and Iranian oil embargoes of that decade. Gasoline at the pump was rationed for the first time since World War II, oil prices quadrupled, inflation ensued, and a recession followed. The United States was seen as vulnerable to having its oil spigot cut off by Middle Eastern oil producers. Also at that time, our culture was experiencing a rising period of environmental consciousness. Combined, the forces of greater environmental awareness, energy vulnerability, and economic sensitivity gave rise to a series of energy policy studies that examined energy more strategically than we had in the past. Strategic energy studies are again in the news and have led to new approaches to how our country, and the world, should think about how to produce, distribute, and consume energy and how to handle its waste.

This chapter will examine the new energy thinking that began more than a generation ago. The next chapter will provide a set of more specific policy prescriptions that follow from the new thinking. To be sure, over the last generation, thoughts about energy policy have evolved. That evolution will be traced here. Briefly, energy policy analysts began their strategic thinking by focusing on specific energy problems such as oil dependence or a nuclear future while occasionally mentioning the environmental consequences of our fossil fuel past. To no small extent, this narrow focus reflected the continued belief in the staying power of our traditional assumptions.

Particularly, we continued to rely on the assumption that economic growth depends on abundant, reliable, and reasonably priced supplies of energy. Traditional energy markets were considered to be of central importance in our economy, whereas the needs of the environment were not contemplated with great urgency. That focus would decisively change by the first decade of the twenty-first century.

The New Narrative of Energy Policy

The narrative of energy policy has significantly changed since the policy studies of the late 1970s and the early 1980s to the extent that today fossil fuel markets must share the stage with carbon emissions. The former divide between energy and the environment has been breached, and that breach spells hope for a truly transformative energy future. Central to that breach was the concept of sustainability, made most famous by the United Nations' report, *Our Common Future*.[2] This report rejected the core assumption that the economy was directly and positively linked to energy alone. Instead, the report argued that the economy is also dependent on a healthy environment and on properly priced natural resources consumed in the production of energy.

In his 1976 article in *Foreign Affairs*[3] and in a later book,[4] Amory Lovins posed what remains the central question for energy policy: Should the country continue to rely on the rapid expansion of centralized high technologies to increase energy supply, or should we promote efficient use and rapid development of renewable energy sources? Lovins, of course, argues that we should choose a low-carbon energy path, and he further argues that such a path can greatly improve energy efficiency, reduce the cost of energy, protect the environment, and create markets to support a sustainable and humane energy future. He argues further that technical fixes, together with a different energy mix, can result in significant capital savings.

Lovins is as much of a realist as he is a visionary. His realist assumption, in addition to faith in workable markets, is that fossil fuels will be transitional fuels; however, as much as possible should be left in the ground for emergency use. Still, his focus lies beyond oil. To the extent that Lovins has a particular political position, his early energy policy analyses were, and continue to be, decidedly anti-nuclear. From the perspective of energy policy, Lovins equated a nuclear generator to a tea kettle – both are used to create steam. He thought it made no sense to use such highly centralized and highly sophisticated technologies to boil water, which, he said, was like cutting butter with a chainsaw. From a socio-political standpoint,

expanding the use of commercial nuclear power in the face of nuclear proliferation threats is senseless to him.[5]

Lovins' analyses serve as the beginning of the last generation of the
new energy thinking. Another *Foreign Affairs* article published in 2003
signifies the conclusion of the last generation and the beginning of the
current one. By the turn of the twenty-first century, energy policy analysts began to accept, then amplify and extend, Lovins' approach. Former
Senator Tim Wirth, superlawyer C. Boyden Gray, and politico John Podesta
lamented the fact that the energy policy dialogue had become stale; that
energy relates as much to homeland security and national defense as it does
to the economy and to the environment; and that energy policy is central
to addressing the problems of globalization, including the growing gap
between rich and poor and the war on terror.[6] No longer can energy policy
be discussed in terms of economic growth alone. Energy policy is inextricably tied to the environment, to our position in the world, and to matters
of peace and security.

Senator Wirth and company were clearly driven by concerns about
energy security in an increasingly fragile world, especially after 9/11.[7] They
were also driven by the growing scientific consensus that global warming
is anthropogenic but, more importantly, that global warming was occurring more quickly than anticipated. We can observe the melting of ice caps,
the retreat of polar bears, the death of coral reefs, and the increasing violence of hurricanes and tidal waves. No less dramatic, though, the authors
drew attention to the stark reality of the energy disparity among the world's
six billion people. Only one-third enjoy access to as reliable and abundant
power as we do; one-third have access to only intermittent power; and the
remaining one-third "simply lack[s] access to modern energy services."[8]
Energy equity, then, is a necessary element for a better world order.

In addition to broadening the ambit of energy policy, the article is
significant for another reason. The authors are from different political
parties. Additionally, the authors come from differing public and private
experiences and perspectives. In short, this article is a microcosm of the
need for a new energy politics, as well as a new energy policy. The new
non-partisan politics of energy will be more fully addressed in Chapter 8.

This chapter, then, describes the arc of energy policy over the last generation. Over that time there were a substantial number of significant reports
and publications, some of which were highly criticized. This chapter will
look at the key reports, as well as their critics, and will argue that there has
been a synthesis over this period and that the need for an energy transformation has now become mainstream. The devil, of course, always resides

in the details. When we talk about the future, we often talk about the next generation. It is the argument of this book that the next generation of energy policy is here, that the necessary set of new energy assumptions are in place, and that the key variables for a sustainable and responsible energy future are available now.

The Beginning of the Generation

Several energy and environmental studies written in the early years of the 1970s inaugurated the idea that the country was ripe for its third energy transition. The studies were driven by different concerns. Some were responding to environmental challenges, others to energy predicaments, others to economic signals, and others to national security issues. In brief time, all of those concerns would coalesce into a consensus energy policy discussed in this and the next chapters.

Limits to Growth

Limits to Growth, the notorious 1972 report from the Club of Rome, employed a computer model called World3 to assess the future direction of population, food production, industrialization, pollution, and the consumption of non-renewable natural resources. The report was variously criticized as unscientific, nonsense, wrong in its modeling, and wrong in its prediction that the world was running out of resources. Many of the criticisms were simply wrong about what the book actually said, and, in fact, the book can be considered the first book arguing for a smart energy policy.

The core argument of *Limits* is that exponential growth can strain non-renewable resources, as well as returns on capital investment, to the point at which a society could collapse. The possibility of collapse is true, the book argues, as a natural consequence of unrestrained growth. In this regard, then, the book posed two central questions: "*Is it better to try to live within that limit by accepting a self-imposed restriction on growth? Or is it preferable to go on growing until some other natural limit arises, in the hope that at some time another technological leap will allow growth to continue still longer?*"[9]

The authors argued that human society has opted for the second course of no limits to growth and has virtually ignored the first course of self-imposed limits. In a real and fundamental sense, these two questions remain to be answered and, increasingly, answers are demanded. At the same time, the questions pose something of a false dichotomy between doing nothing, as if doing nothing means relying on a market that works naturally without human intervention, and heavy-handed government intervention.

As noted in the last chapter, the choice is not between markets or government; it is a choice about how much government intervention is prudent in the face of the market failures of negative externalities and the tragedy of the commons. These market failures provide justification enough to answer the questions posed by *Limits* by saying that self-imposed limits on pollution are not only advisable but they are necessary for further and sustained economic growth.

For all the criticism that *Limits* generated, its conclusions, although important, were not particularly radical. The authors concluded that: (1) if 1970 growth trends continue, then limits to growth would be reached sometime within the next 100 years; (2) those growth trends could be altered to establish conditions of sustainable ecological and economic stability; and (3) chances of success increase the sooner a path to sustainability is pursued.[10] The book then discussed trends, as opposed to predictions, about the future that were generated by its computer model. The concept of exponential growth is at the heart of the analysis. This growth may best be seen relative to world population statistics. Prior to 1800, there were less than 1 billion people on earth. The 3 billion population mark was reached around 1960; and in 1972, when *Limits* was published, world population was approximately 3.9 billion people. At the end of 2009, population topped 6.8 billion. Clearly, the world population curve rises steeply to the right as *Limits* accurately predicted.[11]

What is more significant than the steepness of the population curve is the associated question about the amount of natural resources, with their attendant wastes, needed to sustain that population. In short, are food, water, and energy produced in sufficient quantities and sufficiently distributed to serve the global population? As population increases, demand for resources increases accordingly. The Malthusian risk is that population growth outstrips resource supply and, as a consequence of negative feedback loops, growth is curtailed and economic collapse occurs. The even more daunting problem concerns the fairness in the global distribution of all resources including energy and food, which is currently disturbingly uneven. Additionally, increased economic growth brings with it increased pollution.

Limits accepted the positive correlation between economic growth and energy consumption. However, the authors also warned of the harmful effects of pollution's exponential expansion. In 1970, the atmospheric concentration of CO_2 was approximately 320 ppm, and *Limits* projected that the concentration would reach 380 ppm by 2000. The estimate was not far off. The concentration reached 380 ppm by approximately 2005.

Limits concluded by noting that a sustainable equilibrium is achievable if capital plant and population are kept constant in size and if input and output rates in population and investment are kept to a minimum and are set in the proper ratio.[12] This equilibrium is not intended to create a static state. Instead, it is a dynamic model allowing for growth, changes in a country's standard of living, and technological advances. The impetus behind the model is to draw a portrait of a long-term, sustainable society in which industrial growth does not threaten the health of the planet and the people in it.

Soft Energy Path

Limits was embraced by environmentalists, mocked by market advocates, and had little to say directly about energy policy other than to suggest that nuclear power could slow the expansion of carbon pollution. The direct connection between energy, the environment, and the economy was forcefully made five years later by Amory Lovins in *Soft Energy Paths*. Lovins, a Harvard dropout, pursued a physics degree at Oxford, returned to the United States, and founded the Rocky Mountain Institute (RMI) with its mission of promoting the efficient and restorative use of natural resources.

Lovins' first target was nuclear power, which he saw as a mismatched technology, too costly and cumbersome to boil water and too dangerous to proliferate in a dangerous and contentious world. His nuclear position drew criticism in return – he was out of the mainstream, his thinking was too soft, and his ideas were not economical. Today, Lovins is known as the Sage of Snowmass and, together with his RMI colleagues, advises clients throughout the world on the economical and efficient uses of energy. Their A-list clientele include foreign governments, blue chip firms such as Bank of America, Ford, and Wal-Mart, among many others, as well as the U.S. Departments of Energy and Defense.

Lovins begins by attacking the basic economic assumption that there is a positive and direct correlation between energy production and consumption and economic growth and well-being that can extend indefinitely. Instead, he argues, society must account for all costs of energy, including the cost of its waste as well as the costs of economic and national security that arise from imported oil and an over-reliance on nuclear power. For Lovins, the least-cost option, increased energy efficiency, responds to environmental, economic, and national security problems. Traditional energy sources may contain much potential, but at the point of end use, we can utilize only a fraction of it. That fraction must be larger.

Lovins was brought up with and educated about technology, and he incorporates the lessons he has learned into designing what he calls the soft energy path. He rejects the idea that bigger is better and substitutes the idea that energy production and distribution should be scaled to task. Small-scale and distributed electricity generation is more efficient than running power lines over hundreds of miles from large-scale fossil fuel burning power plants across state lines to residential and commercial as well as industrial end users.

To achieve the transition from the hard to the soft path, Lovins argues that three steps must be taken. Institutional and bureaucratic barriers, which support and sustain the hard path, must be broken. Fossil fuel and nuclear subsidies must be eliminated. And, energy prices must reflect their true costs in order to send accurate price signals into the market for producers and consumers.

Soft Energy Paths promotes energy efficiency in several ways. Technical fixes such as building designs, construction and appliance standards, car mileage requirements, and co-generation can significantly reduce energy consumption. Additionally, soft energy technologies should be adopted to replace the hard path. These small-scale technologies would rely on renewable energy and would deliver energy closer to consumers. Solar heat and cooling and wind power are the primary examples of such technologies. Designing energy systems to be scaled to use contains several economic efficiencies. Residential energy use is a good example. Through distributed generation, capital costs and load losses due to transmission and distribution are greatly reduced. Similarly, maintenance costs of large-scale power plants and the grid can also be significantly decreased.[13]

Like *Limits*, *Soft Energy Paths* recognized the dangers of exponential growth and the damaging impact of increased pollution on global warming and warned that warming would become a concern in a matter of decades, not centuries. In both cases, the predictions were accurate. Additionally, *Limits* and *Soft Energy Paths*, together with the unstable energy events of the 1970s, drew a great deal of attention to our energy future. Double-digit inflation, gas lines, increasing oil imports, a fragile Middle East, and an unpredictable OPEC gave policy analysts much to ponder with the result that numerous studies of the energy future were published before 1980.

Energy Future

In *Energy Future*, a report commissioned by the Harvard Business School, the authors set out to address the matter of oil dependence, arguing that the era of easy oil had ended, domestic production had peaked, imported oil constituted a threat to American power, cheap oil was a thing of the past, and that domestic oil policy needed to change[14] – "U.S. energy policy should give alternative sources of energy, including conservation, an 'equal chance' with the social cost of imported oil."[15] The authors also argued that domestic oil prices must be deregulated and be allowed to rise to the world price in order to spur conservation.

The book also addressed natural gas, recognizing that this was a cleaner fossil fuel, which is largely produced domestically but had suffered under regulation. They applauded efforts of the Carter administration to deregulate gas prices and to eliminate the disparity created by dual natural gas markets. Not only did regulation distort prices, it muddied our ability to estimate recoverable reserves. Natural gas markets are further complicated by the facts that unconventional methods are needed to recover more natural gas; synthetic natural gas from coal can be made available; liquefied natural gas can be imported; and the Outer Continental Shelf (OCS) can be explored for natural gas deposits. All of these alternatives to the traditional recovery of natural gas can be translated into a higher cost resource. Published shortly after the passage of the Natural Gas Policy Act of 1978, *Energy Future* was optimistic about fixing the distorted natural gas market. This optimism was well-placed as natural gas now offers promise as a substantial future energy source.[16]

Coal and nuclear power presented the same conundrums for the authors of *Energy Future* as they do today. Coal's abundance and pollution present opportunity and challenge. During the 1970s, coal was seen as a substitute for oil as legislation attempted to encourage power plants to shift from oil to coal. Additionally, synthetic gas and liquid petroleum might be extracted from this resource. Regardless of its abundance, however, carbon emissions remain problematic. When *Energy Future* was written, coal transportation presented a major cost obstacle because rail transportation was limited, and at the time there was much talk of building multi-state coal slurry pipelines as delivery systems. Slurry pipelines, however, have their own set of economic and political problems. Siting pipelines is expensive, and the pipelines' water requirements bring significant state-based political problems.

Additionally, health and environmental problems affect the coal industry throughout its fuel cycle. Deep pit mining threatens worker health and safety. Open pit mining threatens topsoil and creates erosion; and the overburden, which is deposited in streams and on top of otherwise fertile land, contains its own environmental harms. Transportation is dirty and dangerous; coal burning, with its attendant sulfur dioxide and carbon emissions, fouls the air; and coal ash waste presents hazards of its own.[17] The authors concluded that although "coal may not be a panacea for our immediate energy problems, it will continue to experience steady growth."[18] The authors further concluded that, even with emerging technologies for coal use, its environmental challenges might prove debilitating.

Energy Future concludes its analysis of nuclear power with the observation that "[i]n the United States there is simply no reasonable possibility for 'massive contributions' from nuclear power for at least the rest of the twentieth century."[19] The prediction could not have been more accurate. It was not until after 2000 that utilities began contemplating a return to nuclear plant construction. The reasons for the lack of optimism were familiar. Cost overruns, the accident at Three Mile Island, the unaddressed problem of nuclear waste, and concerns over safety and radiation together spelled the demise of this energy resource. The authors called the debate between advocates and critics of nuclear power a stalemate. Today, the nuclear industry is better assessed in terms of whether or not it can pass a market test and produce electricity at a lower cost than electricity produced from other resources. All indications are that it cannot.[20] Nuclear power is simply too expensive today.

Energy Future promoted conservation as an energy resource. Conservation can be seen in two ways. We can use fewer resources, or we can get more energy out of the resources we currently use. Amory Lovins also based his book, as well as the work of the Rocky Mountain Institute, on improving efficiency and moving to renewable resources. Similarly, *Energy Future* promoted co-generation, motor vehicle fuel mileage standards, and building standards. The authors also recognized the need to align energy prices with their true costs. Overall, conservation or energy efficiency could greatly reduce consumption with no loss in productivity – "In other words, in 1973, the same U.S. living standard could theoretically have been delivered with 40 percent less energy."[21]

Energy Future concluded by arguing that U.S. policy needed to shift from cheap imported oil to a more balanced system of energy resources. They offered a binary choice: import more oil, or accelerate the development of conservation and solar energy. The authors were not sanguine about future prospects, however. They argued that entrenched fossil fuel interests made congressional action unlikely. They were correct. The authors also blamed the public. The public's addiction to cheap energy demanded that society ignore the social costs of production, distribution, and consumption.

When faced with the question of "What to do?" the authors argued that reliance should be placed on the marketplace. A balanced energy economy can be generated when price signals are accurate, the playing field between traditional and alternative resources is leveled, and the social costs of energy production, distribution, and consumption are internalized. To the extent that government regulation can provide financial incentives rather than sticks to stimulate alternative technologies, then tax credits and a realignment of subsidies to encourage domestic production, as well as conservation and renewable energy, should be adopted.

Ford Foundation

Two other major energy studies were published in 1979 by the Ford Foundation and by Resources for the Future. The Ford Study,[22] their third major energy report of the decade,[23] was written by a distinguished group of economists, scientists, and others from the academic and private sectors. The study was comprehensive and was motivated by shocks to the economy as a result of oil disruptions. The central message of the report was: "All the fuels – gas, oil, and coal – will become progressively more expensive."[24] Concomitantly, the country faced an energy "transition." Today, when we speak about transition, we mean the transition from traditional fossil fuel energy policy to a policy that relies on energy efficiency and renewable resources. The Ford Study, however, used the word "transition" differently. The study addressed a transition from cheap, abundant, and reliable energy to costlier, scarcer, and more unreliable supplies. It also intended a transition from conventional oil and gas to a greater reliance on coal, nuclear power, unconventional oil and gas supplies, and renewable resources.

The report was based on several assumptions. The authors did not accept the premise that the world was running out of energy, but they did accept the premise that Middle East oil dependence is likely to continue and contains unacceptable economic and political risks. Higher energy costs were believed to be inevitable, yet prices must accurately reflect those costs.[25] Additionally, the environmental effects of energy must be attended to and conservation, including energy efficiency, must be treated as an energy source. Finally, the authors expected future shocks to our energy supply and to the economy and believed that R&D policy was essential even though no simple technical fixes were likely.

Several themes emerged from these assumptions. Energy prices and resource allocation must be determined by market forces: "In principle, if energy were priced correctly – that is, at a price that equaled its marginal cost of production plus such factors as the cost of depletion, environmental damage, and national security – there would be no need for public policies dealing explicitly with energy conservation."[26] Contingency programs, such as the strategic petroleum reserve, are necessary to soften energy shocks. The political instability in the Middle East adds costs to the price of oil. Energy production should be encouraged as an economic input, not as the sine qua non of economic health. Moreover, environmental conflicts need to be managed. Government research and development policies are necessary to develop a range of information and options but not necessarily promote specific technologies. The study anticipated that by the end of this century, the world would experience "a relative decline in oil and gas use, rising energy costs and prices, and increased efficiency in energy use," regardless of U.S. policy. U.S. policy, however, should anticipate this future and smooth the transition, or expect "a series of perpetual crisis-oriented actions."[27]

The Ford Study was based on an economic model that attempted to treat energy production and consumption as a matter of supply, demand, and price. In 1979, higher energy prices seemed inevitable. Yet, as we have seen, with the exception of price volatility in the oil market, in the most recent years, energy prices have stayed relatively flat. Nevertheless, the Ford Study's emphasis on the centrality of accurate prices was correct and is essential to any future energy policy.

For the most part, the study focused on conventional resources and substitutes for them and was optimistic about the likelihood of their recovery. It also made conservation and energy efficiency central to its analysis.[28] Relative to the environment, the report was cognizant of the threats of climate change and was careful to note the dramatic changes that could come about as a result of continued use of fossil fuels, especially coal.[29] In their acceptance of continued coal use, the authors may have been a bit too accommodating.[30] The problems of the adverse climate effects of carbon dioxide, especially attributable to coal burning, and the possibility of dangerous levels of radioactive waste signal a new type of environmental problem. Unlike simply fouling a river and then cleaning it up, climate change and nuclear waste present problems that are orders of magnitude larger than the typical environmental spillover and will require large-scale, long-term solutions. These environmental problems present risks of scale and potential irreversibility unseen before.

The report then generated a series of recommendations consistent with their assumptions and themes. The Ford Study recommended the decontrol of oil and natural gas prices, which occurred shortly after the report was published. They also recommended that electricity prices reflect the real cost of production through marginal cost pricing, which has also largely occurred. Additionally, the study recommended greater use of science and technology to provide better information about the costs and effects of energy production and to develop future options including cleaner coal and synfuels.

Regarding the environment, air pollution needed to be handled more effectively,[31] solar power needed to be further developed, and conservation and efficiency had to be part of the energy mix. Additionally, whereas market forces were favored over subsidies, the authors acknowledged that subsidies for fossil fuels were greater than those for conservation and renewable resources, and that an economic case could be made for subsidizing newer forms of energy.[32] The report closes with a recommendation for a greater role of science and technology. "Creating new facilities to supply energy, adapting existing technologies to changes in costs or regulation, and introducing new processes that incorporate advances in materials and technology are each aspects of a process that we call research, development, demonstration, and deployment (RDD&D)."[33] Traditionally, research and development was a short-term investment in specific projects. However, our energy challenges require longer-term investments that can be deployed at scale. The role for a new form of R&D will be developed in Chapter 7.

Resources for the Future

The Resources for the Future (RFF)[34] study took a path similar to that of the Ford Study and suggested a general approach to energy policy rather than setting detailed policy prescriptions. The RFF study recognized short-term oil disruptions but recognized the long-term availability of fossil fuel resources. The report held out hope for commercial nuclear power, as well as for substitutes for oil and natural gas. The report also envisioned energy costs leveling as technology contributed to overall cost reductions.

The report expressed a commitment to continued economic growth and estimated that future energy consumption would approximate 115 quads, with an economic growth rate of 3.2 percent per annum.[35] The report slightly overestimated both, but the relationship between energy and the economy was consistent. "Energy use and economic activity are inescapably connected. Energy utilized in productive activity is clearly an essential ingredient in economic growth."[36] The RFF study was less aggressive than Ford about the environmental impacts of energy production, distribution, and consumption. The report stated that "the scientific data and analyses needed to measure and evaluate such impacts are often not well established."[37]

Consistent with other energy studies, the RFF report recognized the need for workable and reliable markets in order to set energy policy. Energy prices should reflect marginal cost and should do so by eliminating subsidies, including all social costs, and removing the existing price regulations. Given the importance of electricity in the economy, and in our daily lives, the report held that large-scale power stations fueled by uranium and coal should play a significant part in our energy portfolio[38] even in the face of growing environmental warnings.[39] Whereas the report assumed that traditional energy resources would play a large role in the future,[40] it also recognized the need to move to conservation and energy efficiency[41] as well as renewable resources, particularly solar and synfuels.[42] In this regard, the study acknowledged that although there is a strong correlation between energy production and economic well-being, there need not be an equivalent rate of growth in each. Conservation and energy efficiency can increase the rate of economic growth at a lower rate of energy production.

The RFF report, like the Ford Study, brought attention to the catastrophic threat of climate change as a result of carbon emissions. The report characterized climate change as a "highly uncertain or at least low-probability event having possible severe impacts on a large number of people."[43] Still, the report was aware of the increasingly rapid buildup of CO_2 in the atmosphere and its expected influence on increasing global temperatures. The study recognized that the problem of climate change was, at the time, controversial, but nevertheless saw this issue as inimical to energy policy design.

The RFF study concluded by contrasting two views of the energy future in what it called the expansionist view of the traditional policy and the limited view as expressed by the Club of Rome. The report argued that both positions have evidentiary support, and it recognized the political reality that a strong consensus was lacking for either alternative. Future energy choices could be made with improved knowledge of the facts, some sense of shared views about the role of energy in our economy, and a reliable decision-making process. The report held out the hope that consensus could be reached on a number of elements including the acceptance of a broad array of energy supply options and the promise of new technologies and practices such as conservation, increased federal involvement in research and development, and a commitment to discovering potential environmental harms of energy production and use. In short, whereas the RFF study did not lay out a transitional approach to the future, it did recognize the central importance of the linkages among energy, the environment, and the economy.

Appraisals and Reappraisals

The 1970s was a defining decade for energy and the environment. President Nixon signed the National Environmental Policy Act on January 1, 1970, and President Carter gambled his administration, which ended with the passage of the Energy Security Act of 1980, on his partially successful ability to change domestic energy policy by waging the "moral equivalent of war" against future oil shocks. *Limits* brought with it an awareness of the fragility of the natural and human environments and the risks of unchecked economic growth and expansion to both. After the oil crises of that decade, energy policy analysts brought with them an awareness of dependence on foreign oil, of the peak in domestic oil production, and of the possible cost consequences of both. Fortunately, through *Limits* and the other policy studies, the country became aware of the pressing needs of both energy and the environment. Unfortunately, energy and environment policies remained separate and unconnected, even in the face of their obvious interrelatedness throughout the fuel cycle.

Looking back on *Limits* and the energy reports following, three lessons emerge. First, the dire predictions about rising energy prices have not transpired. In the same vein, critics of these reports, especially critics of *Limits*, accused the authors of arguing that the country and the world were "running out of energy." These reports did not make such a Chicken Little claim, and the world is not running out of energy.[44] Second, the equally dire predictions of environmental harm, especially global warming, seem to be understated. Indeed, the most salient feature of *Limits* was to connect the world economy with the environment.[45] Yet, with the exception of the Ford Study, most reports did not emphasize the dangers of global warming nearly to the extent that we perceive them today. Third, and in some ways most problematic, the energy analyses of 25 and 30 years ago are pretty much the same today as we approach the end of 2010. In other words, to the extent that those studies were accurate, we have yet to take their lessons under advisement and implement the recommendations with sound energy policy.[46] A review and appraisal of the studies indicate that they were largely accurate with the occasional misstep. We can draw several conclusions from them, but energy policy has not greatly advanced since the reports were published. These conclusions will be further developed in the remaining chapters.

Some Missteps
Nuclear Power. The most notable misstep among all the reports involves the future of nuclear power. All the reports, except Lovins' *Soft Energy*

Paths, anticipated an increased role for nuclear power in our energy future. Lovins, to the contrary, saw its elimination. Today, approximately one-fifth of the country's electricity is generated by nuclear power. Nuclear power plants operate close to capacity, and have increased their share of electricity generation through the realization of greater operating efficiency. However, industry expansion has stalled with some small sign of a nuclear renaissance. Still, it is becoming increasingly unlikely that new units will be constructed in the next few decades largely due to cost.

Energy Prices. This second most notable error in the reports concerns energy prices, which were predicted to increase notably. With the exception of fluctuating oil prices, fossil fuel prices have held fairly steady through 2000. Recent years, however, have witnessed price volatility for crude oil and natural gas with decreasing or level prices for coal.[47] Consequently, although drivers notice an uptick in prices for gasoline at the pump, electricity consumers continue to enjoy fairly cheap electricity. Predictions about increased energy prices are likely to become true in the near and midterms, especially if climate change legislation becomes a reality because of higher carbon prices.

Environment. On the one hand, it may appear curious that although the reports mentioned the negative impact of fossil fuels on the environment, they generally did not make climate change the centerpiece of their analyses. On the other hand, it is understandable that these reports underplayed climate change because they were written during a period in which the dominant model had delivered a healthy economy to the country, and most of the reports, with the exception of the Ford Study, were written as a response to the oil disruptions of the 1970s. From this perspective, then, the reports worried more about fossil fuel independence than about fouling the air with carbon emissions.

Economic Growth. Lovins, most notably, made the argument that the link between energy consumption and economic growth needed to be shattered. The other reports did not address this issue; instead, they assumed continuing economic growth as the country continued its energy productivity. What has become clear, however, is that the direct and positive correlations among world population, industrial production, and carbon dioxide concentration in the atmosphere are increasingly problematic. Hockey-stick projections indicate dramatic growth in world population, industrial production, and carbon dioxide emissions in the atmosphere, thus signaling a need for immediate attention.[48]

World Wealth Disparities. *Limits*, more than the other reports, was driven by concern for global resource and energy needs, not just the needs of the United States. The concern of *Limits* was unavoidable – an increasing world population means either a corresponding increase in per capita industrial production or increasing wealth disparities between developed, developing, and underdeveloped countries.[49] Currently, wealth disparities are increasing as approximately 1.4 billion people live on less than $1.25 per day.[50]

Whereas China and India are the most obvious examples of developing countries with increasing needs for energy, there are too many people in the world that lack fundamental energy resources. Indeed, of the world's nearly 7 billion people, 1.5 billion people lack any electricity at all, 2 to 3 billion more do not have access to reliable electricity,[51] and, for the first time in history, more than 1 billion people go hungry every day.[52] Unsurprisingly, climate change presents greater threats to developing and underdeveloped countries. It has been estimated that if by the end of the century global temperatures rise 2°C from pre-industrial levels, then "[b]etween 100 million and 400 million more people could be at risk of hunger. And 1 billion to 2 billion more people may no longer have enough water to meet their needs."[53]

Oil. Each of the reports was critical of our dependence on foreign oil. The reports diverged, however, regarding what to do about it. For Lovins, the problem is simple – get off of oil completely.[54] The other authors adopted the more traditional approach, arguing that we should do what we can to reduce oil use through such things as vehicle fuel economy standards and that we should build up our strategic petroleum reserve. They further argued that we should also consider developing domestic resources offshore and in Alaska, as well as continue to consider the development of synthetic fuels from oil shale, tar sands, and coal.

To the extent that over the years we have only managed to increase our oil imports, then U.S. oil policy can be judged a failure. Still, there is some light at the end of the tunnel, for even though oil imports have increased, the importance of oil in the economy has declined. "About 50% less oil per dollar of real GDP is consumed in the U.S. today than was the case in 1979."[55]

Markets. Competitive markets can facilitate the transition away from the traditional policy to the smart energy policy of the future. Faith in markets may appear to be a truism; it is also too simplistic. Traditional policy, as noted previously, has greatly favored fossil fuel energy interests with subsidies and other financial supports. Fossil fuel support is a worldwide phenomenon. The World Bank, for example, estimates global subsidies to petroleum products at $150 billion annually.[56] Domestically, it has been estimated that for the period 2002–2008, the federal government has provided $72 billion in subsidies to fossil fuels.[57] In a perversely complementary way, U.S. energy policy provided less support to alternative energy resources with federal subsidies totaling approximately $29 billion over the same period. Whereas $29 billion is not insignificant, it represents time-limited subsidies, and the trend had been to increase fossil fuel subsidies while decreasing subsidies for renewable resources.[58]

Nevertheless, energy policy over the last decades is moving more toward a market orientation. Oil price controls have been lifted. Natural gas price controls have eliminated the dual natural gas market responsible for natural gas shortages. Electricity prices are moving toward marginal cost. Additionally, pollution controls are relying on market-based regulations rather than command-and-control regulations. These are examples of the types of market reliance urged by all of the authors.

Energy Efficiency and Renewable Resources. The reports also agree about the need for increased reliance on energy efficiency and renewable resources. Over the period, energy regulations have fostered efficiency measures by promoting efficiency standards for car mileage, buildings, and appliances. Similarly, all authors argued that renewable resources, predominantly solar and wind power, must play a larger role in our energy portfolio. Unfortunately, the authors also recognized that the use of these resources has remained static over the last few decades and that these resources will not have a larger piece of the energy pie until they become cost-competitive. This idea of cost-competitiveness is directly related to the matter of subsidies as well as research and development expenditures.

Research and Development. All authors acknowledged the need for increased research and development funding, while recognizing the decline in those expenditures in the field of energy. Going forward, R&D must be used creatively and federal funds must be used in a way more akin to venture capital. Federal funding must look not only to funding basic science or the invention of new technologies but federal funding must be seen as an investment that attends to matters of demonstration, scalability, marketability, and the reproduction of technologies across firms and industries. In short, R&D must seek a transformation in energy systems and their regulation instead of hoping for and funding simple technical fixes.

Coal. The reports demonstrated a consensus on coal as well. Coal's abundance, low cost, and widespread use for electricity generation was seen as unlikely to be eliminated from our energy picture. Going forward, the world demand for coal "grows more strongly than all other energy sources except modern non-hydro renewables."[59] Coal's position in world energy economies is increasing sharply in developing countries with their growing need for electricity. Still, coal's contribution to carbon emissions was not ignored, and the need for clean coal technologies was recognized. The failure of those technologies to achieve competitiveness was also recognized.[60]

More than thirty years ago, the nation and the world became aware of the relationship between continued energy use and environmental harm. Additionally, the nation and the world became aware of the relationship between continued economic growth and environmental harm. These three variables – energy, the economy, and the environment – are directly related. Traditional energy policy exacerbated the negative dimensions of the relationships among them. This trend must be reversed. Instead, the nation and the world can enjoy economic growth and a healthier environment with smarter energy use. The energy reports of thirty years ago, however, only brought us halfway there.

The reports, although generally accurate, were too narrow and static and did not go far enough. Still, they contained some hidden good news. From 1970 to 2008, energy consumption per real dollar of GDP declined more than 50 percent from $17.99 to $8.52. As consumers, we are getting more energy per dollar. During that same period, energy consumption per capita stayed relatively flat, dropping from about 359 million BTUs per person in 1978 to 327 BTUs per person in 2008.[61] Further, economic growth has been sustained at roughly 2.5 percent to 3.0 percent annually throughout the period.[62] In short, the country realized lower cost energy together with economic health for a sustained period.[63]

Additionally, the reports achieved consensus on the need to reduce oil dependence, the need to rely on markets, the need to increase energy efficiency and renewable resources, and the growing importance of environmental protection. The reports faltered in two specific ways. With regard to the environment, the reports could have more strongly emphasized the dangers of climate change. More critically, though, several of the energy reports limned the traditional policy model without offering a sound transitional plan for moving to a smart energy future. Although the elements of an alternative energy policy were present, a road map for getting there was not. This transitional road map was to come with the UN report, *Our Common Future*.[64]

The Beginning of the New Energy Thinking

Published in 1987 by the UN World Commission on Environment and Development, *Our Common Future* extended the ideas and analysis in *Limits* and popularized the idea of sustainability. The study defined sustainability as organizing society "to ensure that [humanity] meets the needs of the present without compromising the ability of future generations to meet their own needs."[65] The report recognized the need for economic growth, particularly in light of the growing income and wealth disparities throughout the world, and stated that countries had an essential obligation to relieve poverty. In order to improve the common lot, environmental and energy resources needed to be managed so that they contribute to a positive future, not to a more expensive and challenging one.

Sustainable development is not an anti-growth concept. Instead, it is a path for responsible growth, recognizing the significant interconnections between energy, the environment, the economy, and world population especially in developing and underdeveloped countries. Additionally, the report emphasized the increasing interconnections among global economies, noting that as world economic complexity increases, then threats, not the least of which is nuclear proliferation, to national security increase as well.

Our Common Future, like *Limits*, acknowledged that there are limits to economic growth but that those limits were not absolute ones. Instead, these limits are based on past policies, then existing technologies, and entrenched political and regulatory structures, all of which supported energy and environmental policies that were not sustainable. Sustainable development "is not a fixed state of harmony, but rather a process of change in which the exploitation of resources, the direction of investments, the orientation of technological development, and institutional change are made consistent with future as well as present needs."[66] Hard choices, of course, must be made; and, perhaps, the most challenging choices must be made by political institutions. Political institutions not only need to reassess past policies and practices, but those reassessments must come with a view toward international cooperation instead of continuing to pursue independent national gains.

The report notes that the average person in an industrial economy uses more than eighty times as much energy as an individual in sub-Saharan Africa. People in poverty use minimal amounts of non-renewable and non-reliable energy, and they need energy for economic improvement. Further, it is hardly likely that world population will decline. Consequently, any attempt to relieve the pain of poverty will necessitate significant increases in energy production and consumption. More to the point, energy efficiency is essential for world economic health and must be perceived as transitional to the low-energy and soft-energy paths afforded by renewable energy resources.

Like most of the energy reports before it, *Our Common Future* was unwilling to eliminate nuclear power as a potential climate-friendly energy resource. However, nuclear power can only be considered a significant energy actor after its waste problems are addressed. Similarly, coal use must be circumscribed to the extent it continues its high level of carbon emissions. Together, energy efficiency, renewable resources, nuclear power, and clean coal will all require significant public and private investments in research, development, and demonstration.

Our Common Future made specific energy policy recommendations based on four assumptions. Energy supplies needed to grow at a minimum of 3 percent per capita in developing countries; energy efficiency and conservation is a key element; public health and safety must be taken into account; and the biosphere must be protected. The report's energy recommendations were seen as transitional from an era of unsustainable energy use to a sustainable one. Moreover, growing demand for energy must be made more evenly available throughout the countries of the world. Ideally, energy use in underdeveloped and developing countries would increase while energy use in industrialized countries is reduced through efficiency and conservation measures, thus maintaining total energy production and consumption over a period of time.[67]

Fossil fuels represent a continuing problem. World oil reserves appear to be declining, yet natural gas and coal reserves are bountiful. Nevertheless, each of these resources contributes to global warming, air pollution, and acidification. Increased carbon emissions due to fossil fuel use are aggravated by the fact that as forests are cut down carbon sinks are destroyed. The report was cautious about climate change, recognizing that in 1987 the scientific consensus about global warming was not as accepted as it is today. Still, the report raised the pertinent question, "How much certainty should governments require before agreeing to take action?"[68] More notably, the report recognized the obvious necessity of global cooperation if environmental and economic fairness around the world is to be achieved.

Renewable energy and energy efficiency have the capacity to provide all of the world's energy needs. Yet in 1987, even with growing use of renewable resources, the overall effect on the energy portfolio was negligible and continues to be so. Greater reliance on wind and solar power, biofuels, and geothermal energy all require sustained investments. Additionally, whereas renewable resources generate environmental problems of their own, they are not nearly of the magnitude generated by fossil fuels. Following Amory Lovins' observations, *Our Common Future* recognized that renewable energy systems operate best at small to medium scale, which make them ideally suited for rural and suburban dispatch; this is responsive to the problem of international equity among industrial and developing countries.

Energy efficiency is the centerpiece for a sustainable energy future for two reasons. First, energy efficiency initiatives, such as building and appliance standards, are the most cost-effective of all energy measures. Second, energy efficiency can be implemented in every sector of society and in every country of the world. By way of example, "[t]he woman who cooks in an earthen pot over an open fire uses perhaps eight times more energy than an affluent neighbour with a gas stove and aluminum pans."[69] Energy efficiency can improve the quality of life at the most rural levels. The UN's energy efficiency analysis coincides with the observation made in the other energy reports that efficiency can be enhanced through signals that accurately reflect the full social costs of energy production. Markets are not the enemy of a sustainable energy future, but they must be made to work more effectively. *Our Common Future* notes that within the next fifty years, nations have an opportunity to produce the same level of energy services with as little as half of the primary supply currently consumed. To achieve this goal, however, significant changes must occur in the political and social institutions responsible for energy policy development and implementation.

Our Common Future acknowledged that energy policy is not one thing. Instead, it is a wide mix of products and services, as well as an admixture of public and private efforts. To the extent that a sustainable energy future can be achieved, it is imperative that energy production, distribution, and consumption be more directly aligned with the harmful environmental effects of continuing fossil fuel use. This linkage between energy and the environment breaks the long-standing commitment to the energy-economy correlation. In the United States, our energy policy was based on the idea that we should allow various energy industries to operate independently of and in competition with one another. This commitment to independence is replicated in the world; yet, in both instances, a distorted picture of world markets and independent operators is painted. Instead, traditional fossil fuel energy industries are favored with government finance and support and do not operate as independently as the ideal might suggest.

Domestically and globally new policy directions are required to preserve the environment, sustain economic growth, and alleviate as much hunger and poverty as our global resources allow. Disparities between rich and poor must not be allowed to expand. Energy equity, no less than environmental equity, can coexist with economic health; and we can only achieve that health through cooperative and interrelated efforts. In short, government agencies and institutions responsible for economic and energy policy and planning must do so with an eye toward sustainability. Similarly, individual nations must be aware of and sensitive to the negative consequences of their economic and energy policies throughout the world. No less so, international agencies must likewise be made responsible and accountable for their programs and budgets as they impact a sustainable future. *Our Common Future*, then, urged individual nation states to strengthen their environmental agencies, address the cleanup of existing problems, and commit resources for technical advice, as well as research and development, to underdeveloped countries.

More specific energy policies for the sustainable future envisioned by *Our Common Future* were set out in greater detail in 2002.[70] This later report explored the economic and environmental consequences of energy use in developed and developing countries. Energy is not only directly connected to economic health and environmental harm, it is also directly connected to the design of societies and the development of cultures. The paradox of poverty is that the poor pay a larger share of their income for inadequate access to unreliable energy that, in turn, disadvantages them in terms of health and economic well-being. In order to reverse this cycle, new technologies and smarter energy production and consumption must come about as a result of public-private partnerships around the world.

Our Common Future extended the analysis in *Limits*, and *Limits* repaid its debt by adopting the theme of sustainability in its thirty-year update.[71] Both reports recognized the underlying need to change socioeconomic and political systems as currently structured and to relieve them of their biases. Structural changes must create positive feedback loops as information accurately assesses energy use and environmental impacts and prices them accurately. Further, public and private organizations, together with new technologies and new investments, are all part of the next energy transition. It is not necessary that all of these changes be centrally directed; it is necessary that they all sing from the same sustainable hymnal. "From a systems point of view, a sustainable society is one that has in place informational, social, and institutional mechanisms to keep in check the positive feedback loops that cause exponential population and capital growth."[72] Further, a sustainable future necessitates legal rules and regulations, which promote disclosure and transparency especially in the face of resistance by entrenched political and economic interests.[73]

Instead of a precise blueprint for the future, *The 30-Year Update* offered a set of guiding principles for building a sustainable society. A sustainable society will require focusing on long-term costs and benefits, not on quarterly profits and losses. Improved information, particularly relative to price signals and costs, will better inform society and will better reflect an efficient economy. Neither the United States nor the world can continue to wait thirty or fifty years or more by relying on old policies when new challenges present themselves. Response times to crises must be quicker and more flexible. Non-renewable resource use must be minimized. Renewable resources must be protected. All resources should be used to their maximum energy potential. And, exponential growth of population and physical capital must be slowed or stopped.[74]

The New Energy Thinking

The arc of thinking about energy policy over the last generation began with concerns about the environment in the 1960s, developed through the energy crises of the 1970s, and attempted a synthesis between energy and the economy in the UN's concept of sustainability in the 1980s. Just after the turn of the twenty-first century, energy thinking entered into a significant new phase that attempted to incorporate and expand the various concerns expressed in the previous reports. To be sure, there has been no shortage of energy policy analyses in the last decade. However, two reports stand out among the others, not only for their comprehension and synthesis, but because they were intentionally non-partisan in their approach and they incorporated ideas from business, the academy, and politics.

Energy Future Coalition
Founded in 2001 with foundation support, the Energy Future Coalition was created specifically to examine deficiencies in U.S. energy policy.[75] More specifically, oil dependence, national security, climate change, and economic health, including providing access to energy to the two billion people who lack it, were the guiding principles of the coalition comprised of more than 100 representatives from business, labor, government, the academy, and the NGO community. The Energy Future Coalition engages in policy analysis, coalition building, and political action at the state and federal levels to address the problem areas identified through education and technological innovation. The coalition recognizes the long-term nature of its engagement; the global economic and environmental interconnectedness of energy producers and consumers; and the necessity of U.S. leadership for clean energy to become a reality.

The coalition announced four specific projects. First, energy efficiency is the linchpin of their program to foster clean energy, create jobs, and protect the environment by building lower-cost, cleaner power plants.[76] Next, these efforts will require changes in utility regulation, which are addressed more fully in Chapter 6. Additionally, the coalition, consistent with the

Obama administration's energy plan, has the goal of providing 25 percent of America's energy by the year 2025 from renewable resources.[77] The electric industry can be a major contributor to a more efficient energy future; and to do so the smart grid must be fully deployed, which is the third project of the coalition.[78] The fourth leg of the coalition's energy policy table is clean energy development that is specifically aimed at providing access to modern energy services for the two billion people worldwide who currently lack reliable energy by financing what the coalition calls global development bonds to finance energy investments in developing countries.[79]

The Energy Future Coalition's foundational report, *Challenge and Opportunity: Charting a New Energy Future*,[80] was published in 2003 and was the product of six working groups addressing transportation, bio-energy, coal, the smart grid, efficiency, and international energy matters. *Charting* set specific targets. First, the transportation sector must loosen its reliance on petroleum and reduce its carbon emissions through greater efficiency and biofuels. Second, the electric industry must loosen its reliance on dirty coal by capturing and sequestering carbon dioxide and building new plants through gasification technology. Additionally, the electric industry must increase its efficiency through distributed generation, innovative rate making, and a more intelligent transmission system. Third, private-sector investment was considered pivotal in changing the structure of the entire energy delivery system. Finally, government policies must be aligned with the private sector, and government must participate in investing in innovation and technology.

The Coalition's pragmatic political agenda centered on three core ideas. First, the United States had to become a leader on domestic as well as international energy policies. Second, traditional U.S. energy policy had been largely fragmented, and that policy had been protected by powerful economic and political actors. The grip of those actors had to be loosened. Further, whereas the environmental lobby enjoyed some success blocking proposals harmful to the environment, it enjoyed less success advancing their proposals for cleaner energy. Third, if advances were to be made, they had to be made as the result of a broad-based consensus with support from both sides of the political aisle, from non-governmental organizations, as well as from business and universities. *Charting* also recognized that a sound energy policy was based on sound markets and that blind allegiance to a free-market ideology to the exclusion of government input and support was a recipe for failure.

In its report, the Coalition set out three twenty-five-year goals. Oil dependency must be addressed by cutting U.S. oil consumption by one-third.

Next, carbon emissions must also be reduced by one-third, with the goal of worldwide reduction of two-thirds by the end of the century. Finally, the United States must develop, use, and export innovative energy technologies and institute trade policies to reduce the number of the world's poor, who lack access to energy services and markets. Concern about climate change was central to all of these goals, and climate change was also central to future economic opportunities, which would lead away from past policies and pave the way for jobs in the future.

National Commission on Energy Policy
The National Commission on Energy Policy was created in 2002 and consists of a bipartisan group of approximately twenty of the nation's leading energy experts from industry, government, the academy, as well as labor, consumer, and environmental protection organizations.[81] The National Commission engages in policy analysis and political advocacy, and in 2004 released its comprehensive energy statement, *Ending the Energy Stalemate*,[82] many provisions of which were adopted in the Energy Policy Act of 2005.

Ending the Energy Stalemate was based on, by now, a familiar set of assumptions. Twenty-first-century energy policy must increase oil security and decrease carbon emissions through greater energy efficiency and renewable resources.[83] The report recommended what it considered a revenue-neutral package of initiatives designed to provide affordable and reliable energy supplies. The authors, as with all of the energy reports, accepted as fact the need for energy for a healthy and clean economy. A future energy economy could be delivered through workable markets and with appropriate regulations. The report estimated that federal expenditures of approximately $36 billion for an energy transition would be needed over the next ten years but that those expenditures could be recouped through the sale of emission allowances under a cap-and-trade program.

The report rejected the idea that uncertainties, attributable to the long-term nature of current energy and environmental problems, should be used as a political subterfuge for inaction. Clearly, energy demand will continue to grow. Equally clearly, we cannot continue to satisfy that demand with an antiquated fossil fuel policy, which left the United States with a polluted atmosphere and vulnerable in a fragile world. To the end of greater oil security, the report recommended both increasing and diversifying world oil production and to raising fuel economy standards while investing in hybrid cars.

Ending the Energy Stalemate focused on a cap-and-trade regime rather than a carbon tax to reduce greenhouse gas emissions. Through a permitting process, emissions standards could be established and geared to specific targets for the worldwide reduction of carbon emissions. The permits would be tradable at auction, and markets would be used with the intent of raising revenue and having the price of fossil fuels reflect their social costs. The report estimated that its greenhouse gas recommendations would have only a slight impact on household welfare and on GDP. It was estimated that the plan would cost the typical U.S. household $33 per year and would result in a decrease in GDP growth from 63.5 percent to 63.2 percent between 2005 and 2020.[84] Moreover, also like *Charting*, the report called on the United States to assume national and global leadership roles.

Future U.S. energy supply would depend on: improved energy efficiency as a result of efficiency standards for appliances, equipment, and buildings; targeted efficiency programming by utilities; and efficiency improvements in the industrial sector. Traditional energy resources would be provided by an expanded use of natural gas, including liquefied natural gas, and through advanced coal technologies. Nuclear power was also seen as a future source of supply once waste is effectively managed and advanced nuclear designs have been demonstrated and adopted. Federal funds will be needed to expand the contribution of renewable resources, improve our nation's energy infrastructure, and encourage technological innovations in the private sector.

Conclusion

Reviewing the last generation of energy policy analyses, we can arrive at certain conclusions. First, the analyses are consistent regarding the expected increase in the demand for energy, the expected rise of the world's population, and a concomitant increase in carbon emissions and other forms of pollution. Similarly, there is a significant consensus that energy choices have for too long remained relatively stable, fixed, and unwisely committed to fossil fuels. The reports also acknowledge that past fossil fuel policies have become deeply embedded into our political and economic cultures and that those old cultures are no longer sustainable.

The earlier studies reacted to concerns about domestic pollution and national energy crises caused by oil dependency. The later studies acknowledged the global interconnectedness of energy, the environment, and the economy as well as security.[85] The Energy Future Coalition report succinctly articulated the energy consensus as "the need for change … [presenting] a new vision that linked security, environment, and economics."[86] There are several other points of consensus as well. Energy supply must be based on a wider diversity of resources than currently available, including energy efficiency and renewables. Oil dependency must be broken. Nuclear power remains questionable. Coal must be cleaned. And, whereas markets are the preferred form of social ordering, they cannot be left unconstrained. Therefore, an effective energy transition will require responsible and responsive government regulations. Additionally, federal financial support must move away from fossil fuels and must promote technological innovations for cleaner alternatives. Most importantly, any expenditures that must be incurred should be incurred sooner rather than later when they may be prohibitive. Similarly, when climate change was seen as a potential threat in the 1970s, it is seen as a grave danger today.

The new energy thinking, especially those policy analyses done after 2000, also exhibit an important characteristic. These reports were generated with bipartisan support from the public and private sectors. Further, NGOs are playing a more important role in policy formation and implementation, and organizations such as the National Commission on Energy Policy and the Energy Future Coalition are politically savvy and exist not to have reports that sit gathering dust on shelves but exist to have their policy analyses turned into political action.

The next chapter will examine in more detail the specific policy recommendations that have been developing over the last several years. These recommendations are based on the consensus items that have been identified. These recommendations consider costs and benefits and are not divorced from economic growth. Nor are they divorced from environmental harm.

4

Consensus Energy Policy

Without energy, there is no economy. Without climate, there is no environment. Without the economy and environment, there is no material well-being, no civilized society, no personal or national security. The overriding problem associated with these realities, of course, is that the world has long been getting most of the energy its economies need from fossil fuels whose emissions are imperiling the climate that environment needs.

John P. Holdren[1]

The previous chapters covered the history of energy policy and regulation, the economic and political assumptions behind the traditional energy model, and the emerging critique of that model. Together, these analyses generated a new set of economic and policy assumptions on which to build a new energy policy and a new model of energy regulation. This chapter examines in more specific detail several policy proposals mostly published since 2006. In brief, since the turn of the millennium, a consensus energy policy has been developing from numerous studies, several of which emanate from either bipartisan or non-partisan organizations thus indicating the emergence of a new energy politics.[2] To be sure, the emerging consensus has its direct predecessors, and the path to a new energy policy can be traced back a generation. Still, considering that the traditional policy is over a century old and has entrenched both private- and public-sector interests in multibillion-dollar industries, change comes hard and resistance is neither unexpected nor unpredictable.

Energy – its production, distribution, consumption, and end products – is a massively complex system. Further complicating matters, climate change presents scientific, technological, political, economic, and social challenges and uncertainties that must be managed effectively if the United States is to transition to a low-carbon energy economy.[3] The emerging consensus on energy policy is a hopeful sign of a growing agreement among policy thinkers and of a broader public acceptance of a smarter and cleaner energy future. It now remains to be seen whether the political process will adopt the necessary laws and regulations to promote a secure, environmentally responsible, and economically sound energy policy. The following discussion examines the goals of the consensus policy and the means that legislators and regulators can use to realize the needed energy transformation.

Consensus Targets

What is most apparent from reading the spate of energy reports written in the last few years is the coalescence of ideas about both the goals that a new energy policy must pursue and the means of achieving them. Naturally, debates about goals and means will continue, and interested actors will push their own agendas. Those utilities, for example, with a large stock of nuclear power plants are more likely to promote climate legislation than those with coal-fired facilities. Similarly, wind and solar power firms are more likely to tout the reliability of these renewable resources and seek access to the electric grid as opposed to traditional power plant firms, who own the transmission lines, find it difficult to invest in upgrades, and are unwilling to expand access to these intermittent sources of electricity. Similarly, politics affects policy design as well. A carbon tax, for example, is a better device to price carbon than a more complicated cap-and-trade regime, but any "tax" is anathema to politicians. Nevertheless, as policymakers and private actors think about an energy future less dependent on fossil fuels, a consensus on means and ends has been emerging.

This section reviews two broad goals – reducing carbon emissions and promoting national security. These two broad targets are intended to serve as markers for developing a future energy policy incorporating the several values discussed in the previous chapter. The following section will discuss the various mechanisms for achieving these goals. As we learn more about climate science, we learn better how to set a target for reducing greenhouse gas emissions. To be sure, there is no exact certainty about how much carbon reduction should occur, or the timetable for achieving that reduction, or the precise correlation between increasing carbon emissions and increasing global temperatures. Still, the scientific consensus is strong that anthropogenic carbon emissions play a significant role in global warming. Therefore, a target goal on climate emissions frames the broader policy discussion and, as will be argued, is a necessary element for going forward with a transformational energy agenda.

Are 350 and 2 the Correct Target Numbers?

The atmosphere currently contains 385 ppm of CO_2, which constitutes more than a one-third increase from the 1750 pre-industrial level of 280 ppm and represents a level of greenhouse gases not witnessed in the past 650,000 years.[4] Several recent analyses of climate change have suggested that a safe future level of CO_2 in the atmosphere would approximate 450 ppm.[5] This estimate is based on the partial acceptance of business as usual, which acknowledges a continued reliance on fossil fuels for both transportation and electricity generation. More recent reports, however, argue in favor of a reduction below current levels to 350 ppm, resulting in negative emissions.[6] Consider, then, the implications of *negative* CO_2 emissions. The first implication is that we must move full speed ahead and expand the energy efficiency and renewable resources components of our energy portfolio. The second, and more significant, implication is that we will have to capture all carbon emissions above 350 ppm through such efforts as carbon capture and sequestration (CCS) and the creation of carbon sinks through, for example, reforestation.

There are two further variables to be considered. First, assuming a 350 ppm target, when should that target be reached? Should it be reached by 2100 or by 2200? Second, what are the costs to achieving it? These two variables are related because an earlier date will require higher abatement costs initially. Recent studies argue that both dates are achievable at, of course, varying costs.[7] The cost element is controversial and is seriously debated, yet one consensus point emerges: Start now – future costs (and risks) will only increase.[8]

The second significant number is 2°C.[9] More particularly, the issue becomes whether or not greenhouse gas emissions can be slowed or reduced to the point at which, over the next several decades, the Earth's surface temperature does not increase 2 degrees higher than the pre-industrial level.[10] The scientific record is strong that since the mid-nineteenth century, when modern weather records began, the earth's surface temperature has been rising. More notably, average global temperatures have increased about 0.8°C since then and have been increasing most rapidly over the last two decades, which have been the warmest in recorded history.[11] Today, global mean temperature is the highest it has been since the last ice age roughly 12,000 years ago.[12]

The linear warming trend in the last half of the twentieth century has been accelerating and was nearly twice as high as for the entire century. This warming trend is consistent with the anthropogenic contribution of greenhouse gases into the atmosphere. The human contribution to carbon emissions is the result of burning fossil fuels, deforestation, and imprudent land use policies.[13] At current warming rates, it is predicted that by the end of the twenty-first century temperatures will increase between 1.1°C and 6.4°C.[14] Although the low end of that range may be tolerable, the mid to high ends are catastrophic.

We are able to observe the effects of rising temperatures. Clearly, weather patterns are changing. Cold days and cold nights are less frequent, more intense, and seem to occur out of season. Heat waves are more frequent. Heavy precipitation events have increased. Tropical storms, hurricanes, and cyclones are occurring more frequently and with greater intensity. Sea levels are rising. Ice caps are melting. Permafrost is thawing. Coral reefs are disappearing. Species are becoming extinct at a more rapid rate. Agricultural patterns are changing. Populations are migrating. In short, global warming affects every life form.[15]

We can put another face on global warming. Many climate models show that a doubling of industrial levels of greenhouse gases is likely to commit the earth to a temperature increase of between 2°C and 5°C. A 5°C increase is far outside human experience. Indeed, the earth has not had temperatures higher than 5°C above pre-industrial levels in 30 million to 50 million years.[16] More to the point, the world must change its pattern of energy use if it wishes to keep global warming below the 2°C threshold.[17]

What must be emphasized throughout a discussion of the numbers about carbon emissions and warming trends is that although these numbers may not attain scientific precision, the trend lines are unmistakable and the projected consequences unacceptable.[18] To some degree, the numbers will vary based on any number of assumptions, including society's ability to retard the growth of the emissions through changes in energy consumption patterns; the pace and direction of technological innovation; target dates; estimates of population trends; and choice of climate model. Additionally, the question must be raised as to the relationship between greenhouse gas emissions and temperature. Another label for this relationship is climate sensitivity. Whereas it is clear that there is a direct and positive correlation between emissions and temperature,[19] the proper ratio between the two is less clear.

One way of thinking about climate sensitivity is to imagine a doubling of CO_2 emissions from the current level of 385 ppm to 770 ppm. Analysts vary on the likely consequences of such a doubling. In 2006, the IPCC estimated that if CO_2 concentration doubled, then the Earth will experience an approximate 3°C increase within a range of between 2.0°C and 4.5°C. Other analysts estimate the climate sensitivity of a doubling of CO_2 to be

6°C.[20] In actuality, all of these numbers must be looked on as probabilities and as averages across the whole surface of the planet. In a business-as-usual scenario, carbon levels are projected to reach 610 ppm by the end of the century.[21] If that level is reached, then there is, for example, a 99 to 100 percent certainty that temperatures would increase by between 1.4°C and 2.4°C.[22] Further, at 610 ppm, there is an 82 percent chance of a 3.4°C increase and a 47 percent chance of an increase of about 4.4°C.[23] Ranges and probabilities are an attempt to locate the broader problem of global warming; they are not intended to suggest less risk or to provide data to relieve our most troublesome concerns. Indeed, the import of the numbers is unmistakable. "These are not small probabilities of big changes; they are large probabilities of enormous changes."[24]

There is, however, a noticeable trend in climate sensitivity analyses. Analyses that indicate a climate sensitivity of less than 1°C are decreasing, whereas analyses showing a greater chance of the sensitivity higher than 5°C are increasing. In other words, there is a 20 percent chance that the world can experience a warming in excess of 3°C above pre-industrial levels even if greenhouse gas concentrations are stabilized at today's levels.[25]

Complicating matters further is the existence of feedback loops.[26] The strong possibility exists, for example, that climate change may accelerate future warming by reducing natural absorption and releasing stores of carbon dioxide and methane as the Earth's surface warms. Increased temperatures can also change rainfall patterns, which will also weaken the ability of natural sinks to absorb carbon dioxide. Similarly, the thawing of permafrost regions will increase warming and reduce carbon sinks while releasing greenhouse gases from frozen peat bogs. As the Earth warms, these feedback loops could be triggered and a temperature spike could be 1°C to 2°C higher than the expected estimates by 2100.

To put the issue of 350 ppm and 2°C in context, these numbers are intended to be markers for a broader discussion of changes in energy policy. Climate change is not the only driver for a smarter, cleaner energy policy. However, the reality of increasing greenhouse gas emissions, coupled with the warming of the Earth's surface, provides sufficient reason for moving away from the traditional fossil fuel energy path. Still, one can remain cautious, or even skeptical, about climate change, or, more accurately, about the ability of politicians and administrative bureaucrats to successfully address the challenges while still acknowledging both the science behind climate change and the economic and environmental consequences of business as usual. Assuming, then, that business as usual contributes to global warming, the decision to invest in mitigation measures today becomes an exercise about costs in the future.

Costing out carbon emission reduction investments is deeply problematic along a number of tracks. First, cost-benefit analysis is notoriously difficult in the environmental arena because the costs of action today are more easily quantifiable than benefits to health and reduced health risks tomorrow. In other words, it is argued that the costs and benefits of environmental action are simply incommensurable.[27] Second, whereas economists generally agree that future investments must be discounted to present value, economists do not agree about precise discount rates; others debate whether or not discounting makes any sense in intergenerational contexts.[28] In fact, the debate over discount rates regarding climate change has generated considerable controversy.[29] Third, matters of intergenerational equity and global justice may not be susceptible to an economic calculus and, instead, are based on a variety of political and ethical assumptions about the future.[30] Will future societies be richer or poorer than today's? Will individuals live longer? How long is intergenerational? What types of environmental catastrophe are likely to occur with rising temperatures? Are climate change trends reversible? Will technological discoveries be made that remedy the problem without the need for contemporary abatement investments? Each of these questions involves ethical/political assumptions not susceptible to quantification and cost-benefit analyses.

A useful way to consider the costs of climate change is to contrast a business-as-usual scenario against a policy scenario that incorporates abatement measures. Under this exercise, the trade-off is between: (1) continuing the current energy policy that, we believe, contributes to economic growth while it continues to emit greenhouse gases, which lead to climate damages, and (2) a transformed energy policy with abatement investments at the risk of slower economic growth and lower future climate damages.[31] To put these two scenarios in brief perspective, a recent National Research Council study estimated that business as usual in 2005 cost the economy $120 billion in health-related damages from burning fossil fuels,[32] and a study by the Union of Concerned Scientists estimated that under the low-carbon scenario the economy can realize annual avoided energy costs of $465 billion by 2030 and accumulated savings of more than $1.7 trillion between 2010 and 2030.[33] In short, the environment and the economy can either pay now or pay later. The studies show, however, that it is likely that paying now is less costly than paying later.

Stern places the cost of climate damages under the business-as-usual scenario at between 5 percent and 20 percent GDP[34] and places the cost of abatement at 2 percent,[35] which is the midpoint of a range of studies that place the costs between 1 percent and 3 percent GDP.[36] Assuming an abatement cost of 2.5 percent GDP, a high-end estimate but lower than the 4 percent U.S. annual military expenditure, this is the equivalent of losing one year of growth or waiting twenty-nine years to double the economy instead of waiting twenty-eight years.[37]

Another study in the cost literature reported median estimates that mitigation expenses will reduce GDP by 0.47 percent during the period 2010–2030 and by 0.72 percent during the period 2010–2050.[38] In other words, by 2030 the GDP growth rate under business-as-usual is expected to be 2.86 percent and under a climate change policy scenario to be 2.84 percent, which is a negligible change that can be reversed by adopting other assumptions or models.[39] McKinsey & Company projects a cost to GDP of 1.3 percent[40] that, when positive returns on investment are included, produces a net cost of 0.4 percent GDP.[41] Still, other analysts argue that a low-carbon economy will have net positive economic effects and increase economic growth through the opening of new markets and through job creation.[42] Whereas others still, predominantly business and industry lobbies, predict financial catastrophe with a $669 billion drain on the economy, resulting in $8 a gallon gasoline and the loss of 4 million jobs by 2030.[43] These industry statistics have been roundly criticized.[44]

For our purposes, recognizing limits on carbon dioxide emissions and paying attention to changes in global temperatures serve as useful guides against which changes in energy policy and technology can be measured. Accepting the facts of global warming and greenhouse gas emissions leads to both a reconsideration of traditional energy policy and suggests avenues for the future. Because fossil fuels contribute to both emissions and warming, substitute fuels are necessary. What we find, then, is not only a consensus on climate change goals, we find a consensus on energy policy responses. Most particularly, the consensus is that waiting to move to a low-carbon economy is economically imprudent and environmentally unsound.[45]

The Three Faces of Security

Consensus energy policy recognizes the need for security. Security, however, has three dimensions – national, economic, and energy – that unite around a policy of promoting oil independence and increasing the use of renewable resources.[46] Additionally, there are three concerns surrounding national security that involve annual expenditures of billions of dollars. First, roughly one-half of our imported oil, 4 million to 5 million barrels per day (mbd), comes from OPEC countries, and the United States imports approximately 1 mbd from Saudi Arabia alone. The cost of extracting oil in Saudi Arabia is negligible, and for purposes of discussion we can set that cost at $5 per barrel. In the summer of 2008, oil prices exceeded $140 before receding to between $70 and $80 a barrel in 2009–2010. The math, then, is simple: At $140 per barrel the United States is shipping $135 million to Saudi Arabia every day. At $70 and $80, the United States is shipping $65 million to $75 million a day, respectively. Over the course of the year, U.S. dollar exports range from $24 billion to $50 billion. Again, *to Saudi Arabia alone.* It is further estimated that over the next two decades, the United States will spend more than $5 trillion importing oil.[47]

The staggering amount of money becomes more problematic when we recognize: (1) that we are borrowing the money from China to supply our oil addiction; (2) that the cost of imported oil constitutes a severe trade imbalance; and (3) some of those Saudi funds find their way into the hands of the terrorists who are undermining stability in the Middle East and are actively committing violence against us. Additional consequences for foreign oil imports involve propping up Venezuela and the dictatorship of Hugo Chavez; generally funding petrostates that are notoriously undemocratic;[48] and supporting the quixotic and corrupt government of Vladimir Putin, to name only a few economic/political distortions without mentioning environmental degradation.[49]

If we add the environmental consequences of global warming to our national security concerns, then a second level of adverse consequences follow.[50] Although some countries will benefit from warmer temperatures, masses of people around the world will suffer. Poor and underdeveloped regions are likely to have fewer resources and are unlikely to be able to invest in mitigation measures while adaptation opportunities decrease. Potable water becomes scarcer. Pestilence increases. Famine and poverty escalate. Mass migrations of hundreds of millions of people within nations and across borders, especially in Central America, South Asia, and Southeast Asia, will occur as a result of rising sea levels. Migration will also occur in areas of Africa as a result of desertification, thus exacerbating international tensions and further stretching U.S. military capabilities.

The U.S. military is also stretched and challenged by imported oil. The Department of Defense is the country's, and the world's, largest oil consumer. DOD's annual energy expenditure is $20 billion, and 85 percent is dedicated to oil for transportation and operations. Price volatility dramatically affects the defense budget. For every change of $10 per barrel there is a corresponding budget alteration of $1.3 billion.

Further, military and other foreign policy initiatives, overt as well as covert, involve substantial expenditures and constitute the third area of national security. Just as oil does not include the social costs of pollution, the price of a gallon of gasoline does not include military and foreign policy expenditures needed to bring that fuel to market. Within the last two decades, two Iraq wars; continued Middle East instability; increased incidence of pirates taking oil vessels hostage; and the need to keep shipping lanes open all require investment of military dollars internationally.[51] Other foreign policy expenditures include diplomatic efforts to convince producing countries to maintain oil flows and price stability; the exploitation of new oil producing regions; and actions to stabilize oil producing, and sometimes despotic,[52] regimes.[53] Cost estimates are, not surprisingly, difficult and vary. Still, military expenditures for supporting oil imports range from $27.3 billion to $98.5 billion per year,[54] and overall foreign policy costs are in excess of $1 trillion in wealth transfers to foreign nations during the last decades of the twentieth century.[55]

The second dimension of security is economic security. More specifically, and more troublesome, the concern is about the economic insecurity of price volatility in oil markets. Here, as elsewhere, history is instructive. During the oil crisis of the 1970s, the price of oil quadrupled from roughly $10 to $40 per barrel. Given the decline of domestic oil production, substitutes for conventional oil were sought in tar sands and oil shales in the western United States. After hundreds of millions of dollars of investments in these alternative fuels, world oil prices declined and, because oil from tar sands and oil shales was not cost-competitive with world oil, these efforts collapsed completely when the price of oil receded to historic levels. Parenthetically, the same phenomenon occurred with less drastic consequences in the first decade of the twenty-first century, and we continue to monitor the price of oil as compared to the price of unconventional substitutes to gauge when, and whether, to move energy investments into oil substitutes. As attractive as domestic substitutes may be, price instability prevents investments from being made in those sectors and in renewable resources.

Oil shocks affect the economy by raising prices, which are notoriously inelastic in the short run. It has been estimated, for example, that a 1 percent change in oil supply can have a 5 percent to 10 percent effect on price.[56] In other words, demand, as well as alternative supply, are slow to respond to price hikes, which have a regressive effect on consumers who must pay a larger percentage of their household budget for energy. Additionally, higher prices can negatively affect the country's productivity. In the 1970s, the oil disruptions contributed significantly to double-digit inflation and high unemployment. Economists have argued that the three oil shocks of the 1970s and early 1990s preceded three economic recessions. In the 1970s, it was estimated that GDP fell by between 0.6 percent and 4.5 percent as a result of high oil prices. Further, it has been estimated that the oil shocks cost the economy on the order of $1 trillion to $3 trillion since 1973.[57]

Oil markets have been changing, and not necessarily for the better. Although OPEC power may be slightly lessened because of new market actors, it is still a force that can dramatically affect prices. Today, oil markets are not controlled solely by cartels such as OPEC and by state-run oil companies; oil markets are now subject to speculation by a diverse set of investors including hedge funds and investment banks.[58] Consequently, oil prices have become harder to predict and, therefore, difficult and expensive to rely on. Oil, like all commodities, is priced based on supply and demand. World projections are that demand will continue particularly from India and China regardless of efforts to move to non-fossil fuel-based energy economies elsewhere. Supply, however, is troublesome and subject to price volatility, primarily due to continued OPEC cartelization and geopolitical disruptions, as political instability continues to threaten supplies.

The third dimension is energy security. Put most simply, if the United States had the capacity to generate all of its energy needs domestically, then economic security follows and national security is improved substantially. Indeed, the arguments for energy independence are similar to those for energy security with one caveat. In a peaceful world with competitive markets, there would be no need for a country to produce all, if any, of its energy domestically. Instead, market transactions would take place at the cheapest (most efficient) point. However, in a contentious world with unfriendly cartelized markets, which is to say the oil sector, energy independence is highly desirable and the consensus exists for a future energy policy to achieve as much oil independence and, therefore, energy security as possible.

Consensus Energy Policy

The core element of the consensus energy policy is that private-sector investment is the single most important component; it is the engine that will bring innovative technologies to market, create new jobs, and build the new energy economy.[59] However, given the country's history of fossil fuel favoritism, government has a necessary role to play in facilitating the investment in the transition. Just as there will be debates around defining precise targets for carbon emissions and about how best to reach energy security, there will be debates around specific policy measures. Nevertheless, a strong consensus exists as to the overall goal of moving to a low-carbon energy economy by replacing fossil fuels with energy efficiency and renewable resources even though interested actors will debate about what measures are most promising and available to achieve future gains.[60] Mostly, the debate about means is a debate about how heavily to invest in particular measures and which sector of the energy economy is more likely to prove to be successful. Carbon capture and storage, for example, is a crucial technology for reducing carbon emissions, but there is a lack of large-scale demonstration due to the enormous costs involved. Should, then, CCS be initially developed by the public or private sectors or some combination of both? Another example of debate about future energy policy is: Should investments be made in oil substitutes or in nuclear power? If either or both, then should the investments be made by public or private actors? This section lists six strategies and mechanisms about which there is large agreement for successfully transforming energy policy.[61]

Energy Efficiency in Buildings, Transportation, and Industry[62]

Energy efficiency promises the most significant opportunities for a sustainable and clean energy future at the least cost in buildings, transportation, and industrial processes. McKinsey & Company reports that once barriers are removed and efficiency improvements are fully deployed at scale, then energy savings can reach $1.2 trillion, which comfortably covers the estimated $520 billion needed through 2020 for up-front efficiency investments.[63] It is also estimated that by 2030, the United States can realize a 30 percent reduction of energy use in these sectors. Further, new energy efficiency markets are opening and expanding. In the residential sector alone, the energy retrofit market can run to $400 billion; and the addition of commercial buildings doubles that figure,[64] resulting in substantial job creation.

Energy efficiency moves along two fronts. First, we can use less energy and use less costly energy through conservation. Second, and more importantly, we can increase productivity from the energy that we consume, and we have been doing so for several decades. The United States has, since 1973, reduced its energy intensity by 45 percent. In other words, we use less energy per unit of GDP today than we did more than thirty years ago.[65] Going forward, McKinsey & Company reports that energy productivity can reduce energy demand by 22 percent by the year 2020.[66] Furthermore, investments in energy productivity are beneficial not only for the environment, they benefit the world economy as well with projected internal rates of return on investment of 17 percent, resulting in savings of $900 billion annually by 2020.[67]

The Obama administration recognizes the gains to be made through energy efficiency both in the Stimulus Bill, which has dedicated $24.4 billion to energy efficiency,[68] and in the proposed climate and energy legislation, which contains energy efficiency provisions for each of the sectors discussed here.

Notably, for the most part, efficiency technologies in this sector already exist and, also for the most part, are cost effective.[69] An investment of $2,500 in such things as caulking and insulation for a single-family residence, as an example, can be recouped in three to four years.[70] Indeed, a new home today consumes 30 percent less energy than one constructed prior to 1970, and today's refrigerator consumes less than half that of a twelve-year-old model.[71] Even seemingly small gains reap dramatic payoffs. The EPA's Energy Star Program, for example, reduced electricity usage by 4 percent, which translates into a $12 billion savings and is equal to avoiding greenhouse gas emissions in 23 million cars.[72]

There are more than 115 million buildings in the United States that consume roughly 40 percent of all energy in the United States and consume more than 70 percent of all electricity produced. The consensus is emerging that between 20 percent and 30 percent savings against projected consumption can be realized over the next twenty to twenty-five years. More ambitiously, the nonprofit organization Architecture 2030[73] has set a target for 2030, at which time all new buildings will be zero-carbon emitters while generating 4.5 million jobs and more than $1 trillion in construction revenue.[74] Savings in new buildings will come from building design, green standards, construction techniques, and improved materials. Savings for both new and old buildings will also come from existing technologies in space heating and cooling, water heating, and lighting. Advanced lighting and cooling techniques can achieve even larger gains.[75] The National Academy report shows that a cumulative investment by public and private funders of $440 billion in *existing technologies* between now and 2030 can produce annual savings of approximately $170 billion in reduced energy costs.[76] Advanced technologies such as solid-state lighting, advanced cooling, and smart metering promise further gains. A report from The Center for American Progress and the Energy Future Coalition estimates that through 2020 up to 1.25 million jobs can be created with consumer savings of up to $64 billion.[77]

Barriers, however, have prevented broader adoption of a full range of energy efficiency measures.[78] One problem is known as the principal-agent problem. Consider the situation of a contractor or landlord and the owner or tenant. Although an owner or tenant will benefit from energy efficiency improvements to construction, neither the contractor nor the landlord have incentives to raise the cost of construction or rental unless competitor contractors and landlords are also making the same improvements; otherwise they lose sales. Further, imperfect information about cost savings makes consumers reluctant to invest in efficiency improvements until they feel assured that their investments will be returned with interest.[79] Consequently, labeling programs, such as Energy Star or LEED certification, are efforts to provide information and public education to consumers about energy efficiency.

These and other barriers can be overcome and efficiency improvements can be realized with existing technologies and with little or no direct public expenditures. Instead, technological innovations can be facilitated through smart building and design standards, favorable rate treatment by public utility commissions for efficiency expenditures made by private utilities, favorable tax treatment such as credits for improvements, and accounting practices that can favor new technologies. Capital investments, for example, are often depreciated over the thirty-year life of a building. If those investments were treated as expenses, then they can be deducted in the year in which they are incurred, thus serving as an incentive for increased energy technology expenditures. These sorts of government regulations, then: (1) attempt to pull these innovations into the market; (2) rationalize the market through uniform standards; and (3) disseminate information about efficiency innovations. In other words, there is no need for direct public funding; instead, indirect funding is intended to bring current technologies to market scale.

The transportation sector will require a different set of regulations and investments. Transportation is almost solely reliant on petroleum and produces about one-third of U.S. greenhouse gas emissions. Energy savings can be realized through improvements in fuel economy standards; fossil fuel use can be reduced through hybrid vehicles and biofuels; and, in the longer term, fuel cell vehicles and battery electric vehicles will be developed. Improvements in car design, engine efficiency, and diesel engines can also improve vehicle efficiency. Additionally, to realize greater savings, improvements in battery life and fuel cell technologies will be necessary.

In the transportation sector, both direct and indirect investments are needed on both the supply and demand sides. On the demand side, vehicle fuel efficiency standards have recently been raised from 25 miles per gallon to a target of 35 miles per gallon, which will improve fuel efficiency. On the supply side, additional improvements in vehicle design and engines will be necessary to move into the new energy economy. The potential is significant for a deployment of a variety of vehicles, including hybrid electric vehicles (HEV); plug-in hybrids (PHEV); battery electric vehicles (BEV); and hydrogen fuel-cell vehicles (HFV). Currently, hybrid electric vehicles are on the streets and plug-ins are emerging. BEVs and HFVs, however, will require technological advances particularly in the area of longer storage batteries, the greater use of lithium technologies, and reduced vehicle weight.[80] Consequently, investments must be made throughout the innovation stream from basic science, such as advanced batteries, to the development of transportation alternatives and to the construction of supporting infrastructure to bring these technologies up to market scale.

The industrial sector should realize energy efficiency savings of approximately 20 percent by 2020, and the Rocky Mountain Institute reports that efficiency measures can increase industrial electric productivity alone by 8 percent.[81] Energy-intensive industries such as chemicals, petroleum, steel, and cement should be able to capture efficiency gains through combined heat and power production, better facilities and machine design, and improved manufacturing processes. In the future, advanced materials that resist corrosion and degradation during performance at high temperatures should improve manufacturing processes and reduce energy costs. Further automation and process improvements can likewise result in savings. The industrial sector has been slow to adopt these measures because of short-term thinking when designing new processes and facilities; failure to consider and set energy and resource efficiency targets; and a reluctance to make up-front capital investments even if they promise reasonable returns, among other barriers.[82] Federal energy innovation policy can fund basic science and technology research for materials, and tax and accounting measures can be used for investments in manufacturing processes and combined heat and power (CHP) applications.

Again, innovation investments can be made along a spectrum from direct funding of basic science involving advanced materials to simply expanding the utilization of CHP through either tax credits or other back-end incentives. It should be noted that as a particular technology, or new application of an existing one, moves closer to market, the more important and greater the role of the private sector. Indeed, innovation policy must treat new energy technologies along a continuum with the endgame being private-sector ownership. Government has a necessary role in stimulating technological energy innovation but has little role in operation and ownership of the technologies developed through R&D investments, unlike its historic ownership and operator roles in defense and space exploration.

Alternative Transportation Fuels[83]

The transportation sector consumes roughly 14 million barrels of oil per day, most of which is imported. To further energy security, technologies that reduce dependence on imported oil by domestically producing alternative liquid fuels to replace gasoline and diesel are a high priority investment. The United States has abundant coal and biomass resources that will require cost-effective technologies to convert those resources to liquid fuels.

Biofuels present an opportunity to move toward oil independence, develop the agricultural sector, retrieve energy from waste, and create new markets and jobs.[84] President Bush, in his 2007 State of the Union messages, proposed that biofuels should replace 20 percent of our gasoline consumption by 2017. Today, the Departments of Energy and Agriculture are moving forward to meet that goal through projects that both increase the supply of alternative fuels and improve vehicle efficiency. The DOE has funded a variety of programs from research on forest waste to algae and from development and demonstration to infrastructure.

Biomass fuels for the production of ethanol have substantial potential. Currently, gasoline contains about 4 percent alcohol to reduce smog, and it has been estimated that cars can be built to use about 85 percent ethanol at an additional cost of between $50 and $100 per vehicle.[85] Yet a choice must be made between corn, which is the current primary biofuel feedstock, and cellulosic fuels.[86] It would be an unwise energy policy to rely on corn ethanol for two reasons. First, corn is only about 12 percent energy efficient and, second, it has a negative effect on food prices. Cellulosic biomass from dedicated energy crops, agricultural and forest residues, and municipal solid waste, however, has minimal impact on food prices and is more energy efficient.[87] At today's prices, however, producers will likely need financial incentives to grow biomass instead of other crops.

Additionally, conversion technologies play a necessary role in realizing full deployment of biomass resources. Currently, biochemical conversion of starch from grains to ethanol has been demonstrated but remains in the early stages of commercial-scale deployment. Additionally, a transportation and distribution infrastructure will be needed because ethanol cannot be transported in oil pipelines. It has been estimated that if the necessary conversion and distribution infrastructure were in place, then biomass could reduce oil consumption approximately 1.7 mbd to 2.0 mbd.[88]

Coal and biomass, either separately or together, can be converted into a liquid transportation fuel. The United States, as noted, has substantial coal reserves and has the technological capability to deploy conversion to fuel liquids. Fifty thousand barrels of oil per day can be derived from 7,000,000 tons of coal. In other terms, 100 plants of this size could produce the oil equivalent of 5 million barrels per day by utilizing 700 million tons of coal. This constitutes a 70 percent increase in current coal consumption. To achieve this goal, substantial investment in plant and infrastructure will be necessary. It has been estimated that a combination of biomass-to-liquid fuels and coal-to-liquid fuels can displace more than 5 million barrels of oil equivalent per day by 2035. However, this large-scale deployment will not occur absent aggressive demonstration in the short term.[89]

Technologies currently exist for the indirect liquefaction of coal to produce transportation fuels. These technologies are commercially deployable; however, without carbon storage, these technologies would generate twice as much carbon emissions as current petroleum-based fuels because carbon is emitted in the conversion process as well as during burning for consumption. Coal-to-liquid and biomass-to-liquid fuels can significantly reduce our dependence on foreign oil. However, the technologies with CCS are not proven and are not commercially viable. It is estimated that coal-to-liquid fuels without CCS can be produced at roughly $70 per barrel of gasoline equivalent. Biomass fuels without CCS can be produced at $115 to $140 per barrel. In other words, these alternative fuels present three problems: (1) even without CCS, they are not commercially viable; (2) they fail to reduce carbon emissions; and (3) with CCS, they are projected to be even more expensive and less economically viable.

To become competitive with current fossil fuels, a carbon price needs to be imposed that will make these alternative fuels cost-competitive. Additionally, demonstration projects for conversion at scale and a distribution infrastructure are necessary. Further, if energy targets are to be met within the next generation, these projects must be undertaken immediately. Carbon capture and sequestration, then, is a necessary energy technology that achieves several purposes including providing liquid fuels as well as generating low-carbon-emitting electricity. In developing alternative fuels, several challenges present themselves. All of these challenges can be overcome through targeted innovation policies to facilitate these programs. First, agricultural policy must attempt to provide sufficient and sustainable biomass products without adverse environmental impacts. Second, CCS must be shown to be both technologically and commercially feasible as well as economically and politically viable. Large-scale demonstrations and procedures for operation and monitoring of CCS must be actively pursued if solid-to-liquid conversion is to make a notable contribution to a new energy policy. Additionally, standards must be set on carbon emissions for both old and new power plants.[90] Third, cellulosic ethanol is in the early stages of commercial development, and demonstration plants are on the near horizon. Fourth, a distribution infrastructure will be necessary to utilize these fuels. Fifth, the costs of alternative fuels are currently dynamic and fluctuate as a result of crop availability, labor, and government policies.

In addition to these resources, multiple technologies exist for deployment beyond the year 2020. Many of these technologies involve bioengineering from yeast, algae, and bacteria. Further, hydrogen fuel cell technology has progressed rapidly yet remains too expensive for deployment. Thus, direct public funding is needed along the alternative fuels spectrum from basic science to large-scale development and deployment, and either direct or indirect funding will be needed for full-scale commercialization.

The Obama administration announced three measures to promote the production of biofuels and reduce oil dependence.[91] The Environmental Protection Agency has finalized the rules to achieve a long-term renewable fuels standard of 36 billion gallons by 2022.[92] Second, the Department of Agriculture has proposed a rule to provide financing to increase the conversion of biomass to bioenergy, and the administration released a report setting out a strategy to advance the development and commercialization of a sustainable biofuels industry.[93] Finally, the president issued a memorandum creating an Inter-Agency Task Force on Carbon Capture and Storage to develop a comprehensive and coordinated federal strategy to speed the development and deployment of clean coal technologies. In that memorandum, the president calls for five to ten commercial demonstration projects by the year 2016.[94]

Renewable Energy[95]
The primary renewable energy resources are solar, wind, geothermal, hydropower, and biomass. For the most part, we will depend on these resources to generate cleaner electricity. For decades, renewable resources have played a negligible role largely because they have not been cost-effective. However, in 2008 investments in renewable resources exceeded fossil fuel investments by $30 billion, and new installed capacity from renewable resources exceeded that of nuclear power and fossil fuels combined.[96] Thus, using renewable resources to produce electricity presents an opportunity for dramatic change in energy policy. Unlike coal, natural gas, and

nuclear power, renewable resources can be decentralized and can operate at significantly reduced scale. Additionally, a wide variety of different types of renewable resources can be used to produce electricity, thus increasing consumer choice and expanding competition while opening new markets to new entrants.

The current electricity sector is problematic for a number of reasons. It is responsible for 40 percent of the country's carbon dioxide emissions. Moreover, because the country is relying increasingly on coal, emissions from power plants have grown by more than 33 percent since 1990. Coal-burning power plants constitute the country's largest source of carbon emissions, representing more than all of the elements in the transportation sector combined.[97] Consequently, low-carbon electricity-producing technologies play a necessary role in a clean energy future, and their potential is promising enough to provide all of the nation's electricity needs. Indeed, a series of comprehensive reports concluded that the United States can achieve carbon emissions goals by pursuing both energy efficiency and renewable resources.[98]

Given this promise, the question arises as to why we have not deployed renewable resources to any substantial extent in the last fifty years. Several barriers remain in addition to cost-effectiveness. Conflicts over land use, higher short-term costs, difficulties in scaling these technologies, access to transmission, and public acceptance have slowed the development of renewable resources for electricity production. It is notable, though, that the problem of technologies is absent from this list. Indeed, it is a strength of renewable resources that most technologies are currently available and technological improvements will be made that have the effect of further lowering costs.

In recent years, renewable energy, most notably wind power, has grown in importance as an electricity provider. Although estimates vary, non-hydroelectric renewables could provide 10 percent of the U.S. electricity by 2020 and 20 percent or more by 2035. Wind has installed more new electricity production than any other source, other than natural gas, and is responsible for approximately 35,000 jobs in 2008.[99] By the end of 2009, the United States had more than 35,000 MW of wind power installed, which makes it the world leader in wind capacity.[100] Since 2000, wind capacity has increased an average of 30 percent per year. Still, actual contributions are marginal as wind provides less than 1 percent of the nation's electricity.

Wind power is also notable for its diversity in size. Smaller turbines can be used by homes and farms to generate 100 kW or more. Larger turbines can be used for utility-scale generation from 500 kW to more than 3 MW.

The DOE estimates that wind alone has the technical potential to provide ten times today's U.S. electricity needs and that it can provide up to 20 percent of U.S. electricity by 2030 without affecting the reliability of the power supply.[101] If wind is to continue to play an increasing role in electricity generation, state and federal renewable portfolio standards and federal production tax credits will continue to be needed.

Solar electricity is generated from either concentrating collectors (CSP) or photovoltaics (PV). The primary difference between these two technologies is that CSP can operate as a large-scale power producer that needs to be connected to a grid, and PV technologies are more flexible and can be used at various scales down from individual buildings to wrist watches. Like wind, solar power has the potential to generate all of the country's electricity needs through PV installations affecting less than 1 percent of U.S. land area or through CSP installations affecting a 100 square-mile area.

Currently, solar contributes 0.01 percent of the total electricity in the United States, yet both solar markets are beginning to expand. The PV market grew 62 percent in 2007. Moreover, a study of the wind, solar, and geothermal technologies shows that the cost of those technologies fell between 50 percent and 90 percent from 1980 to 2005.[102] These gains have been made as a result of technological advances, growing sales, economies of scale, and government support through tax incentives and R&D. Future projections indicate that costs will continue to decline. Further, to the extent that a carbon price is imposed on fossil fuel electricity, then renewable resources become not only more cost-competitive, they can be priced cheaper than fossil fuel electricity. The most significant challenges to large-scale adoption of solar power lie in the need to bring existing technologies to commercial scale with advanced technologies and to bring the power to market through grid access.

Government has a role to play in increasing the percentage of renewable resources used to produce electricity. Although the federal government has yet to adopt renewable portfolio standards (RPS), more than half of the states have a renewable resource requirement that has the effect of expanding the use of these alternatives. Renewable markets can be enhanced through more comprehensive use of tradable renewable energy credits, ideally on a national, rather than the regional, market.[103] RPS requirements, together with investment tax credits and production tax credits, will help expand the markets for solar and renewable resources, especially if these requirements are set for a number of years going forward instead of for two or three years at a time as in past legislation. Additionally, transmission grid improvements together with interconnection and access policies must be settled. All of these measures will help expand the market, thus lowering costs.[104]

The argument is often made that wind and solar power are intermittent resources and, therefore, present problems of reliability. Whereas it is the case that different sections of the country have variable access to both resources, it is also the case that even coal-fired power plants have downtime and, therefore, are intermittent. Central power stations must align supply and demand not only for reliability but for efficiency as well. There is, then, no technical reason why smaller wind and solar producers cannot be plugged into the grid and have power dispatched both reliably and efficiently.[105] In short, utility dispatchers will change the way they do business from operating a small number of large power plants to operating a large number of small and diverse producers while maintaining grid reliability.[106] Because of the clean potential of renewable resources for the production of electricity, the smart grid becomes all the more important and will be discussed further in this chapter and in Chapter 6.

Fossil-Fuel Energy[107]

In addition to dominating our energy economy, it is also noteworthy to recognize that fossil fuels have segmented markets. Oil supplies 98 percent of the transportation sector; coal generates more than 50 percent of our electricity; and natural gas provides 74 percent of non-electric energy use in commercial and residential sectors. At the moment, there is little crossover or substitution among these resources. Future energy policy, particularly through the use of renewable resources and alternative transportation fuels, will witness a breaking down of these sectoral barriers. Most obviously, biofuels and electricity will be used to a greater extent for transportation.

Fossil fuel markets are coming under increasing pressure not only because of climate change. Demand from developing countries is growing significantly, thus aggravating the international security tensions mentioned previously and further tightening markets. The world currently consumes approximately 85 million barrels of oil per day with estimates of future consumption in excess of 105 million barrels per day by 2030[108] against proven reserves of 1.2 trillion barrels.[109] We can, therefore, project at least another century of oil reliance at current, and expected, rates of production and consumption.

Because fossil fuels have served segmented markets for so long, a large fossil fuel infrastructure has developed at enormous expense. The existing stock of vehicles, heating systems, power plants, pipelines, and filling stations represent extraordinary investments, not to mention the supporting political and bureaucratic structure in place. Neither private investors nor their supporting political actors will quickly walk away from this energy system. Although it may be unrealistic to assume that the country's energy portfolio can divest itself completely of fossil fuels in the short term, even in the transportation sector, it is not unrealistic, indeed it is necessary, to redeploy these resources and to expect them to burn cleaner in the midterm. Further, whereas it is also realistic to assume continued exploration for oil and natural gas, as well as to plan for technical improvements in enhanced recovery techniques to extract more oil from old wells and to recover new oil from new places such as deep water offshore fields, it is not unrealistic to also seek cleaner substitutes with environmental safeguards. Nevertheless, even assuming technological advances, economic feasibility, and access to new oil, the DOE estimates that domestic production will not increase more than marginally from our current production rate of 5.1 mbd to no higher than 6.4 mbd under the best case scenario, especially given the fact that the United States has only 2 percent of proven world reserves.[110] In other words, the United States cannot drill its way to energy independence. Additionally, although crude oil from shale and tar sands may be available, its economic viability depends on the world price of oil and our tolerance for even dirtier burning fuels.

Natural gas, however, can be produced domestically and may well serve our current and future demand. Domestically, the United States produces 86 percent of the natural gas that it consumes, and large recent discoveries, particularly from shale, have been made.[111] Shale gas is a promising source of a lower-carbon-emitting fossil fuel. In 2000, shale gas contributed 1 percent to our nation's gas supply. In 2010, it contributed 20 percent and is estimated to contribute 50 percent by 2035.[112] Although promising, the United States has less than 10 percent of world reserves of natural gas; and unless the new discoveries are sufficient to meet demand, LNG imports will be necessary. The key trade-off between business as usual for coal-fired electricity and more efficient or cleaner electricity is the price of carbon. As carbon continues to be priced at zero and demand stays flat or increases, it is unlikely that any pulverized coal plants will be retired or modified even though few new ones will be built. As new electric plants are built, however, natural gas combined cycle plants will compete with coal; they release half of the carbon emissions of coal plants.

Whether or not these new natural gas plants are competitive with coal will be based on the cost per million BTU. At $6 per million BTU, new natural gas plants are competitive; at $16 per million BTU, they are not. A range of $7 per million BTU to $9 per million BTU can serve for planning and discussion purposes. Over the last few years, natural gas prices have fluctuated from $4 per million BTU to $13 per million BTU, thus introducing a significant financial risk to the planning process. Natural gas will be competitive with coal if and when significant and sustained domestic discoveries are made and are efficiently brought to market.

Carbon capture and storage is, of course, the key determinant for future clean coal projects. Whereas such technologies have been demonstrated at commercial scale for purposes such as enhanced oil recovery, no large electric plant captures and stores carbon today. Storage is possible in oil and gas reserves, deep formations with salt water, and deep coal beds. Sites would have to be selected and operated safely and environmentally soundly, and the geology must be suitable for injection and capture without migration of CO_2 underground. Current surveys indicate that a sufficient number of sites are available. However, the projected need for carbon storage would require not only substantial investment but significant engineering. The chicken and egg problem for CCS is that reliable cost and performance data are needed before the private sector will invest and such data can only be obtained by construction and operation of full-scale demonstration facilities. Additionally, it is posited that there should be a diversity of demonstration projects, which will require substantial public investment up to the point of bringing CCS plants to scale.[113]

Coal-fired electricity represents a substantial capital investment, and most of the units can produce electricity for the next twenty-five years or more. Nationally, coal-fired generation exceeds 300 GW and emits one-third of total U.S. carbon emissions and 7 percent of the world's emissions. Coal-fired electricity is not only dirty, it is notoriously inefficient. Pulverized coal plants produce nearly all of the electricity at 34 percent to 38 percent efficiency. Efficiency may be increased up to 44 percent, yet technological improvements would be required.

Three alternatives are available to reduce emissions from coal-fired power plants: (1) improve generating efficiency to reduce CO_2; (2) retrofit existing plants with CCS;[114] or (3) retire units and replace them with CCS-equipped plants.[115] The trade-off among these options is that the first option is less costly but will not have a significant impact on carbon emissions, whereas the last two options can significantly reduce emissions but at a much higher cost. It is estimated that whereas carbon capture up to 90 percent is technologically feasible, the cost to build or retrofit a plant with CCS would approach the cost of the original plant.

Current discussions about future coal policy raise the issue as to whether or not plants coming online can be "capture-ready." The term itself is deeply ambiguous given the fact that we have neither put a price on carbon through a tax nor have we set carbon emissions limits through cap-and-trade regulations, which would also have the effect of pricing carbon. Consequently, plant owners would be undertaking extraordinary financial risk without assurances that future regulations will not render those investments worthless. This financial risk exists even if plant owners make use of the most currently available efficient technologies and leave room for future technological improvements.

In light of financial uncertainties, utilities have historically chosen to build natural gas plants, which can be constructed in three to four years versus four to eight years for coal plant construction. Construction costs, then, are lower. Still, natural gas has been subject to financial risks because of fluctuating gas prices that are affected by global markets and because of the uncertainty of reserves. Additionally, even though natural gas emits carbon at half the rate of coal-fired electricity, carbon capture is still necessary for natural gas plants and presents the same cost patterns.

Coal-fired electricity will play a sizable role in the country's energy future until limits are placed on emissions or carbon is priced. Without such regulatory actions, coal plants will be retired at a very slow rate, having no notable effect on reducing emissions. Nor is there any financial incentive to put expensive carbon-capturing plants online. Business as usual will change with either a strong CCS policy or with the successful implementation of new technologies such as transforming CO_2 into useful building materials.[116] Even then, a demonstration period lasting until 2020 will be necessary to develop confidence in CCS technologies as well as develop appropriate and complementary regulatory policies.

Assuming that regulators price carbon correctly, that is, at a rate to induce the adoption of CCS, then if CCS is commercialized by 2020 it will still take two or three decades before the current dirty coal fleet is replaced.[117] The U.S. coal and traditional electricity industries can maintain their market niche, or even improve them, under this scenario for the next half-century or more.

The United States can achieve a cleaner energy future in the fossil fuel sector by pursuing the following policies. First, and most cheaply, energy efficiency must reduce electricity demand. Second, either emissions standards or carbon pricing must accelerate the retirement of dirty plants. Third, the greater use of natural gas for electricity generation will cut into carbon emissions. Fourth, new power generation technologies, as well as

CCS technologies, must be funded through innovative R&D funding in order to develop a suite of technologies that can be brought to demonstration for testing and then to commercialization.

Nuclear Energy[118]

Nuclear energy poses several intriguing problems. On the one hand, it has a minimal carbon footprint, which together with growing public acceptability suggests that nuclear power is an attractive option for our energy portfolio.[119] In addition to low-carbon emissions, nuclear power appears attractive because of the estimated growth in electricity demand as well as the volatility of natural gas prices in the past. On the other hand, waste disposal and management and high construction costs undermine its cost-competitiveness with other resources. Additionally, concerns are regularly raised about weapons proliferation; and a National Research Council report, among others, unanimously recommended that the DOE not go forward with its program for fast-neutron reactors and fuel reprocessing precisely because of proliferation risks.[120] Nevertheless, over the last few years a number of articles and reports discuss a nuclear renaissance.[121] In 2010, there were approximately thirty nuclear units in some stage of planning,[122] yet because of long construction times we are unlikely to see nuclear-generated electricity within the next twenty years. The key to a nuclear renaissance lies in the ability of the industry to construct prototype plants that can provide the necessary learning to construct plants more cheaply in the future. Right now, though, affordable nuclear plants are ideas on the drawing board only; they are not real world fixtures.[123]

Historically, the United States has not duplicated a reactor design in any of its 104 nuclear plants, thus making construction costs notoriously high. Construction cost uncertainty is partially the result of more than three decades of inaction in this sector, which raises the costs of the learning curve for building a new nuclear plant fleet. Domestic estimates indicate that nuclear construction costs are rising faster than other technologies, and the European experience has been that nuclear construction costs remain high as well as variable.[124] Construction costs can be reduced through standardization, innovation, learning-by-doing, and government support. The Energy Security Act of 2005 provided financial support for a first generation of new nuclear plants, and the Obama administration announced its support of the industry with more than $18 billion of loan guarantees.[125] Nevertheless, high construction costs and continued concerns about waste disposal keep the new nuclear fleet mothballed.

Technological innovations are available in the nuclear industry with the hope that structural costs will decline, bringing nuclear power closer to the cost of coal-fired electricity. Technological innovation in the nuclear sector involves improvements to existing plants as well as new reactor designs and new fuel processes. Upgrades to existing plants are substantially less expensive than adding new plants, and several plants are adding capacity. Nuclear plants continue to operate more efficiently, thus reducing downtime. Over the last two decades, the cost of operating nuclear power plants has reduced the price of electricity as these plants operate at near capacity and as their useful life is been extended from forty to sixty years with discussions about another twenty-year extension, which should only be made if it can be shown that the reactors do not pose a substantially high risk of catastrophic radiation release.[126] In other words, existing plants offer the potential to increase their power output through higher power fuels; new plant components; capacity additions; extending their operating lives through license extensions to sixty or eighty years instead of the anticipated forty-year life; and reducing downtimes. These improvements are much less expensive than constructing new plants and can be implemented in the short term.

Additionally, two new nuclear power technologies are being considered, both of which are intended to improve safety, security, liability, and efficiency. First, technologies that improve on existing light-water reactors are referred to as evolutionary reactor designs, and these technologies are ready for deployment without additional R&D. All reactors in the United States are collectively known as light-water reactors, which use ordinary water to cool the plants and control neutron reactions.[127] An example of an evolutionary reactor design is one that allows a reactor to operate at higher temperatures, thus offering process heat in addition to the production of electricity.

Second, new technologies can also provide alternative reactor designs, and many of these technologies are currently being developed and will need further R&D as well as demonstration before deployment. A current example of an alternative reactor design ready for deployment is the small modular reactors on the order of 10 MW to 140 MW, or one-hundredth to one-tenth the size of the largest current plants, which are substantially cheaper and quicker to build.[128] Further, siting for small reactors is easier and more flexible, and some designs would use depleted uranium, run for 100 years, and would not require refueling.[129] Additionally, alternative fuel cycles, instead of the once-through cycle now in use, will allow fuel to be reused. Reprocessed uranium lowers fuel costs and reduces waste but can achieve weapons-grade status, thus increasing proliferation risks.

The nuclear industry faces a serious conundrum. Even if license extensions are granted for a total of eighty years, the first generation of nuclear plants will begin to be retired, thus raising the question of whether or not the rate of new plants can keep up with the retirements of old plants let alone generate additional electricity. Other barriers include length of the licensing process, continuing uncertainty about waste disposal and management, continuing concerns about safety and security including nuclear proliferation, and the availability of adequate quantities of uranium. Given the slowdown worldwide in nuclear plant construction, extraordinary measures will be needed if nuclear power is to make a substantial contribution to carbon reductions by 2050. To achieve notable reductions, an average of twenty-five U.S. plants per year for the next forty years will have to be constructed. Moreover, if nuclear power is to play a significant contributing role by the end of that period, thirty to fifty reactors will be needed annually worldwide.[130] Still, existing regulations such as loan guarantees, production tax credits, liability limitations, regulatory lag insurance, and new plant subsidies favor nuclear power even in light of negative market signals.[131]

Used fuel continues to be a problem insofar as there is no permanent storage facility, and Yucca Mountain has been scrapped as a possible site after decades of study. President Obama has appointed a Blue Ribbon Commission on America's Nuclear Future specifically charged with examining the backend of the fuel cycle and recommending alternatives for waste storage, reprocessing, and disposal.[132] It is estimated that on-site storage can continue for several decades. Nuclear plant expansion, then, is technologically feasible, and the barriers that confront the industry are: *financial* – relative to the cost per kilowatt hour generated and for waste disposal; *regulatory* – relative to licensing processes and the time to review and approve applications; and *practical* – because of questions about the adequacy of sufficiently skilled personnel after years of a moribund industry.

Electricity Transmission and Distribution[133]

Investment in infrastructure – a smart infrastructure – is necessary to achieve any or all of the previously mentioned gains in energy efficiency, demand response, emissions reductions, and energy productivity. The electric grid has the capacity to allow renewable resources to be sold to consumers; to provide cleaner electricity; to reduce fossil fuel dependence; and to promote economic growth. The grid can do this by diversifying the resources used in the production of electricity and by sending more accurate price signals and other information to both producers and consumers.

The current grid is comprised of transmission and distribution. The transmission system consists of approximately 130,000 miles of high-voltage transmission lines and the distribution system of millions of miles of lower voltage lines that bring electricity to consumers. The United States is comprised of three transmission systems; all the generators within an interconnection are synchronized with one another but not between the three synchronized systems. Over the last several decades, investments in the transmission system have dropped off dramatically. The drop-off is principally attributable to uncertainty about earning a return on investment in the grid. As a consequence, reliability is threatened as the system operates near its physical limits, thus increasing the risk of cascading disturbances.

A modernized transmission and distribution system would incorporate a variety of new technologies not only to distribute electricity more efficiently but would allow for two-way communications between producers and consumers. Advanced communications can provide information about pricing, reliability, and stress. Additionally, a modern grid will accommodate a diversity of generation options including efficiency and renewable resources; will be self-healing to avoid widespread blackouts; provide high-quality power in a digital society; and will provide more security in the event of human error, natural disaster, or terrorist attack.

Today, however, the nation's electric grid is over-extended, under-invested, and in need of significant upgrade and improvement. It is highly unlikely, for a large number of reasons, to expect public utilities to increase significant investments without supporting federal and state policies geared to facilitating the construction of a smart and environmentally responsive grid.[134] For the most part, advanced technologies are available and breakthroughs are not necessary. High-voltage transmission lines can be constructed, cost-effective electric storage is available, and intermittent renewable energy sources can be accommodated. As the grid becomes more intelligent, information flows among providers and consumers can be used to produce and consume electricity more efficiently, reliably, and safely.

Modernization and expansion can be accomplished at a cost of $225 billion for transmission and $640 billion for distribution. These investments are substantial; and although it will not necessarily require direct financial support from government, rules and regulations by which utilities are regulated will have to change to encourage that investment and to allow the development of new rate-making processes.[135] Most notably, FERC must develop a legally effective cost allocation formula for interstate transmission investments, and state public utility commissions must develop policies to allow utilities to recoup smart grid investments in their rates.[136] These regulations are discussed more fully in Chapter 6.

Conclusion

In recent years, there has been a multiplicity of reports from interest groups, non-governmental energy organizations, trade associations, private-sector actors, and the current presidential administration all addressing our energy future. What is remarkable about these reports is that they evince a significant consensus on a large number of policy issues regarding goals and the means of achieving them. The goals include environmental sensitivity and the three faces of security. A secure energy future is one that does not leave us vulnerable to international markets and international economic disruptions. A secure energy future is one in which abundant, reliable, and affordable clean energy is available. Additionally, a secure energy future is one that limits threats to our national security.

Our energy future no longer resides in fossil fuels; it resides in a substantial ramping up of energy efficiency and renewable resources. Energy efficiency is an abundant, virtually untapped, and cheap energy source. Our economy can not only survive but thrive by deriving more productivity from using less (and cleaner) energy. For over a century, fossil fuels have dominated our energy profile and have been advanced by economic and political support structures at the expense of new entrants and new energy sources. Solar, wind, distributed generation, alternative transportation fuels and vehicles, new building designs and standards, and innovations yet unnamed will change the way we do our energy business.

The continuing open question about our energy future remains nuclear power. Operating nuclear power has become increasingly affordable as the cost of the first generation of nuclear plants has been recouped and as the operational costs have decreased. Nuclear power leaves little or no carbon footprint certainly as contrasted with coal-fired electricity. However, the construction of new nuclear plants presents economic, political, and waste concerns that undercut its promise of providing clean and abundant fuel. Similar open questions remain about the midterm use of fossil fuels. Coal's abundance and oil's essential role in transportation, together with the substantial infrastructures surrounding each resource, make it appear unlikely that we can wean ourselves from either in the short term. This thinking must be rejected and must be replaced by immediate commitments to redesign energy policy to shift financial incentives and supports away from fossil fuels to efficiency and renewable resources.

On the one hand, the challenges of climate change may span the time frame of a century. Moreover, the new energy thinking is only beginning to emerge and be felt in private capital and investment markets after over a generation. On the other hand, we must act now, well within a generation, or the costs may be beyond our ability to pay them. Again, the question: How do we move away from fossil fuels and achieve energy independence?

The next two chapters will explore in more detail an energy policy that reduces reliance on fossil fuels in both the transportation and electricity sectors. The remaining chapters address a new politics of energy and proposals for changes in government regulation intended to move us to a safer energy future.

Fossil Fuel Future

Coal, oil, and natural gas will remain indispensable to meeting total projected energy demand growth.

National Petroleum Council, 2007[1]

On March 21, 2010, President Obama proposed opening offshore waters along the Atlantic Coast, the eastern Gulf of Mexico, and the north coast of Alaska to oil and natural gas exploration.[2] The area has preliminary estimates of 3.8 billion barrels of oil and 137 trillion cubic feet of natural gas.[3] Was this announcement political heresy and a sellout of environmentalism or a hard-nosed realism consistent with a pragmatic view of our energy needs and of our energy politics? Oil exploration and drilling, in the face of cries for energy independence and environmental protection, is seen by some as a betrayal of a smart energy future. To others, though, it is unrealistic to assume that we will be driving electric vehicles powered by renewable fuels to any significant degree within the next generation, and it is equally unrealistic to assume that renewable fuels will supplant coal for electricity generation in the near term. Indeed, projections about the growth of electric vehicles reveal no significant market penetration for nearly a generation. Moreover, projections for electricity generation from renewable resources remain marginal, ranging from 9 percent in 2008 to 17 percent in 2035. Consequently, as attractive as a non-fossil-fuel future may be, a transition period appears to be most likely in which oil, natural gas, and coal play prominent roles. Yet problems persist.

April 2010 may have been the cruelest month for fossil fuels. It was crueler to human lives and to the environment. On April 5, 2010, the Upper Big Branch coal mine in West Virginia exploded, killing 29 workers and constituted the worst mining disaster in forty years.[4] Fifteen days later, on April 20, BP's Deepwater Horizon oil rig, with 126 workers on board, exploded then caught fire in the Gulf of Mexico.[5] After having been on fire for more than a day, the rig sank on April 22 with 11 workers missing. In addition to the loss of life, the rig continued to spew in excess of 5 million barrels of toxic crude into the Gulf, becoming the worst oil disaster in the nation's history.[6] And, before the end of the month, a roof collapse in a Kentucky coal mine with a history of safety problems left two miners dead.[7] None of these events can be described as anomalies or extraordinary events. Instead, they reveal a pattern of human health and safety neglect, ecological insensitivity, and environmental callousness born of our traditional fossil fuel policy. On the smart side of the energy ledger, on April 28 an event, cutting against both traditional energy policy and Not In My Backyard (NIMBY) opposition, occurred as Interior Secretary Ken Salazar announced federal approval for the Cape Wind Project, the first U.S. offshore wind farm, off the coast of Cape Cod.[8] Energy policy may well be changing, but it is changing too slowly.

The failure of our nation's oil and coal policies is, in significant part, a market failure – the failure to properly price these resources. As fossil fuels, oil and coal have, for nearly a century, doubly benefitted. In the first instance, the true costs of health, safety, and environmental protection have not been included in the price of these products, thus oil and coal have been underpriced. In the second instance, a century of traditional energy regulation has financially and politically supported the growth and stability of both industries thus subsidizing these fuels. Even if we assume that a transition away from a transportation sector fueled by oil and an electricity sector fueled by coal is a generation off, today we must, nevertheless, treat fossil fuels noticeably differently than from the past. These industries can no longer continue to rely on historical favoritism; subsidies and supports must be removed; and to the extent that they are transitional resources in energy markets, they must compete with new energy entrants.

Introduction to Fossil Fuels

World energy use is expected to grow at an annual rate of 1.5 percent between 2007 and 2030, which translates into an overall increase in energy consumption of 40 percent during this period.[9] Fossil fuels play a continued dominant role in that expansion. At 20 million barrels per day (mdb), the U.S. consumes nearly one-quarter of world oil output of 85 mbd at an annual cost in excess of $500 billion, which constitutes approximately 40 percent of our trade deficit.[10] World output is expected to reach 100 mbd by 2030, and U.S. consumption is projected to increase modestly to 21.5 mbd.

Relative to future global supplies, productive capacity is expected to grow through 2030 with no immediate peak in global production even in the event that world demand rises to 115 mbd.[11]

Coal consumption in the United States is 1.1 billion short tons (bst) with 1.0 bst dedicated to electricity production. This level of consumption is twice the amount of coal consumed in 1950, and current projections are that coal consumption should increase to 1.3 bst. Worldwide coal consumption is projected to increase at an average rate of nearly 2 percent per year, which is more than any fuel source other than growth in non-hydro renewables. China's coal demand has increased noticeably as it has begun to increase its imports from the United States appreciably.[12] Natural gas consumption is also expected to increase 1.5 percent per year globally, and U.S. consumption is expected to grow to 24.3 trillion cubic feet (tcf) by 2030 from 23.2 tcf in 2008. Fossil fuel consumption, with its attendant carbon emissions, is projected to increase over the next two decades unless energy policy forces a change.[13]

There are, however, some positive signs even in the face of predicted demand growth. First, today, the United States uses about half as much oil to produce a dollar of GDP as it did in the mid-1970s. This reduction has occurred as a result of increasing fuel economy standards, virtually eliminating oil as a base fuel for electricity generation, and from moving to a less energy intense information services economy from a heavy manufacturing one. These gains in energy efficiency have slowed in recent years, and now we must re-calibrate goals for further reductions in oil intensity throughout the economy. Recommendations have been made that the United States can further reduce intensity by 50 percent by 2030 and by 80 percent by 2050.[14] Other positive signs are that utilities and state Public Utility Commissions (PUCs) are slow to build new coal-fired power plants precisely because of polluting carbon emissions and are projecting significantly higher capacity additions from wind and natural gas.[15]

For nearly fifty years, U.S. politicians, albeit with low levels of commitment, have decried the U.S. dependence on imported oil. There are three chapters to the oil independence story. Starting in the 1950s, our first concern was that domestic production failed to provide all our demand and, therefore, we began importing foreign oil. As noted previously, in response to imported oil, the Eisenhower administration sought to protect domestic firms by putting a quota on the amount of imported oil that could come into the country, thus propping up domestic producers. The reaction of foreign producers to U.S. protectionism was the creation of OPEC in 1960. We have been wrestling with OPEC ever since.

The second chapter occurred in 1970 when domestic production peaked as predicted by geophysicist M. King Hubbert who hypothesized that oil production can be plotted along a bell curve and that after a maximum point, production will peak and then begin to decline. Hubbert correctly predicted that U.S. production would peak sometime between 1965 and 1970.[16]

The third chapter came about as a result of the oil crises of the 1970s, which intensified around 1975 when oil imports exceeded domestic production. The cries for oil independence have continued, with little to no effective responses, as the gap between domestic production and imported oil only continues to grow. The world problem for oil production is the increasing demands placed on producers by the rapidly growing and modernizing economies of India and China. This increased demand for oil by developing and modernizing nations again raises the question about peak oil; this time, however, on a global scale. There are those, such as Matthew Simmons, who argue that we can extend Hubbert's Peak theory worldwide and that the globe has hit its peak and oil production will soon begin to decline with disastrous economic dislocations.[17] The prevailing view, though, is that production will grow with increasing demand; that non-OPEC countries have peaked or will soon peak; and that the OPEC countries will meet that demand for two or more decades.[18]

We now find ourselves writing yet another chapter to the oil saga. On July 11, 2008, benchmark crude hit the price of $147.27 per barrel. One year later, the price dipped to $59.87, having hit a bottom of $32.40. In the summer of 2010, oil climbed back to $80 per barrel. In short, the new world of oil is one in which price volatility attracts our attention and concern. Price volatility is important not only to speculators and hedge fund managers, it is important to ordinary consumers as well as businesses of any size because the price of energy impacts all budgets. Price volatility is also a matter of concern in the oil futures markets, which daily trade more than ten times a single day's consumption.[19] Price volatility means higher risk to investors and instability for consumers, thus threatening energy markets and world economies. In addition to growing world demand, the new oil chapter also involves increased security concerns, renewed environmental awareness, and the continuing search for new sources.

After the BP oil spill in the Gulf, offshore oil exploration and drilling will be scrutinized at least in the near-term.[20] Still, the public consensus, as well as government plans, are to not take offshore oil off the table. In addition, the United States looks to Canada for importing synthetic crude from its oil sands. Although there is no chance of a deep water oil spill like

BP, oil sands pose other environmental risks including toxic sludge ponds, greenhouse gas emissions, and destruction of forests. Still, Canadian oil sands are plentiful, and the Obama administration is reviewing a Canadian company's request to build a 2,000-mile underground pipeline running from Alberta to the Texas Gulf. It is estimated that Canadian oil sands can become America's top source of imported oil, surpassing the combined imports from Saudi Arabia and Kuwait[21] and reaching 37 percent of U.S. imports by 2035.[22] Synthetic fuels do not solve the problem of carbon emissions. Instead, they address matters of national security. The potential of oil sands is directly related to price volatility. It is estimated that oil sands become economically competitive in the range of $60 to $85 per barrel, and price volatility in recent years has not afforded a reliable business plan for oil sands investments.[23]

Traditional policy is a paradigmatic example of how an idea – cheap fossil fuel energy – has captured U.S. politics and markets to the benefit of industry interests and to the detriment of environmentally sensitive and economically valuable alternatives. Fossil fuel market rhetoric has framed the debate about energy, and consequently about ensuing government fossil fuel policies, in terms of sustaining a healthy economy and protecting national security. Government should support, and lightly regulate, fossil fuels, the argument goes, as good (and necessary) for our country to have and enjoy a healthy economy. More importantly, in a dangerous and troubled world, a sound fossil fuel policy is also necessary (and good) for our country's domestic and global security. This argument, and the free-market ideology supporting it, would be unobjectionable except for three things. First, the idea itself is flawed in theory and flawed in execution. Fossil fuel markets are neither free nor competitive; they are heavily subsidized and ignore the extraordinarily inefficient social costs generated by pollution. Second, fossil fuel favoritism hampers the development of preferable, less costly, and more promising energy alternatives that will strengthen the economy and make our energy mix more self-sustaining. Third, the political and financial support enjoyed by the fossil fuel industry retards the development of a necessary transformation of the country's transportation system. In short, fossil fuel ideology, as embraced by the industry and the regulators and politicians who support it, is insensitive to the economic dynamics that can stimulate innovation, open new markets, and bring new technologies online all to the point of responding to the challenges of climate change and developing a smart and secure energy policy.

Fossil fuels have been the bedrock of U.S. energy policy for more than a century. Protection of oil, natural gas, and coal interests during that time has resulted in increased greenhouse gas emissions while limiting the role of energy efficiency and renewable resources. The economic ideas and the policy assumptions behind fossil fuel protection not only support those industries, they also thwart efforts to enact meaningful measures addressing global warming. Our nation's energy ideology focuses on solving energy problems by increasing the supply of fossil fuels and keeping the price of fossil fuels low through subsidies and other measures including under-enforcement and outright neglect of health, safety, and environmental protection laws already on the books. The pursuit of oil in ever more difficult places at ever higher prices, as well as the reluctance to price carbon and develop meaningful CCS technologies, demonstrate how ingrained the ideology is in our regulatory policy and politics.[24] Traditional energy policy is, thus, dependent on a well-worn path that produces dirty energy, discourages cleaner alternatives, and disadvantages the emergence of new markets and new technologies by tilting the economic playing field its way.

Fossil Fuel Favoritism

The United Sates consumes 100 quadrillion BTUs of energy annually, and 85 quads come from fossil fuels.[25] The total costs of fossil fuels should not be underestimated. Nor should the amount of financial support enjoyed by these industries at taxpayer and consumer expense be underestimated. Historically, regulators have promoted the use of oil and coal for both military and civilian uses; politicians have created institutions to promote these markets; and the oil and coal industries have expanded to become among the most highly centralized and profitable businesses throughout the economy. Whereas unaccounted-for pollution costs underprice oil and gas, other subsidies abound. The price of gasoline at the pump does not include national security costs estimated at $750 billion for 2008 alone[26] and in excess of $1.5 trillion for the years 2004–2008.[27] These costs involve conducting foreign policy, increasing risk by limiting U.S. strategic options, strengthening foreign adversaries, and aggravating geopolitical competition for oil resources.[28] Similarly on the coal side, we underestimate, and underprice, the cost of coal throughout its fuel cycle. Each day, for example, 260 million gallons of water are used for coal mining. Further, 120 million tons of solid waste are produced by burning coal each year. Ninety million gallons of waste slurry are produced annually. Moreover, mountaintop removal has either buried or polluted 1,200 miles of streams in addition to the air pollution, health problems, and deaths affecting coal workers.[29]

In short, slogans such as "drill, baby, drill" cannot resolve the problem of our dependency on foreign oil. Nor does a domestic petroleum market ensure price stability because of the power exercised by global oil. Intentional supply disruptions, price volatility, weak growth in domestic production, geopolitical instability, and the fact that 90 percent of the world's oil is controlled by state-owned oil companies preclude competitive markets from sending reliable price signals. The lesson to be learned from the political and economic reality of world oil should be clear: "Policies and technologies that bring fuel choice to transportation represent the only long-term solution to America's oil dependence."[30]

The U.S. energy profile can be roughly divided between oil for transportation and electricity for heating, cooling, and lighting. Today, oil and electricity are not substitutes for one another. The key lesson for the energy future is that these sectors must be integrated as electricity assumes the predominant role in the transportation sector. Indeed, the single most important breakthrough technology that has the capacity to radically reform energy policy and wean us from oil is electric vehicles. Consequently, any program for reducing carbon emissions must address the amount of oil consumed by vehicles and the amount of coal burned to generate electricity. Neither task will be easy because of the dominance of fossil fuels in our economy and in our politics.

Fossil fuel energy policy exemplifies the old idea of a static, rather than a dynamic, energy economy. Fossil fuels have established a supporting regulatory structure and bureaucracy; and, as public choice theory tells us, interest-group politics make it difficult to change policy direction as incumbents enjoy the competitive advantages that access yields. In short, fossil fuel favoritism is firmly entrenched in our political and regulatory cultures. A contemporary snapshot of U.S. energy policy proves the point through the following observations. Today, we find oil prices reaching historic highs with no drop in demand;[31] electricity prices rising as demand increases;[32] plans accelerating for the construction of liquefied natural gas ports; a relaxation of environmental and other regulations of coal mining;[33] and a revised enthusiasm for synfuels development to extract oil and gas from coal liquefaction, oil shales, and tar sands. All of these developments continue to rely on a worn out energy policy, and all require a greater, and more expensive, use of fossil fuels.[34]

This protectionist policy is a true tragedy of the commons as increased fossil fuel use means increased carbon emissions. Instead of taking climate change seriously, treating it as a warning about the future and accepting it as a stimulus for rethinking traditional energy policy, policymakers continue down a well-worn path. Climate change requires us to revamp energy policy by incorporating environmental concerns. Yet, even though energy and environmental policies are *beginning* to be thought about and discussed together, they are still treated as separate regulatory regimes because fossil fuels dominate renewable and alternative energy resources.

Renewable and sustainable alternatives to fossil fuels are attracting increasing attention and increasing venture and investment capital. Nevertheless, alternatives such as solar, wind, biofuels, and other, perhaps more exotic, fuels such as hydrogen and fusion, remain problematic. Markets are slow to develop for these alternatives, and their projected shares of the energy economy remain small. Nearly thirty years ago, for example, solar power was projected to supply 20 percent of all energy needs.[35] Today, however, the Department of Energy estimates that *in twenty-five years*, solar power will account for, at best, 2 percent to 3 percent of our electricity needs whereas now it accounts for less that .01 percent.[36] Thus, traditional U.S. energy policy imposes a double barrier to reducing global warming. On the one hand, traditional policy promotes fossil fuel use. On the other hand, it fails to promote alternative and renewable resources on an equal basis.

Dirty Energy Subsidies

The claim is often made, especially when oil profits hit historic highs,[37] that oil companies enjoy abundant government subsidies and thus enjoy an unfair competitive advantage in the marketplace. Simply as a matter of foreign defense, it has been estimated that military expenditures to protect the flow of Mideast oil range from $20 billion to $90 billion per year.[38] Preliminarily, there are two things to say about oil subsidies. First, quantifying the dollar value of subsidies is extremely difficult, if not impossible. There are direct and indirect subsidies that do not lend themselves to a common metric and often remain hidden.[39] There is no single source, government or private, to find accurate and reliable data, although the Pew Charitable Trusts has started Subsidy Scope to build a comprehensive source for collecting subsidy data including data on energy subsidies.[40] Further, it is often the case that the data that is available is biased either for or against the industry. A study of oil subsidies published in 1998, for example, calculated that oil subsidies in 1995 amount to between $15.7 billion and $35.2 billion.[41] Another study of oil subsides concluded that the oil industry is not a net beneficiary of government support and that it is dubious to suggest that oil companies do not pay their fair share of environmental costs.[42] A reader might be forgiven her skepticism upon recognizing that the first report was commissioned by pro-environmentalist Greenpeace and the second by the conservative, free-market Cato Institute.

The second preliminary observation is that subsidies are a real part of the political economy. In true Madisonian fashion, "the interests" will lobby government for favors and advantages. It is in the nature not only of a political economy in which government and business often work together, it is in the nature of a pluralistic democracy. Oil companies will and have successfully lobbied for a great variety of advantages and so have countless other interests. Whereas a strong theoretical, economic case can be made for the elimination of all subsides, no good, realistic political case can be made. More to the point, subsidies are being used and will be needed to combat climate change. We cannot and should not throw out the subsidy baby with the dirty energy policy bath water. Oil politics will not soon go away; therefore, oil interests will continue to lobby for political favors. Available data and solid examples provide ample evidence of significant government support for fossil fuels. Still, that support runs counter to generally accepted market principles and counter to sound environmental and energy policies.

Most simply, a subsidy is some form of government expenditure that makes production, and therefore the cost, of any good cheaper.[43] A direct cash payment to farmers to produce wheat or an aggressive accelerated depreciation accounting standard on oil company property both operate the same way and have the same effect – oil and wheat are cheaper to produce. The farmer will plant more wheat because it is cheaper to do so, and the oil company will produce more oil because their tax liability is lessened. There is another way to perceive subsidies – they are economically perverse because they do not follow the rules of competitive markets. Subsidies distort markets and create surpluses. More of a subsidized good is put on the market than would otherwise be placed there under competitive conditions.[44]

So subsidies are a part of our politics and a part of our economy even though they have perverse economic effects. The argument against fossil fuel subsidies is that they give the fossil fuel industry an unfair competitive advantage in the political and economic market places. That argument is correct. Energy subsidies must be redirected away from dirty energy policies that distort economic choices and toward a clean energy future. One example of redirected subsidies would be to cut tax expenditures for coal, oil, and natural gas, which currently enjoy the lowest taxes of any industry even in the face of historically high prices and profits, and then provide tax breaks for energy efficiency and renewable resources.[45] After a brief overview of fossil fuels subsidies, two case studies of recent fossil fuel catastrophes highlight the government favoritism enjoyed by the oil and coal industries.

Taxes and Accounting[46]

Energy tax policy, unsurprisingly, follows energy politics. For most of the twentieth century, tax policy favored the exploration, production, and distribution of oil and natural gas resources and did nothing to encourage conservation or renewable resources. As reported, "[A]n examination of the American tax code indicates that oil production is among the most heavily subsidized businesses, with tax breaks available at virtually every stage of the exploration and extraction process."[47] The most notable tax incentive for the oil industry has been the oil depletion allowance, which allows an oil company to deduct all or part of its mineral investment. The depletion allowance was based on the idea that because oil is a wasting resource, a tax break can be used to compensate for, and encourage, that investment. The counter argument is that other businesses treat such investment as an understandable business risk that is built into their price structure and investment policy.

The rules operate in such a way that under the proper circumstances, depletion deductions could exceed the original cost of investment. Either way, the depletion allowance reduces tax liability for oil companies in a greater measure than allowed for other business firms. Another financial incentive allows oil and gas producers to treat drilling costs, such as labor, materials, supplies, and repairs, as expenses. With this treatment, the expenses can be deducted in the year in which they are incurred instead of capitalizing them and writing them down over a period of years as other businesses do. In this way, these "intangible drilling costs" are recouped in the first year, leaving more capital available for other investments. As a consequence of favorable accounting treatment, capital investments by oil companies are taxed at an effective rate of 9 percent whereas all other businesses pay an overall rate of 25 percent.[48] Such devices reduce the tax liability of oil companies to such an extent that, combined, these tax subsidies have resulted in little or no income tax for much of the petroleum industry.[49] Tax laws can be changed to eliminate the expensing of exploration and development costs and to eliminate percentage depletion and alternate fuel production credits, which would raise approximately $4.1 billion a year while cutting carbon emissions.[50]

Oil companies also enjoy the financial benefit of tax havens. By moving corporate operations out of the United States, firms can pay little or no taxes.

Transocean, the company operating the Deepwater Horizon oil rig, moved its corporate headquarters from Houston to the Cayman Islands and then to Switzerland; in the process, it reduced its tax bill by $2 billion.[51]

Unsurprisingly, oil company responses to charges of favoritism have been predictable. We pay our fair share. Subsidies provide jobs, and cuts in subsidies will result in job losses. The economy needs fossil fuel productivity.[52] However, the U.S. Treasury reports that a cut in subsidies will have a negligible effect on productivity and might well make the economy more efficient.[53]

In the 1970s, domestic oil policy shifted away from encouraging production and toward conservation of oil to promote the substitution of burning coal in power plants. The period after 1970 saw the introduction of a "gas guzzler" excise tax, the Crude Oil Windfall Profits Tax, price freezes on oil, and other energy legislation also promoting conservation and greater use of coal. This seemingly dramatic switch from oil favoritism to oil taxation was less formidable than it first might appear. As noted earlier, the Windfall Profits Tax did not last and did not collect the hoped-for revenue; price controls were evaded, further reducing the hoped-for revenue collection;[54] and, until recently, vehicle fuel standards have been stuck in neutral for more than thirty years.

The Royalty Treatment

Royalty payments are a major source of non-tax income to the United States; nearly $50 billion was collected between 1982 and 1990. Curiously, though, the federal government was content to rely on lessee's royalty calculations rather than have them calculated by government agents. Such reliance was a boon to the oil companies and a bust to the Treasury. It has been estimated that oil companies underpay the Treasury approximately $500 million annually.[55] In response to those estimates, Congress passed legislation that established the Minerals Management Service (MMS) within the Department of Interior to oversee royalty payments on oil and gas taken from public lands.[56] The MMS, however, has come under congressional and public criticism, which resulted in an internal investigation revealing lax royalty enforcement, among other practices of questionable ethics and legality, to the benefit of the oil industry.

There are tens of thousands of oil and gas leases on public lands both on shore and offshore that account for about 35 percent of the oil and 26 percent of the natural gas produced in the United States.[57] In exchange for the opportunity to develop the land and produce oil and natural gas, the lease holder is obligated to pay royalties, sometimes as much as 18 percent, into the U.S. Treasury. Royalty payments, however, have been less than reliable. In 1995, for example, Congress suspended royalty payments as a way to encourage deepwater drilling. Payments were to start after oil reached $28 per barrel, but the law was written so poorly that oil companies continued to evade that obligation past the benchmark price, resulting in losses of billions of dollars to the Treasury.[58]

Other royalty abuses existed. In the mid-1990s, by way of example, petroleum engineers discovered a conspiracy by several oil companies to depress the value of oil and then understate the total amount of royalties owed to the U.S. A *qui tam* whistleblower lawsuit was filed under the False Claims Act against a dozen oil companies in Texas. The complaint alleged that the oil companies underpaid royalties from 1980 through 1998. The prosecution of the suit resulted in payments in excess of $450 million.[59]

In recent years, the Office of the Inspector General of the Minerals Management Service released a report on False Claims Act allegations that were filed by employees and former employees of MMS against their own department.[60] The report chronicles the allegations by, as well as the claims of retaliation against, the employees for their whistle-blowing. The Inspector General found that the Mineral Management Program, within MMS, is "fraught with difficulties" including "the bureau's conflicting roles and relationships with the energy industry," which contributed to a "profound failure" to hold "together one of the Federal Government's largest revenue producing operations."[61] By way of one example, the MMS IT system was not programmed to automatically calculate interest owed on the royalty payments, relying instead on manual calculation of interest by the oil companies. No one was watching the regulators.

Even when employees attempted to police their own agencies, fossil fuel companies won. For example, in law suits filed by MMS employees, who were frequently auditors in the field, the employees alleged that oil companies under-reported royalties owed and underpaid interest payments. One lawsuit claimed that an oil company would sell oil to a marketing company at a reduced price in exchange for marketing services. The oil company, in turn, would use the reduced price without including the value of the marketing services to calculate the royalties owed to the United States. In this suit, a jury found that the company owed $7.5 million, but the judge dismissed the suit holding that the employee could not sue under the False Claims Act. A week after the lawsuit became public, the employee lost his job because of a departmental "reorganization."

In another case, the employees charged that the oil companies routinely under-reported interest payments and that when the MMS employees reported the under-payments to their supervisors, the supervisors refused to take action. The employees then filed lawsuits charging that senior officials refused to demand the additional interest payments from the oil companies because it would be a hardship for the companies to do the interest calculations. These complaining employees were removed from their jobs at MMS and were given below-entry-level positions at the Bureau of Land Management, also in the Interior Department.

In both instances, the Inspector General's investigation can be best described as Kafkaesque. The IG's report shows a complete lack of communications within MMS. Different auditors and different investigators took different positions as to the propriety of the oil companies' behavior. Worse still, the complaining employees were exposed to criminal prosecution for using government documents to bring their lawsuits in an effort to perform their assigned jobs and collect royalties due to the United States.

The Inspector General's report did not calculate the total amount of royalties possibly lost. More importantly, though, the report exposes the agency's failure to clearly define and fulfill its mission. An earlier government investigation suggested that the government might lose about $10 billion over a decade just because of a legal mistake in oil and gas leases that had been ignored.[62] Consistent with the Inspector General's report was a 2006 Inspector General audit that revealed that MMS did not have an effective compliance system, which prevented them from fully determining the costs and benefits of compliance reviews.[63] Given both reports, it is clear that MMS is dysfunctional and that the dysfunction benefits fossil fuel interests. There is more about MMS later in this chapter.

Assessing Subsidies

Subsidy estimates are subject to a number of methodological choices and problems. In its most recent estimate, DOE's Energy Information Administration (EIA) reported a subsidy valuation of $16.6 billion for all energy sources.[64] This figure is in contrast to another estimate of $75 billion reported by a non-governmental organization.[65] The disparity in estimates between DOE and non-governmental sources has been attributed to the EIA's lack of rigorous peer review; a restricted set of data sources such as no references to NGOs; and a narrow research mandate.[66] In other words, politics as usual. Still, fossil fuel subsidies are substantial.

Even though subsidies are a part of our political economy and although it is unrealistic to imagine that they will disappear in an unchecked spasm of pure competitive free-market thinking, oil subsidies run contrary to the public good for economic and social reasons. Further, oil subsidies are not insubstantial. The United Nations Environment Programme, for example, estimates that global energy subsidies are nearly $400 billion annually, with the vast bulk going to fossil fuels.[67] Domestically, a 2009 study shows that fossil fuels continue to receive the majority of $11 billion of annual tax expenditures.[68] Another 2009 study revealed that during the period 2002–2008 subsidies for fossil fuels totaled $72 billion whereas subsidies for non-fossil fuels totaled $29 billion over the same period.[69] Further, an April 2010 report estimates that oil subsidies will cost the U.S. government $3 billion for 2011 and $20 billion over the next five years.[70] Still, we are beginning to see a shift away from fossil fuels to renewable resources as the Obama administration reduces subsidies to fossil fuels and shifts them to low-carbon initiatives.

Economically, oil subsidies distort markets and do not correct market imperfections. Instead, they exacerbate them. As private energy firms, oil companies need no additional support to compete in energy markets. Assuming that we can take the claim that oil companies do not exhibit market power at face value, then we should hold oil companies to a single standard. If they do not have market power, then they are competitive and should be allowed to operate in markets without additional government backing. They claim, however, that in order to compete special subsidies are necessary. A case for a subsidy might be made if an oil firm can demonstrate some market imperfection. This is not a case that an oil company can make. Instead, they do not suffer under any market imperfection that may require, let alone justify, government support. Indeed, oil companies do not have any of the classic indicia of market imperfection such as information asymmetries or rival market power that would put them at a competitive disadvantage.[71]

The second, and related, problem is that subsidies to oil companies have severe social costs, and they simultaneously reduce the public good of environmental protection. Oil companies do suffer from a different sort of market imperfection that should be addressed by regulatory controls, not by subsidies. Carbon pollution is a social harm that must be addressed if oil markets are to work more efficiently. Consequently, on the basis of market principles alone, oil subsidies cannot be justified. Economically, then, subsidies overproduce oil at prices lower than a competitive market would set and without the added social costs of pollution.[72]

Simply, fossil fuel companies do not pay their way. Subsidies send bad price signals into the market with the result that they reduce the real price of oil and coal and increase consumption above competitive levels, thus disadvantaging energy competitors. Subsidies to fossil fuel firms retard environmental health, contribute to global warming, and impede the development of an alternative energy policy that promotes renewable resources and energy efficiency. Subsidies to oil and coal companies, therefore, have the dual negative impact of over-producing the product to the disadvantage of superior competitors and disproportionately imposing costs on society. Fossil fuel subsides simply do not promote the public good.

Realigning fossil fuel subsidies, incentives, and financial assistance is the first step in formulating a new energy policy. Although our domestic and foreign energy policies have been greatly supportive of fossil fuel interests and although we will continue to use oil for transportation, it is imperative that we level the playing field for an alternative energy policy that acknowledges the need for more fuel-efficient vehicles, transportation fuels based on resources other than fossil fuels, paying the real cost of oil and coal, and truly weaning ourselves away from Middle East oil exporters. The political will needed to accomplish these goals, as well as consumer tolerance for the costs likely to be involved, are not inconsiderable. Nevertheless, the conversation must begin in earnest. We must identify not only the short-term costs but the long-term risks to our economy, to our environment, and to our security.

We do not have the luxury of low-cost, abundant domestic oil. We cannot continue to assume that foreign oil will flow freely. Nor can we continue to ignore our weakening position in global markets. We can regain our economic leadership through an energy strategy that moves away from old markets and into new ones; that invests deeply in domestic education and technology; and that drills deeply into our most valued resource – innovation. Our old fossil fuel policy retards new growth and the momentum to change and must, therefore, be rejected.

Fossil Fuel Subsidies – Two Case Studies

Two recent events, one involving the Gulf oil disaster by BP and the other involving the coal mine explosion by Massey Energy, reveal the power, and the tragedy, of traditional energy policy. These two studies contain remarkable similarities. In the first instance, both involve the primary fossil fuels that the United States relies on for transportation and electricity. Second, the case studies reveal how entrenched these resources are in the country's traditional energy policy. Third, and most importantly, the studies demonstrate the political power, bureaucratic participation, and financial support by government of that traditional policy. Further, the studies reveal a lack of regulatory oversight and control, resulting in environmental degradation and human fatalities. Finally, the studies underscore the pernicious effects of the combination of economic and political power on the environment and on human life in the name of fossil fuel energy.

Deepwater Horizon

The explosion and sinking of the Deepwater Horizon oil rig in the Gulf of Mexico is the worst oil spill in U.S. history. The rig operator, BP, formerly British Petroleum, has over the last decade attempted to reshape its image and refashion itself as a green oil company that goes by the slogan "Beyond Petroleum." Yet the company's reality is far different. BP has been responsible for several recent oil disasters, and the CEO's first response to this event was to deny liability and blame other participants. Worse, perhaps, was that after three weeks of spewing tens to hundreds of thousands of gallons of oil a day into the Gulf the CEO, Tony Hayward, claimed that "the Gulf of Mexico is a very big ocean. The amount of volume of oil and dispersant we are putting into it is tiny in relation to the total water volume."[73] Oil spill estimates have ranged from BP's low of 5,000 barrels (or 210,000 gallons) to federal estimates of up to 19,000 barrels (or 798,000 gallons)[74] per day of oil and hundreds of thousands of gallons of toxic dispersants. BP's estimates were significantly lower than others', and by the time the well was capped, it was estimated that 5 million barrels (or 210 million gallons) of oil had been released into the Gulf.

Denial and blame-shifting are not responsive to the severity of the spill. Nor are they responsive to the need for a smarter energy policy even though denial and blame are consistent with the dominant energy ideology. More disturbing, though, is that big oil, just like big coal, has been the recipient not only of direct and indirect government financial favoritism, they have greatly benefitted from conscious government neglect, the neglect to enforce the very laws intended to protect the lives, health, and safety of fossil fuel workers and the environment in which they work.[75] As a result, neither government nor industry was prepared to address a spill of this magnitude nor was appropriate clean-up technology available.[76] In fact, clean-up technology had not advanced since the Exxon Valdez oil spill in 1989.[77]

Offshore oil leasing regulations have been notoriously lax. Their purpose is two-fold. First, the regulations are intended to protect the fragile ecosystems in which exploration and drilling take place. Second, the leasing program is intended to generate revenue for the U.S. Treasury through bidding and then through royalty payments. The federal agency most responsible for its not-so-benign neglect of monitoring and regulating the oil leases is the MMS within the Department of the Interior. The MMS is now notorious for its favorable treatment of oil companies even to the point of trading sexual favors and drugs in exchange for turning inspectors' heads away from enforcing compliance and collecting revenue.

Most simply, MMS had created a culture of corruption. In 2008, the Inspector General for the Department of the Interior issued the results of a two-year investigation that documented the widespread malfeasance and ethical transgression of MMS employees. The list of legal violations and ethical breaches is disturbing. In some instances, senior employees left government service to work for the companies that they were charged with regulating and, in violation of conflict-of-interest rules, helped assure that lucrative contracts were awarded to their new employers. In other instances, the investigators simply let the oil companies fill out the required forms themselves in pencil and then the investigators completed them in pen.[78] The report also revealed that nearly one-third of the staff of one program accepted numerous gratuities from oil and gas companies. In one example, two employees received gifts and gratuities on at least 135 occasions from four major oil and gas companies. The Inspector General also concluded that MMS had developed a culture of substance abuse and promiscuity in which employees engaged in illegal drug use, alcohol abuse, and sexual relations with subordinates and with industry contacts. The Inspector General concluded that although a relatively small group of individuals engaged in misconduct, "management through passive neglect, at best, or purposeful ignorance, at worst, was blind to easily discernible misconduct."[79]

In the case of Deepwater Horizon, MMS gave BP permission to drill without obtaining the necessary permits required by the National Oceanic and Atmospheric Administration and by the National Environmental Policy Act to protect the environment and the species within the Gulf despite that agency's warnings about likely dangers. It has been further reported that the MMS even overruled its own staff scientists and engineers who warned of health, safety, and environmental concerns over this very project and that it continued to exempt BP from environmental permits after the April 20, 2010, explosion.[80] By way of example, it had been assumed that the Gulf spill occurred because of the absence of a "blowout preventer" and inadequate well casings. These deficiencies were apparently known by BP to be substantial risks to the safe operation of the well.[81] Although MMS warned of the necessity of this equipment, the agency did not enact regulations to address the need for this protection. Additionally, BP knowingly took riskier options in casing the well because it was cheaper; this most likely led to the blowout.[82] Regulator and regulatee looked the other way on safety and environmental protection in the Gulf.

The Obama administration's response, initially, was not especially strong. On the surface, the administration declared a moratorium on offshore exploration and drilling and vowed to dismantle MMS. Most simply, the agency would be divided into three offices. One office would be devoted to public safety and environmental enforcement, another to licensing, and the other would be responsible for leasing and revenue collection.[83] Slightly beneath the surface, however, the moratorium was to last three weeks, and it was partial and did not end the type of exemptions that benefitted BP's Gulf project.[84] In the weeks following the sinking of the rig, the Department of the Interior issued environmental waivers and seventeen drilling permits for the type of work performed by Deepwater Horizon about the time of the explosion. Government officials responded that a moratorium was meant to apply to drilling permits for new wells and not to existing projects.[85]

The waivers have been particularly problematic for environmentalists because the provision authorizing the waivers is intended to limit projects that pose minimal or no risk to the environment. Waivers that were granted after the spill, however, were designated for drilling projects at lower depths than Deepwater's nearly 5,000 foot well and extended to wells drilled to a depth of more than 9,000 feet. The administration's declared moratorium, together with its continuing the process of granting offshore permits, evinces both the pull of traditional fossil fuel policy and the administration's approach to trying to move away from it.

After the spill continued for more than a month and was determined to be the worst spill in history, President Obama stepped up his approach to the crisis. After admitting that the administration's response should have come sooner and more forcefully, the President suspended virtually all current and new offshore drilling pending a full safety review including the newly opened areas; chastised the MMS for its "cozy and sometimes corrupt" relationship with oil companies; and accepted the resignation the MMS director.[86]

Upper Big Branch Mine

Massey Energy's Upper Big Branch mining disaster occurred on April 5, 2010, and resulted in twenty-nine miner deaths. The mine exploded as a result of excessive methane levels. As unfortunate as this tragedy was, Massey Energy is no stranger to the U.S. Mine Safety and Health Administration (MSHA), which since 1995 had cited the coal company for more than 3,000 mine safety violations. Although the sixth largest coal mine operator, Massey was responsible for more deaths since 2000 than any other operator.[87] Further regarding violations, in 2009 alone the agency issued double the number of violations issued in any previous year. As recently as March 2010, fifty-three safety violations were issued including citations regarding the mine's ventilation plan; and, on the day of the explosion, a violation was issued for failing to provide escape maps and for failing to properly insulate electrical cables.[88] Massey's intentional ignoring of health and safety regulations simply continued its aggressive pattern of overworking the miners, union busting, and profit making above all else.[89]

Tragically, the history of safety violations and the politics of fossil fuel favoritism are deeply embedded in mining culture, including its regulation. Coal mining in general is subject to an array of health and safety regulations; unfortunately, too often those regulations are woefully under-enforced to the direct detriment of mine workers and to the surrounding environment. To the mine's CEO, Don Blankenship, safety violations are part of doing business, and he was reported as having said that "We don't pay much attention to the violation count."[90] More callous still, before Congress, Blankenship testified that in the previous decade his company had experienced twenty-three miner deaths, thus implying that the twenty-nine deaths at Upper Big Branch were simply an ordinary cost of doing business. In response to that claim, Cecil Roberts, the president of the United Mine Workers, testified that he could find no other coal company that had a fatality rate of twenty-three deaths in ten years.[91]

Blankenship has been deeply involved in partisan politics. Sitting on the boards of the U.S. Chamber of Commerce and the National Mining Association, Blankenship has been an active Republican fund-raiser, notably anti-union, and notorious for having tried to buy at least one seat on the West Virginia Supreme Court and influence another. A West Virginia jury issued a $50 million award against the Massey Coal Company for a variety of tortious actions. During the pendency of the appeal, Blankenship entertained Justice, later Chief Justice, Elliott Maynard in Monte Carlo.[92] Further, realizing that the West Virginia Supreme Court would likely hear an appeal of the award and that the court was closely divided, Blankenship contributed $3 million to elect Brent Benjamin to the court. The state supreme court heard the case with Justice Benjamin refusing to recuse himself and reversed the $50 million jury verdict. The case was ultimately heard by the United States Supreme Court, which ruled that recusal was required.[93]

A second level of political influence involves the failure of the regulators to regulate. More specifically, the U.S. Mine Safety and Health Administration had been notorious for its lack of oversight and for its willingness to go lightly on mine owners. The agency preferred volunteerism to heavy-handed enforcement and, thus, relied on the mine owners and operators to police themselves.[94] After the Sago mining disaster in 2006, also in West Virginia, Congress passed the Mine Improvement and New Emergency Response Act, amending the 1977 Federal Mine Safety and Health Act, to upgrade mine safety enforcement. In carrying out the legislative mandates, MSHA began to increase its enforcement efforts yet fell behind training its inspectors, resulting in continued lax enforcement.[95] MSHA continues to be subject to criticism as a weak agency that has been reluctant to issue significant fines, close mines, or close repeat offenders. Instead, the fines are often small and also go uncollected for several years.[96] An analysis of fines levied by the MSHA revealed that since 2005, $123.4 million of fines have been assessed, yet only $10.2 million have been paid.[97]

The Upper Big Branch Mine, and its operator Massey Energy, were the beneficiaries of lax enforcement. In addition to the several hundred cited violations, Massey Energy contested many of its violations, which hand-tied the agency from taking more stringent steps. After the explosion, federal officials noted that the mine's safety record was suspect enough to declare the mine as having a "pattern of violations," which would have authorized MSHA to increase oversight even to the point of shutting down the mine in the event of finding a significant violation. The agency has rarely used its power to find a pattern of violations, even though it has had that authority for decades. It was not until 2007, however, that the agency even began warning mines that a pattern of violations existed and has cited a mine for finding a pattern only once. To date, the mine owner's ability to contest citations has effectively stayed the hand of the agency, allowed mines to operate as usual or pay minimal fines,[98] and, it appears, may well have led to the Upper Big Branch explosion.

The coal mine lobby is neither without resources nor political clout. During the Bush II administration, both coal and oil interests played prominent roles on Vice President Cheney's Energy Discussion Group, which produced the administration's energy policy in May 2001. Neither the vice president's office nor the president's office has ever released the names of the persons on the Task Force despite litigation to force open disclosure. Those efforts resulted in a Supreme Court opinion that kept that information secret.[99] Still, it has been reported that representatives of Exxon Mobil, Enron, British Petroleum, and the American Petroleum Institute, as well as other industry officials, met personally with either Vice President Cheney or key staffers. From the coal and electric utility side, Duke Energy, the Constellation Energy Group, and the National Mining Association, among

others, met with the vice president's energy group.[100] Further, the CEO
of Peabody Energy, a major coal company, served on Bush's EPA transition team; two other senior Peabody officials served on the Department of
Energy transition team; and still others served on the transition team for
the Interior Department. Peabody fared well in the administration, perhaps
after having given $846,000 to federal campaigns, 98 percent of which went
to Republicans.[101]

In recent years, the coal industry has expanded its lobbying efforts to
include more than 100 lobbyists with a budget of $14 million, increased
from $2.5 million in 2003. As a result, the lobby has successfully rebutted
proposals to install safety seals to contain methane leaks; safety measures to
contain fires and explosions; and environmental protections against mercury, mountain top removal, and new source pollution while also attacking
renewable energy and climate change legislation.[102]

Coal industry lobbying is based on certain fundamental premises. First,
coal is abundant and cheap and has been the mainstay of U.S. industry for
over a century. Second, coal, like America, demonstrates strength and economic growth. Third, any additional regulations are not only harmful to
the industry but ultimately harm miners, their families, and the U.S. economy. Fourth, climate change regulations, as well as additional environmental regulations, tend to be hysterical reactions to isolated incidents and are
neither well-thought-out nor in the best interest of the country. Whereas all
of these claims are sacrosanct to mine owners, none are true.

Although coal is abundant and cheap, it is only so because coal has never
been properly priced. Further, although coal has been the mainstay of U.S.
industry, other fuels can be easily, and beneficially, substituted. Similarly,
big coal has symbolized U.S. strength and economic growth, particularly
during periods of industrialization and heavy manufacturing. Today, the
U.S. economy is being transformed into an information and digital economy with less reliance on either heavy manufacturing or industries such
as steel and large equipment. Third, additional regulations, particularly as
they pertain to miner health and safety as well as environmental protection,
are intended to preserve the economy, provide jobs, and serve workers.
Finally, new environmental and energy regulations, including climate
change initiatives, are anything but knee-jerk reactions to a series of cascading environmental hazards such as those involving Deepwater Horizon and
Upper Big Branch. Instead, the need for new regulations for a low-carbon
economy is based on over a generation of serious policy analysis and study;
is responsive to a growing complexity of needs including economic and
environmental protection as well as security; and is long past due.

Breaking the Addiction

At its root, a transformed energy policy requires moving away from fossil fuels to a low-carbon energy economy. To do so, we must break our addiction to fossil fuels, and throughout this book a series of proposals to do so are made. If we believe that energy policy is primarily about national security and if we believe that our addiction is addiction to oil from petrostates, then one possibility is to increase domestic production. However, that strategy is unlikely to succeed because our domestic resources cannot supply all of our needs. Even if we were to tap all available resources, they would yield only a few years' more of oil at the current rate of demand. Similarly, we can search for oil substitutes in the tar sands and oil shale of the western United States and Canada. These efforts, however, simply substitute one liquid fossil fuel for another and will continue or increase carbon emissions.

To the extent, then, that we move away from substitutes for foreign and domestic fossil fuels, an array of options is available. Primarily, we must electrify the transportation sector. Nuclear power generators must be constructed more cheaply and its waste disposed of more safely. Renewable resources must be promoted, especially solar and wind power. The electric grid must be upgraded substantially. We must develop a range of biofuels including ethanol as well as algae. We must also develop coal-to-liquid technologies for the transportation sector. Additionally, given the abundance of coal, advanced clean coal technologies as well as carbon capture and storage become necessities. Federal investment in innovative energy technologies must be increased substantially, and innovation policies must proceed as a partnership between the public and private sectors. Finally, energy efficiency and smarter consumption behaviors must also be a part of a strategy to break our oil addiction.

Each of these strategies is discussed throughout the book. Here, two essential strategies – electric vehicles and carbon capture and storage – are discussed as essential components of moving away from traditional oil and coal resources. The transportation sector must be transformed radically if we are to achieve oil independence. That transformation will occur through the use of electric vehicles, as well as through alternative transportation fuels, as the transportation and electricity sectors of the energy economy merge.

The transition from fossil fuels to a low-carbon energy economy has encountered substantial resistance over the last four decades, and it continues to meet resistance today as fossil fuel companies fund anti-climate change efforts and promote bad science as good policy even while paying lip service to the need for new energy strategies. Since 2009, for example, the fossil fuel industry spent $543 million in lobbying expenditures to shape or kill climate legislation, and Exxon alone spent more than the entire pro-environment lobby.[103] The arguments against transition are cast in terms of economics, but all are, at bottom, political. If there is any lesson to be learned from this book it is that energy policy has never operated in a competitive "free market." Instead, energy policy has been the product of industry and government cooperation and mutual support. It is equally clear that projections about economic costs and benefits of the transition are difficult to make with a high degree of confidence. Nevertheless, modelers have made economic projections that show net positive economic gains with a smart energy policy.

A study by the Energy Security Leadership Council, a group comprised of nationally recognized business and military leaders, employed a computer model to assess the economic consequences of the type of energy policy outlined previously. The findings in that study are impressive. The study, for example, finds that individual household income would increase nearly 2.1 percent annually. Cumulatively between 2009 and 2050, there will be a net addition of $13.9 trillion in aggregate household income as a result of a new energy policy. The study also projects the creation of more than 3 million jobs.[104]

Regarding our dependence on oil, the report says that we can reduce oil imports by 6.6 million barrels a day, which translates to 60 billion fewer barrels of foreign oil from 2009 to 2050, while cutting off billions of dollars from flowing to unstable regimes and, in some instances, to our enemies. The effect of reduced world demand will have a positive effect on our trade deficit; lower the world price of oil; and, as a consequence of higher levels of income, produce federal net revenues in the range of $1.46 trillion, thus promoting national economic security while also reducing carbon emissions by greater than 19 percent by 2050 under a business-as-usual scenario.[105] To be sure, these gains, as well as the costs of transition, will continue to be debated, and that debate must take place. The caveats should be clear: Debate should be open and vigorous, leaving all ideas on the table for discussion; hard science, as well as hard economics and sound economic modeling, must inform the conversation and not warp it; and new energy policies cannot continue to be held captive to old ideologies.[106]

Electric Cars
By conservative estimates, there are in excess of 200 million cars and light trucks on our nation's highways, virtually all of which run on the internal

combustion engine (ICE), consume fossil fuel products, and constitute 70 percent of the demand in the oil sector.[107] If the United States is to have any success in breaking its oil addiction, the transportation sector must be transformed from reliance on oil to being powered by electricity. The promise of an electric transportation sector is enticing. Even using the currently most expensive forms of electricity, solar and wind energy, it has been estimated that clean energy can supply the entire car fleet of the country for the next fifty years at the amount of money it costs to import crude oil for one year.[108] Electricity is notably cheaper than oil.

In order to accomplish this transformation, three elements are necessary. First, electric vehicles (EVs) must be built. Second, alternative fuels, liquid as well as electric, must be developed. Third, the distribution infrastructure must be constructed. Within each of these three building blocks, existing technologies demonstrate the possibility of an electric car future. Still, advanced technological developments are highly desirable, and perhaps even necessary, for a successful transition. By way of one example, advanced batteries will greatly facilitate this transition. Most importantly, a transformed transportation sector will only occur when all of these technologies are successfully brought to commercial scale.[109]

In brief, cars must become more efficient. Currently, the "tank-to-wheel" efficiency of a car is a fraction of the energy contained in a gallon of gasoline. More precisely, the ratio of energy output to input is 1:8. In other words, the standard ICE uses approximately 12.6 percent of the energy contained in a gallon of gasoline. Much of the energy is lost to friction, heat, and idling, and a small fraction is lost to operate accessories such as the radio, lights, and air conditioning.[110] Even without significant technological inventions, notable gains in efficiency can occur with the introduction of new aerodynamic car designs, lighter vehicles constructed with new, safer materials, and increased engine efficiency.[111]

EVs grow in importance once we realize that addressing oil dependence in terms of imposing effective carbon emission regulations is not promising. EVs can become the preferred method of transportation if they are cheaper to purchase and operate than ICE vehicles. As a matter of simple market dynamics, they can become cheaper if gasoline prices rise and stay above $8 a gallon. At that level, imports can be reduced to 5.7 million barrels a day. Also at that price, there will be a reduction in vehicle miles traveled (VMT).[112] Yet, sustained $8 gas is not on the horizon. Congress, for example, is not considering legislation that prices carbon to that level either through cap-and-trade rules or through a carbon tax. Similarly, given the price volatility in world oil markets, and given past history of unsuccessfully moving money from conventional oil to the development of synfuels, it is unrealistic to assume that world oil prices will rise to and sustain the price level needed for a dramatic rollout of EVs based on high gas prices alone.

For EVs to reach market scale, they must be seen as desirable by consumers. Part of that desirability lies in the EV serving as an effective substitute for conventional vehicles. To be sure, the United States is a car culture. We depend on personal vehicles for commuting to work, extended vacations, and casual trips for shopping and socializing. With the exception of a few large cities, public transportation is unavailable to most Americans; consequently, cars are essential. Consumers will substitute an EV for an ICE, not only when the EV is cost-competitive to buy and operate, but when it has the range and acceleration, the space and weight, and the price and refueling convenience of ICEs.

In Chapter 4, a variety of electric cars were described. In brief, plug-in hybrid electric vehicles (PHEV), such as the Chevy Volt, contain both batteries, which can be recharged from an external source, and internal combustion engines. The PHEV will contain a larger battery that can recharge at rest. These vehicles also have the capacity to allow the engine to recharge the battery. These cars have the capacity to run completely on electricity but generally travel short distances in the range of ten to forty miles before the need for a recharge. Hybrid electric vehicles (HEVs) run on both electricity and petroleum, with the engine recharging the battery without the capacity for external recharge. Finally, the last major category consists of electric vehicles (EVs), such as the Tesla Roadster, which operate on electricity alone and will require significant advances in battery technology.

The most dramatic difference between PHEVs and EVs is that PHEVs continue to require fossil fuels for operation and therefore continue carbon discharges and oil dependence. EVs, in comparison, use fuel resources more efficiently; much of the infrastructure (i.e., the electric grid) already exists; they are beginning to be produced affordably and promise to be able to be produced at scale; are cheaper to operate; and emit no carbon. The downside to EVs, however, is that they have high upfront costs largely due to the cost of the battery, and some have a limited range. The Tesla Roadster, for example, projects a new Model S with a 300-mile range and acceleration of 0 to 60 mph in 5.6 seconds. The Roadster, then, is comparable to many ICE sports cars but comes with a base sticker price of $49,900.

Currently, there are significant differences among alternative vehicles. Some PHEVs, for example, are relatively affordable and reduce the cost of fuel per mile traveled but have limited acceleration and range. The first generation of PHEV, with a range of ten miles, has been estimated to cost $5,500 to $6,300 more than a mid-size ICE, and a PHEV with a forty-mile range has been estimated to cost $14,000 to $18,000 more.[113] EVs are more costly still because of higher-cost batteries. However, they also reduce fuel price and have acceleration and range comparable to the cars we drive today. In both cases, however, projections for market penetration are relatively modest, ranging from a best case of 40 million PHEVs on the road by 2030 to 1.5 million units worldwide by 2020.[114] These projections depend on several factors such as price of oil, advances in battery technology, grid reliability, and consumer education. Through all of that, the internal combustion engine will continue to dominate the transportation market for the next generation.

The transformation of the transportation sector also requires alternative fuels to supplant oil and reduce dependence on foreign imports. In addition to electricity, new transportation vehicles will also be driven by liquid fuels ranging from bio-algae to liquid from coal and to various ethanol products. The National Academies report, for example, states that liquid fuels made from biomass (plant and waste) and coal are promising transportation fuel technologies. These technologies are deployable within the next ten to twenty-five years and can become cost-competitive with petroleum, also reducing carbon emissions and improving our energy security.[115]

Electricity, then, promises to be the most efficient fuel. Even the most expensive electricity, solar power, costs approximately $.02 a mile for electric vehicles. In other terms, at that rate an electricity-fueled vehicle would run on the equivalent of $1.50 per gallon for an ICE with oil selling at $25 per barrel, a price we are unlikely to see again. In still other terms, it costs roughly $.15 a mile to operate an ICE compared to $.08 per mile, currently, to operate an EV with cost projections declining to $.04 by 2015 and $.02 by 2020 due to scale economies and declining production costs for EVs.[116]

Once the vehicles have been designed and manufactured and once the fuels have been developed, a distribution infrastructure is necessary. The current understanding is that the existing electric grid, with a few modifications, can sustain the current growth in EVs because we will use off-peak energy at night to recharge batteries.[117] As market penetration increases, though, projections indicate that approximated thirty-eight large new power plants will be needed.[118] Additionally, private-sector actors, such as Better Place, are developing a recharging infrastructure so that electric vehicles can be utilized with the convenience of the cars we currently drive.

Better Place is a well-capitalized corporation listing assets of $1.25 billion with plans for commercial-scale projects in Denmark and Israel.[119] The

Better Place solution is to make the battery a separate, not fixed, component to the car so that batteries can be switched at stations installed along highways in less time than it takes to fill a tank of gas. In addition to switching stations, recharging stations will exist in various shopping malls, business parks, and the like.[120] The continuing improvement in the production of EVs must take place alongside the development of alternative fuels and the infrastructure. Consumers will not purchase in quantity unless their "range anxiety" is reduced or eliminated and they can depend on a reliable recharging system for long-range trips.

Whereas Better Place is a for-profit smart energy company, Amory Lovins' Rocky Mountain Institute (RMI) is a nonprofit think tank that has been involved with transforming our energy economy for over a generation. Its initiatives include developing and marketing PHEV vehicles with a consortium of private companies through Bright Automotive. In addition, RMI has been developing the idea of the "smart garage." Smart Garage is more metaphor than physical space, and its intent is to integrate transportation, the electricity grid, and buildings to enable the broad-based adoption of electric vehicles. Through such integration, energy efficiency can be maximized as the vehicles can serve to help stabilize generation and increase the use of solar and wind power while also increasing customer control and choice.[121] A profitable by-product of developing Smart Garage is to stimulate a new industry in ancillary businesses and services involving such things as vehicles, energy storage, charging, metering, and energy systems for buildings, as well as marketing, financing, and associated software programs.[122]

The Better Place and Rocky Mountain Institute stories demonstrate that for-profit and nonprofit sectors are moving forward to bring new energy policies to the market through the inventive use of science, technology, and business models. Still, for transition to electric transportation to occur on a wide scale, government support will be necessary for R&D expenditures, especially in battery technology, and through market-creating incentives such as loan guarantees, tax credits, and the like. Currently, the Energy Independence and Security Act of 2007 has provided manufacturing loan incentive programs for developing advanced vehicles and other components. The act also authorizes nearly $300 million a year for six research and development programs focusing on energy storage and $90 million a year for plug-in electric drive vehicle demonstration programs. The Emergency Economic Stabilization Act of 2008 provides tax credits of up to $7,500 for plug-in electric vehicles, and various state programs also offer tax incentives for electric vehicles.

Car manufacturers are becoming increasingly invested in the electric vehicle market. Toyota recently entered into an agreement with Tesla Motors to provide funding for the development of cars, parts, production systems, and engineering support.[123] Additionally, Toyota has been a leader in developing EVs by offering its RAV4 and Prius PHEV and by planning to develop an urban commuter EV by 2012. Ford has been dedicating $450 million for its EV plan as well as using $550 million to transform a Michigan plant from an SUV to an EV manufacturing facility.[124] Ford continues to develop its EV lines including releasing its Ford Focus in 2011, which has a 100-mile range per charge. The Nissan LEAF is to be released in 2011 as the world's first all-electric, zero-emission car designed for the mass market. The LEAF will be a five-seat passenger car also with a 100-mile range that can be charged in less than thirty minutes with special equipment at a cost of between $25,000 and $33,000.[125] General Motors' PHEV Volt is to be released in late 2010 and will allow consumers to drive forty miles on a single charge and longer with a range extender.[126]

EVs are symbolic of the types of changes needed to successfully accomplish an energy transition. Whereas public-sector policymakers play an important role in helping to implement change, alone they do not constitute the type of leadership that is necessary for a successful transition. Instead, the public and private sectors must work in tandem to take the innovative potential of new energy science and technology and bring it to scale. EVs also show the interrelatedness of future energy policy. In the future, the oil/transportation and electricity segments of the energy economy will no longer be separate. Instead, one sector may well fully replace the other as electricity replaces oil more cheaply, more responsibly environmentally, and more promotive of national security.

Carbon Capture and Storage

The U.S. coal-fired electricity industry is experiencing some conflicting indicators. DOE projects that the country will increase its coal consumption; yet, as of 2010, there has been a de facto moratorium on the construction of new coal plants.[127] At the same time, there is a consensus that clean coal technologies are necessary, yet none are commercially available. Although there may be somewhat of a (temporary) slowdown in coal use domestically, there is a notable increase worldwide, especially in India and China.[128]

There are in excess of 600 coal-fired power plants in the United States that generate half of our electricity. Given the low cost, abundance, and existing multibillion-dollar infrastructure for coal, this resource is not likely to be eliminated from our energy future. The United States has one-quarter of the world's coal reserves, and the energy content of those reserves exceeds the energy content of all the world's known recoverable oil.[129] However, it is our dirtiest fuel and the most significant source of anthropogenic CO_2 emissions. Consequently, for coal to continue to provide electricity, especially as our energy economy moves from oil to electricity for transportation, coal generation must become cleaner. Carbon capture and sequestration (or storage) (CCS) is the process of capturing CO_2 from coal-fired power plants and other industrial users, transporting it to suitable geologic formations, then injecting it into that formation for thousands of years to reduce carbon emissions.[130]

CCS technologies can occur either prior to combustion or after it. Although pre-combustion technologies can remove carbon more cheaply, the construction costs are more expensive. Concomitantly, post-combustion strategies are less efficient but hold the potential of being cost-effectively installed in existing power plants through retrofits. Consequently, to reach any level of carbon emissions reductions, existing plants must either be retrofitted with clean coal technologies or rebuilt. Thus, there are two basic approaches to dealing with existing plants. Retrofits or rebuilds can be undertaken to either increase efficiency, effectively decreasing emissions, or directly decrease emissions.[131] Efficiency retrofits will only realize modest emissions reductions; however, they will be attractive to older and less efficient plants. Rebuilds can realize greater carbon reductions and may promote larger efficiency gains; however, they are more costly.

Currently, several CCS projects are operating around the world and are capturing millions of tons of CO_2. Yet, none of the projects are ready for commercialization, which will require the sequestration of billions of tons annually. Projections are that widespread commercialization can start as soon as 2020.[132] The federal government has been supportive of developing CCS technologies with the last two administrations dedicating more than $8 billion to pilot projects. Moreover, the DOE estimates that sufficient storage capacity exists in the United States for up to 500 years.

CCS technologies face significant hurdles including: (1) increased cost of plant construction and operation;[133] (2) reduced output from electricity plants in order to capture the carbon, also known as the "energy penalty"; (3) the need for significant infrastructure, especially pipelines, to transport the captured carbon dioxide to storage sites; (4) finding adequate storage capacity;[134] and (5) protecting against environmental risks including seepage into the air or into underground water supplies. In short, CCS must find answers to several cost problems. To put energy costs in some perspective, we can currently generate coal energy at a cost of between $1 and $2 per million BTUs as compared with oil or natural gas energy at $6 to $12 per million BTUs.[135] Then the question becomes: At what cost can electricity with CCS be generated?

In an important study titled *The Future of Coal,* MIT researchers concluded that CCS is the "critical enabling technology that would reduce CO_2 emissions significantly while also allowing coal to meet the world's pressing energy needs."[136] The researchers also considered the cost issues and concluded that CCS can become economical as long as an effective carbon charge is imposed on carbon emissions. The exact amount of that charge is subject to debate, and the MIT study examined a high-price scenario of $25 per ton increasing at a 4 percent rate per year and a low-price scenario of $7 per ton increasing at a 5 percent annual rate. These two scenarios were compared with business-as-usual with the result that even under the high-price scenario less coal will be consumed relative to business as usual, but coal use overall will increase regardless. In other words, for CCS to be economically sustainable, the carbon price must exceed $25 per ton. Nevertheless, to attain a low-carbon energy economy, CCS is a necessary element to reduce emissions in light of increased coal consumption with the goal of reducing carbon emissions to less than half of today's level by 2050.

The pollution problem can be addressed through clean coal or advanced coal technologies such as integrated gasification combined cycle generation (IGCC). Most power plants use pulverized coal, which involves grinding the coal and burning it to turn turbines to generate electricity. IGCC, by contrast, converts coal to gas, runs the gas through a turbine to generate electricity, and uses the excess heat from that process to generate additional electricity, which forms the "combined cycle."

IGCC has several advantages over pulverized coal. With this process, CO_2 is easier to separate and then capture. Although IGCC is more expensive than pulverized coal, it removes more pollutants, including nitrogen oxides, sulfur dioxide, mercury, and particulate matter, and uses less water than pulverized coal operations. IGCC can dramatically reduce particulate matter and capture in excess of 90 percent of carbon emissions depending on the particular IGCC technological configuration.[137] Carbon dioxide can be captured and then pressurized and transported for underground injection and storage. The energy penalty with this process consumes roughly 25 percent of the power generated by a facility.[138]

IGCC technologies are commercially available but only partially deployed. On April 30, 2010, General Electric announced the shipment of its second advanced high-efficiency turbine to be installed at Duke Energy's Edwardsport, Indiana, power plant, which is expected to be in operation in 2012. This technology converts coal to gas, a synfuel, while removing pollutants such as nitrous oxide, sulfur oxide, mercury, and particulate matter. Coal is then used to generate a new, cleaner burning fuel that is then used in a gas turbine to produce electricity as well as directly power the turbine to create electricity.[139] This technology has the ability to capture carbon dioxide prior to combustion. Once captured, the carbon then requires storage, which is currently cost-prohibitive and may necessitate regulatory requirements.

The major issues facing clean coal technologies in general, even the most promising IGCC projects, involve commercial scalability. The recommendation in the MIT study, and in others,[140] is that increased R&D, which currently stands in excess of $8 billion, be used to build several demonstration projects using various clean coal technologies. The demonstration projects should test the energy and economic effectiveness of CCS technologies from capture to storage. In addition, to successfully implement CCS strategies, a regulatory system that imposes strict standards on site selection, project design, operation, and long-term monitoring is necessary.[141]

During the Bush administration, the Department of Energy began a clean coal project known as FutureGen that has been retooled under the Obama administration to build a first-of-its-kind coal-fueled, near-zero emission power plant. FutureGen is a public-private partnership intended to eliminate environmental problems associated with coal-fired plants utilizing IGCC technologies. It is anticipated that the federal contribution will be in excess of $1 billion, most of which has been earmarked in the Stimulus Bill, and that about $500 million will be provided through a consortium of private companies.[142]

Additionally, the DOE is continuing other public-private ventures, including the Clean Coal Power Initiative, which will finance technologies to cut sulfur, nitrogen, and mercury emissions from power plants.[143] Still, more needs to be done to reach commercial scale for CCS. To achieve such a goal, several steps are needed. First, carbon must be priced either through a tax or through a cap-and-trade requirement. Second, funding must be available for early demonstration projects and retrofit technologies. Third, large-scale capture and storage projects must be undertaken. Additionally, the risks of large-scale geologic sequestration, including issues of liability and property rights, must be better known.[144]

The legal obstacles are not insubstantial. A regulatory regime will be needed to both fund and monitor CCS projects. Jurisdictional disputes between states and the federal government over pipeline siting, as well as licensing of storage facilities, must be worked out. In this regard, the potential for conflict between clean energy advocates and environmentalists is significant. CCS can be analogized to onshore and offshore oil and gas drilling and exploration. The onshore analogy compares the ownership of underground geologic formations for storage to the surface owners, and through some sort of unitization regime, property rights can be determined. The offshore analogy compares a liability for leakage or other environmental damage to the type of environmental harms experienced with offshore oil and gas extraction. In short, environmental NIMBYism has the potential to slow the transition to a smart energy economy not only for CCS but for any number of new energy technologies and applications.

Conclusion

A new politics and a new economics of energy is needed to combat the challenges of climate change, to realize a healthier environment, and to maintain a robust energy economy. At the same time, the domestic and international need for energy from fossil fuels continues to grow, and the legal and regulatory structures in place continue to favor these fuels. Whereas investments are being made in alternative and renewable energy resources, investments are also being made in seeking larger quantities of traditional fossil fuel resources. A new energy strategy cannot be blind to the old ways. Instead, the new energy strategy must take established institutions from Washington to Wall Street into account to successfully make the transition to smart energy. The regulatory histories of, and recent disasters in, the oil and coal industries more than amply demonstrate how legislation and regulations, politics and policy, have been used in the past to justify a fossil fuel economy at significant human and environmental cost. Rethinking those assumptions is a necessary prelude to adopting a new regulatory scheme to address global warming.

Fossil fuels dominate our energy economy because of history and politics, not only because of the great wealth these industries generate. Entrenched energy policies and the markets that support them will not be soon displaced. Still, leveling the playing field by redirecting subsidies away from fossil fuels to alternative and renewable resources is a necessary approach to incorporating the new energy thinking. The old ways must change, and dirty energy must be cleaned through new technologies such as CCS. Additionally, new markets, new technologies, and new energy strategies, such as EVs, are attracting private investment. As energy costs continue to rise, as global warming is perceived as more of a threat to our lifestyles and livelihoods, and as smart alternatives gain a growing share of the market, fossil fuels will lessen in importance. For that transition to occur, though, both political and market forces must be aligned so that not only our economy but the global economy can fully address the challenges of climate change and we wean ourselves from dirty energy.

6

Electricity Future

Energy business-as-usual is not a viable option for the United States.
CNA Military Advisory Board[1]

Introduction

Moving away from energy business as usual has a direct and significant impact on our electricity future, particularly as electricity plays a larger role in transportation. To get there, however, the oil and electricity sectors will require different regulatory regimes. In the past, the two sectors were regulated significantly differently. Except in times of crisis, the oil industry was largely immune from price and allocation regulations. Instead, the oil industry was free to set its prices and, except for the beginning of the twentieth century, was treated as a competitive non-monopolistic industry. To the extent that government intervened in the oil industry, it did so through subsidies and financial supports, some of which were direct, some of which were indirect, and some, such as military support, were hidden. Such financial support subsidized, and therefore underpriced, the cost of gasoline at the pump. By contrast, the electricity industry was heavily regulated as a natural monopoly, which had the intended effect of supporting its expansion and capital development.

As our energy future evolves, the case can easily be made for the removal of government price supports and subsidies to the oil industry. The case can further be made for the necessity of pricing carbon, which will affect the price of oil, bringing it closer to its true cost. Still, with the exception of carbon pricing and other environmental protections, the oil industry can be left to its own devices and can be left unregulated regarding prices and allocation. The same, however, cannot be said for the electricity industry, and government regulations are necessary for two reasons.[2] First, the electricity industry continues to exhibit natural monopoly characteristics, especially in its transportation and distribution segments. Second, if we are to move to a low-carbon economy, then not only must a carbon price be set for the coal and natural gas feedstocks used in electricity production, but regulations are also necessary to generate electricity more efficiently and are necessary for the greater production of electricity through renewable resources. In short, government regulations can no longer sustain fossil fuels; instead, they must sustain alternative resources and efficiency.

Our electricity future, then, requires a transformation in: (1) the way the industry is regulated; (2) how it is funded; and (3) how the privately owned utilities run their businesses. More specifically, the regulatory scheme for the electric industry must promote a low-carbon energy mix as well as design a smart infrastructure to deliver it. Further, electric utilities must change their business model to accommodate the sale of efficiency and renewable resources. In other words, electric utilities will be "selling" conservation and energy efficiency and will move away from their dependency on coal. This chapter explains how both the electricity industry and its regulation must be transformed to accommodate the consensus energy policy.[3] Through a coordination of policy, technology, and investment, a low-carbon electricity future can become permanent.[4]

In significant ways, our electricity future is now. Utilities such as Duke Energy, Portland Gas & Electric, and Pacific Gas & Electric are transforming into iUtilties. State policies are capitalizing on energy efficiency. California, for example, has kept electricity consumption *per capita* flat since the mid-1970s, cutting against the national trend of increasing consumption. Additionally, California emits about half of the carbon per dollar than the rest of the nation, and the state generates significantly more electricity from non-hydropower clean sources than any other state, with a goal of 33 percent by 2020.[5] Vermont, as another example, has created an energy efficiency utility that pays its citizens to use less electricity. Still, to bring these individual efforts to scale will require national leadership, policy coordination, a shift in public funding and private financing, and a new regulatory regime, among other next steps.

Industry Overview

U.S. coal-fired power plants are responsible for 10 percent of the world's CO_2 emissions.[6] Consequently, any strategy aimed at reducing carbon emissions must address the electric industry. The electric industry, in turn, must reform itself, and the single most significant reform involves a change in an electric utility's mission. Traditionally, electric utilities were driven by one concern and one concern only – sell electricity. The mission was simple and understandable. Electricity consumption supported the economy: QED the more electricity that is sold and consumed, then the stronger the economy. The traditional mission served the country well, delivering cheap, reliable electricity.

The untold part of the story is that the electricity industry, except in its very nascent stage, never operated in a free, competitive market. Rather, it has been supported by a regulatory compact between government and industry. The regulatory compact, and the parties to it, must change their relationship to meet the demands of a new energy economy. Government regulation must change to promote the public good of a cleaner and more efficient energy. Utilities must change their businesses to ones that sell energy services and products – not just electricity. Moreover, consumers must be prepared to make smart, more informed choices. The transition from a traditionally structured utility to a modern, smart utility will change the nature of an electric utility's business model from its product mix to its pricing policies; will change the nature of the government regulation of that industry; and will change consumption patterns.

The iPhone and the iPad serve as models, and as metaphors, for the new electric utility – the iUtility. The iUtility no more resembles a smoke-belching power plant than the iPhone resembles the heavy black rotary dial telephone, or even the Princess phone, or the iPad resembles the clunky desktop and tower computer of only a few years ago. Rather, the iUtility, like the iPhone and the iPad, is technologically sophisticated, offers various services and products at various prices, and can be customized, decentralized, and personalized as distinguished from the large-scale, centralized, fossil fuel power plants that, once constructed, were intended to last a generation or more.

Most importantly, the iUtility must capitalize on its new business model by adopting analogous forms of network effects exhibited in the iPad economy. Just as the iPad creates social networks for the exchange of information and for new patterns of consumption, the iUtility too can create social networks for providing information about demand, supply, and prices to promote more competitive transactions as well as better inform producers and consumers. The iUtility social network can contribute to grid reliability; encourage new entrants; and promote innovation, all through real-time information systems.

The electricity industry has roughly $300 billion in annual sales with more than $1 trillion in assets.[7] There are approximately 4,000 electric power suppliers, approximately 203 of which are known as investor-owned utilities (IOUs). IOUs are privately owned and generate 72 percent[8] of all electricity, while public entities, both federal and local, produce the remainder.[9] The industry has its unique characteristics. Consumers expect that electricity will be abundant, affordable, and literally available at the flip of the switch. Similarly, producers expect that large-scale electricity production can achieve significant economies of scale (and profits) as they have invested billions of dollars in plant and infrastructure.

Electricity has its own unique characteristics. Electricity cannot be effectively stored. Just think of how quickly the battery in your laptop or cell phone runs down. Electricity is also a completely fungible or homogeneous good. There is no quality difference between electricity from one source or another. Electricity moves literally at the speed of light; consequently, a purchaser cannot identify its point of origin or its consumption path. In other words, consumers and producers do not identify specific goods for sale. Instead, producers place electricity onto the electric grid, and consumers purchase what they need. Although all electricity performs exactly the same functions, electricity service can be more or less reliable, and therefore electricity *service* varies in price and quality. Further, the Second Law of Thermodynamics dictates that as resources are burned to generate electricity and as electricity moves over distances, entropy will result in a loss of energy. There are, thus, efficiency limits in burning the resources that produce electricity as well as in electricity transmission. We will see, though, that we have not reached those limits and that efficiency improvements in the industry can be made.

Given unique industry and product characteristics, electricity regulation itself has developed its own unique structure that appears resistant to significant change after nearly one century. Fortunately, the industry has enjoyed great growth and profits throughout the twentieth century with the help of government regulation. Unfortunately, the regulatory scheme has failed to fully account for the social costs of pollution, and this failure presents a significant barrier to meeting the challenges of a new energy future. More unfortunately, regulators have been stuck in neutral for nearly a generation. The traditional regulatory structure must now confront the twenty-first-century reality of energy transition.

The electricity industry and its regulation worked well until the latter third of the twentieth century when the industry began to change. In approximately 1965, the marginal cost of electricity began to exceed its average cost, an economic event that affected consumers and producers alike. From an industry standpoint, it appeared that a technological plateau had been reached as economies of scale did not continue to be realized. Although utilities continued to invest in new plants, those plants (especially nuclear plants) were more expensive to build and contributed to excess capacity, which, in turn, raised the price of electricity. From a consumer's standpoint, the price rise meant that rates would neither continue to stay flat nor decline as they previously had for decades. As a further consequence, the electricity industry became more politicized both in the federally regulated wholesale market and in the state regulated retail market.[10]

One group of economists estimated that by 1970, the real price of electricity was 2.5 percent of the cost of what Edison charged his first customers. However, the era of cheap electricity is over as electricity prices rose 50 percent from 1970 to 1975[11] and continue to rise, although not as dramatically.[12] Since the Golden Age of electricity post–World War II,[13] both industry actors and regulators have been trying to reform themselves in response to technological and market changes but have met with little success.

The post-1965 era for the electricity industry has been troubling. Not only did plants cost more, in the 1970s all energy firms confronted higher costs, and the electricity industry seemed particularly hard hit as nuclear plants were cancelled or converted to coal-burning plants.[14] Congress wrestled with oil independence by trying to encourage utilities to switch to coal even as they recognized the adverse environmental effects of such a change.[15] In addition, in the mid-1970s, initially through the efforts of President Carter, Congress began to "deregulate" all network industries including energy industries such as oil, natural gas, and electricity; but the electricity industry resisted and continues to resist significant change.[16] Still, as a result of federal legislation, we learned that efficiency gains are possible in the electric industry because of the presence of new producers who can generate lower-cost electricity if only they can get their product to market.[17]

Policymakers and regulators were well aware of the significant and underlying changes in the electricity industry and, over the last three decades or more, state and federal regulators have been trying to restructure the industry so that old, expensive, dirty, traditional utility electricity can be first supplemented, then replaced, by cheaper, renewable, and alternative electricity sources. The difficulty, and sometimes inability, of new and alternative producers to enter the market, however, is a direct consequence of the century-old scheme of regulation that not only shaped the industry but supported the sales of low-cost, dirty electricity by traditionally structured utilities and constructed an infrastructure to do so.

As the electric industry moved from a local competitive industry to a federally regulated one, the firms within that industry, relying on the traditional form of regulation, developed their own corporate structure. In brief, both state and federal regulations encouraged electricity firms to integrate vertically and to serve local markets. Firms were granted government-backed monopoly status through what is known as the regulatory compact. In reliance on that compact, firms undertook a service obligation within an exclusive territory.[18] Utilities were given the incentives to sell as much electricity as they could and had an obligation to serve their local customers, thus giving the local utility a captured (and profitable) market. The government would protect that service territory from competition and would effectively ensure that privately operated firms would earn a reasonable return on their capital investment. In other words, the more generation that the utility built, the more it earned for its shareholders. It also meant that the utility could invest in transmission and distribution for their own customers, privately own those wires, and earn returns on those investments while avoiding competition from other providers.

The consequences of this regulatory design should be apparent – utilities were encouraged to sell more and more electricity; electricity costs rose once the infrastructure was built; local customers were preferred because profits are made within the service territory; the grid and its interconnections were jealously guarded because they were privately owned by the vertically integrated firms; utilities served as many customers within their service territory as they could; and the cheapest and most abundant natural resource, that is, coal, was relied on to generate electricity. Today, dirty fossil fuels and expensive nuclear power account for greater than 90 percent of the electricity that is generated, whereas the renewable resources of solar and wind power account for less than 1 percent.[19]

The regulatory structure, then, rewarded traditional, vertically integrated, privately owned utilities for building fossil fuel plants rather than investing in alternative or renewable resources. Further, the entrenched regulatory design has directly and negatively affected grid modernization because full access to the grid by cheaper and alternative power producers has not been achieved despite federal and state efforts to "deregulate" wholesale and retail electricity markets. Transmission line owners preferred their own profits to providing service to competitors, and state regulators preferred to keep rates low for their citizens. Why, for example, would the Arizona Corporation Commission grant approval of a transmission line proposed by a California utility to service its California territory?[20] The simple, and obvious, answer is that it would not. The failure to promote grid modernization is directly attributable to the regulatory incentives favoring local privately owned distribution and transmission facilities, which allow traditional utilities to maintain control for its customers thus protecting their shareholders while discouraging competitors. The failure to modernize has also led to under-investment in needed grid upgrades, leading to grid failure and power outages.

Traditional Utility Regulation

Traditional utility regulation is based on two complimentary ideas – the economic idea of natural monopoly and the regulatory idea of a compact between government and industry in the name of the public good. Together, these ideas enabled the industry to grow, to nationalize, and to produce cheap electricity for most of the century. The problem with this combination of ideas is that when they reached the end of their useful lives, unwanted consequences followed and the public interest suffered. Traditional regulation constructed a utility industry that became increasingly costly to maintain, that hampered new entrants, that generated millions of tons of GHG annually, and that is still resistant to change. Before moving to the iUtility and its regulation, fundamental utility concepts will be examined.

Natural Monopoly

Industry consolidation in the early part of the twentieth century revealed a central fact about the electricity industry – it constituted a natural monopoly. Monopolies are economically perverse as prices are set above competitive levels. Under monopoly conditions, consumers suffer losses that they would not suffer in competitive markets; some competitive producers are prevented from putting their products on line; and society does not maximize the use of its resources. Left unchecked, electric monopolies could, and did, set prices above and reduce supplies below competitive levels, thus causing inefficient social losses. But was there any other way to run a utility industry? Perhaps not, as utilities were seen as *natural* monopolies.

The easiest way to understand a natural monopoly is to imagine competing electric firms laying competing sets of transmission and distribution lines throughout the same geographic area.[21] It is the characteristic of a natural monopoly that one provider can provide the service more cheaply than multiple providers because multiple providers with multiple facilities are simply wasteful. Natural monopoly, then, became the justification for regulation in the form of the regulatory compact between utilities and the government to the great benefit of the industry and its customers for several decades.[22]

The Regulatory Compact

The government response to the market imperfection of the natural monopoly in the electricity industry was to regulate the industry. Ironically, regulation came in the form of a government imposed monopoly. Simply, a private monopoly was replaced by a government supported one. The government monopoly, through the regulatory compact, set prices at competitive, not supra-competitive, levels and did so through the process known as rate making. The compact is based on a quid pro quo. Privately owned utilities are subjected to government price and profit controls, in exchange for which the utility undertakes a service obligation to its captive customers. Under the compact, consumers avoid monopolistic prices and are entitled to electric service. The utility recoups its reasonable expenditures and earns a profit on its prudent or useful capital investments. The compact worked well for decades as both producers and consumers benefitted. Producers made a profit and, as utilities continued to enjoy economies of scale, consumer prices fell. In addition, public utility commission (PUC) cases were generally non-controversial, and as a consequence utility regulation fell below the political radar screen – that is, until energy prices rose in the 1970s.

Traditional Rate Making

Rate making is the device that drives the regulatory compact.[23] The fundamental idea behind rate making is to mimic the market; that is, set prices at efficient or competitive levels. The primary objective is to enable a private utility to operate as a competitive business able to attract capital investment and provide a return to its investors. Rate making serves other objectives as well, such as controlling demand through promoting or discouraging consumption. However, providing a return on investment to utilities and maintaining abundant electricity dominate the traditional reasons for rate making.[24] Electricity was to be universally available, reliable, and abundant, which could best be accomplished through large, centralized power plants. Those objectives were achieved through the use of the traditional rate formula as utilities built more and larger power plants and as they constructed a national electricity grid. In the early years of electricity regulation, as the industry expanded and prices declined, these goals were largely satisfied. Today, the continued use of the traditional formula comes at an economic cost of distorted market incentives and retarded competition and at the social cost of environmental pollution.

The heart of rate making is the utility's revenue requirement, which is the amount of money that the utility can collect from its customers to stay in business.[25] Regulators set the *rate level* that enables utilities to realize their revenue requirement. In essence, the traditional rate formula is a type of cost-plus contracting. A utility will recoup all of its prudently incurred expenses and will receive a government determined rate of return on its capital investment. The amount of capital investment is also known as the *rate base*. The traditional rate-making process had the effect of encouraging capital investment because the larger the rate base, the greater the utility's profits. Therefore, the rate formula rewarded utilities for building plants and selling electricity.[26] A continuing pattern of such investment, however, became less than optimal because the formula led to excess capacity, which means higher costs and, therefore, higher priced electricity. Further, this old pattern led to lower quality services, especially as investments in infrastructure declined threatening service reliability.

In addition to the rate level, traditional electricity regulation was also driven by *rate structure*. For most of the twentieth century, the most prevalent rate structure was known as the declining block rate. The rate charged for the first period of consumption would be higher than the rates set for the second and subsequent periods. As a consequence, utilities were able to rely on high rates to recover their fixed costs in the first period and would recover their variable costs in subsequent periods. The price signal sent to consumers, however, indicated that the more energy they purchased, the cheaper that energy would be, thus encouraging consumption. The declining block rate structure is also known, more descriptively, as a promotional rate.

The traditional rate-making formula caused distorted electricity markets through excess capacity, high prices, and pollution while encouraging coal use and discouraging the use of alternative and renewable energy resources. A new compact must reverse all of these trends by increasing efficiency, sending more accurate price signals, and moving away from coal. The new compact will continue to enable the utility to earn its revenue requirement. However, the local utility must now see itself in the business of selling efficient and clean energy services and products. The new utility's service obligation will be refocused from providing electricity to local customers to participating in a dynamic market space of clean and efficient energy. The regulation of the utility of the future will use a rate formula that transforms the utility to the iUtility and enables the iUtility to advance a low-carbon consensus energy policy. A new regulatory compact and a new application of the traditional formula can accomplish all of these objectives.

The Future of Electricity

Regulatory reform in the electric industry has been ongoing for more than three decades; and reform efforts, particularly at the retail level, have stalled as a result of a series of setbacks including the Enron debacle in 2001, the California electricity crises of 2000 and 2001, and the August 2003 Northeast blackout.[27] There may well be something of a silver lining with these stalled efforts as the industry confronts climate change. Instead of simply concentrating on retail price deregulation, now industry transformation efforts must pay attention to a broader range of energy and environmental demands. Regulators must confront renewable portfolio standards (RPS); the need for renewable feeder connections to the grid;[28] potential cap-and-trade obligations; grid modernization and expansion; the need for new demand response and energy efficiency regulations; the development of common standards for energy efficiency and reliability;[29] and continued improvements in creating regional transmissions organizations and independent system operators, which have been effective in lowering the price of electricity due to increased competition and improved electricity markets.[30] Additionally, new lines of federal-state jurisdictional authority must be developed. The realignment will require significant planning[31] and coordination among state and federal governments as well as significant amounts of public education.[32]

A future electricity policy will be based on new assumptions. First, we can assume that consumers prefer affordable, reliable, and *clean and efficient* electricity without a significant disruption in lifestyle. This assumption means that the public good of energy regulation is no longer cheap and available energy regardless of source. Instead, it recognizes that significant market imperfections occur where energy and environment converge and that a clean and efficient environment is the preferred public good. This assumption acknowledges that private competitive markets cannot provide clean energy at the optimum level. Second, private investment will continue to seek and realize market opportunities. Competition should become more robust with an increase of new entrants for electricity production and with an influx of technological goods and services to provide clean electricity and energy efficiency. Combining these two assumptions, government regulation can and should be used to correct the market imperfections that damage the public good of environmental quality, and government regulation should promote more competition in this sector. Third, less regulation is preferable to more. More particularly, market-based regulations are preferable to heavy-handed or command-and-control regulations. Fourth, we can also assume that every policy has its costs and trade-offs.[33] Substituting greater use of nuclear power for coal to generate electricity is an example of such a trade-off. So is ramping up renewable resources to displace the position of coal. Finally, we must assume that business as usual is unacceptable if the country is to craft an energy policy that is secure, economically sustainable, and environmentally responsible. At the heart of any sound future electric policy is the need for a new, renegotiated regulatory compact based on these assumptions.

The traditional business model of the electric utility must be replaced with a smarter version – the iUtility. The traditional model of utility regulation must be replaced with a smarter one as well. Where the old model encouraged consumption, the new model must encourage conservation; where the old model fostered economic inefficiency, the new one must foster the efficient use of electricity; where the old model was content with capital-intensive centralized power production, the new one must promote distributed, small-scale power production; and where the old model was satisfied to burn dirty fossil fuels, the new model must expand the development, production, and consumption of alternative and renewable resources. Much of these gains can be realized through a renegotiated regulatory compact. Nevertheless, full gains will only be realized as utilities redesign their business models. The next sections discuss both the public and private requirements for a cleaner energy future.

The iUtility

The model for the new utility, the iUtility, looks at new businesses such as e-Bay, Amazon, Dell, Wikipedia, and any number of social networks and innovative manufacturing and distribution processes.[34] The traditional utility was a capital-intensive, centralized monopoly connected to a large regional/national grid and had a single mission to sell electricity. The iUtility, by contrast, is less centralized;[35] encourages more consumer choice among various energy products and services, including information as well as conservation and efficiency, at various prices; is more competitive; and is continually looking at innovation instead of continuing to reproduce the plant and equipment just constructed.[36] Where the traditional monopoly enjoyed its monopoly status free from competition, the successful iUtility will thrive in a more competitive and innovative environment. "Instead of being proprietary, monopolistic, and large-scale, energy could become interchangeable, competitive, and personal. Moreover, intelligent generation could fundamentally shift the business model of energy companies from commodity sellers to value-added service providers."[37]

The reward structure for the iUtility, through a renegotiated rate formula, will be based on *energy* sales instead of sales of electricity to customers. The iUtility can sell, as examples, energy efficiency services such as audits or energy planning; efficient products such as energy-saving appliances; green energy from renewables such as solar and wind power; and energy efficiency through either conservation or superior performance through technical improvements. The problem, then, for the iUtility should be apparent. The traditional utility earned revenue based on the gross volume of its electricity sales. The iUtility will earn its revenue, at least in part, by reduced electricity sales through either increased efficiencies or conservation. How, then, should the new regulatory compact, and its rate structure, be designed to keep the iUtility in business if electricity sales will be declining?

The New Regulatory Compact

On the surface, the new regulatory compact will look like the traditional one. Consumer prices will be regulated and kept to competitive, not monopolistic, levels, and producers will earn a return on their investments. However, the new compact will be applied much differently and will encourage efficiency and renewable resources while discouraging fossil fuels. Under the new regulatory compact, the iUtility will sell energy services and products under a rate formula that gives the iUtility the necessary revenue to promote environmental protection while earning enough money to continue to attract investors.

The genesis for the new regulatory compact emerged as the industry changed in the 1970s and as regulators attempted to manage that change. Deregulation and restructuring efforts at the time were based on the idea that old electricity had gotten too expensive and that new producers could sell cheaper electricity if only they could get it to customers. In the first wave of restructuring, utilities were asked to unbundle generation from transmission and distribution so that access could be given to new producers and self-dealing by incumbents could be reduced or avoided. It was also believed that grid reliability could be improved if utilities organized themselves regionally while participating in regional transmission organizations (RTOs) run by independent operators. Further, FERC believed that it could design electricity markets in which prices would be competitive and price signals would be clear.

None of those goals have fully come to pass. Prices have risen above the anticipated rate due to the high cost of the resources used to produce electricity, especially natural gas and coal. Retail rates have also increased as rate caps have been lifted after the first round of restructuring activity. Additional pressure on prices has been caused by the need for additional generating plants and the great need for transmission investment. Finally, price pressure has also been brought about by increased concern for the environmental consequences of generation.[38] Redesigned electricity markets have met with mixed success. The RTO/ISO structure is realizing some efficiency gains and other benefits, but FERC has abandoned its proposed rules to redesign regional electricity markets.[39]

As a result of these factors, utilities began to address ways of acting more efficiently and more environmentally responsible by diversifying their fuel mix and by including cleaner burning resources as well as engaging in conservation efforts. In order to address those concerns, utilities developed, along with a regulatory nudge, Demand Side Management (DSM)[40] and Integrated Research Planning (IRP) programs. Both programs were intended to reduce demand during peak periods and to: (1) promote conservation by consumers (DSM) and (2) encourage utilities to have a mix of resources that were more environmentally sensitive including conservation (IRP). Although these programs have not been touted as widely successful,[41] they were the beginning of a merger between energy regulation and environmental protection as state regulatory commissions began to explicitly recognize the link between energy, the environment, and the economy.

Regardless of these efforts, the electricity industry continues to face severe problems. The demand for electricity continues to grow; prices continue to rise; grid reliability is suspect; carbon emissions are unacceptable; new entrants and new technologies are slow to emerge; and coordination of new producers, such as wind and solar, into the old system remains troublesome. The response to these problems lies in part on the transformation of the utility away from its traditional business model to the model of the iUtility and in part to a need for a new regulatory compact among the utility, its regulators, and its customers. Another dimension of the changing electricity industry is the expansion of non-utility power providers; that is, firms that generate and sell power but are not hampered by the regulatory constraints of public utilities.[42]

The full realization of the new regulatory compact can be broken down into two key components. The first component involves the particular charges and obligations that the iUtility will incur by regulatory order. The second component involves the wider variety of energy services and products sold by the iUtility. Both components work together as the new obligations are directed to investment in smarter and cleaner energy resources while moving away from traditional dirty energy. The new regulatory compact will involve new rate designs that promote conservation, energy efficiency, smart consumption, and technological innovation instead of encouraging both consumption and capital expansion in traditional fossil fuel resources. To be sure, the demand for electricity will continue. And new plants will need to be built. Regulators now, however, will be given a broader array of policies, tools, and objectives with which to enter the electricity future.

A note of caution must be struck about the renegotiated regulatory compact and the electricity future that is reminiscent of the enthusiasm for nuclear power in the late 1950s and early1960s. During the promotional years of commercial nuclear power, the comment was made that commercial nuclear power would be "too cheap to meter"[43] because the cost of nuclear fuel was significantly below the cost of coal, oil, or natural gas. That prophecy never became true as construction costs escalated beyond any reasonable estimates.[44] Today, the hoped-for economic gains in energy efficiency and a better environment through renewable resources must fully consider costs and risks in setting estimates and in imposing new obligations and new rate designs. The electricity future will not be "too cheap to meter."[45]

The costs and risks of a new regulatory compact must take into account a realistic assessment of supply-side needs not only from new resources but from traditional ones as well. Additionally, just as the nation is designing a set of uniform reliability standards[46] and renewable portfolio standards,[47] it should also develop uniform metrics and protocols for energy efficiency that can be applied across regions and across utilities.[48] Standards will need to be developed to evaluate and report energy and capacity savings of energy efficiency programs. Additionally, standards must be developed to identify and quantify net GHG emission reductions and accurate estimates of avoided costs, as well as make allowances for any increasing consumption due to savings in energy.[49]

Finally, we must be clear about the costs and risks to the iUtility. Declining sales, of course, mean declining revenues, which is a financial risk to the utility. Those risks can be offset with new services and products. Nevertheless risks remain. In addition, because growth in consumption will continue, we will continue to require new sources of supply that may make the traditional sources of nuclear power and coal appear attractive. The regulatory compact must be sensitive to encouraging smart energy while neither paralyzing nor favoring traditional sources. Utilities will then require the capture of revenue and will need to maintain cash flow as well as returns on investment.[50] The new regulatory compact must address revenue, return, and stability as it stimulates innovation, opens new markets, and invites new actors to the energy future.

iUtility Energy Obligations

Under the terms of the renegotiated regulatory compact, the iUtility will continue to have its rates set by regulators under a differently applied rate formula. However, instead of selling just electricity, the iUtility will sell energy services and products either in a protected geographic market or to a protected set of customers. The rate formula will then provide the iUtility with the necessary revenue to transform their business model while keeping rates just, reasonable, and non-discriminatory. Renewable Portfolio Standards (RPS), Renewable Energy Credits (RECs), disclosure requirements, surcharges, and decoupling are examples of the new terms in the new compact as regulators impose new obligations on the newly restructured iUtility.

Utility investments in renewable and alternative resources can be stimulated through government financial incentives such as: (1) production tax credits; (2) investment tax credits; (3) a more stable and reliable timeline for both types of credits; and (4) loan guarantee programs for green energy investments.[51] State regulation can also assist in green investment by including such investments in the rate formula either as expenses or in the rate base or by providing tax incentives of their own. Additionally, states may have to rethink their recent unbundling policies so that a utility can invest in generation, transmission and distribution, energy efficiency programs, and distributed generation[52] all with the purpose of becoming a full-service energy provider.

RPS. Under a renewable portfolio standards (RPS) requirement, the iUtility is required to provide a specific percentage of its electricity from renewable energy sources such as wind, solar, or bioenergy by a certain date. The obligation is on the utility to purchase the power in the market, thus reducing its dependence on fossil fuel-generated electricity while stimulating new markets. The two essential variables of these policies involve the percent of electricity that is to be distributed from renewable resources and the nomination of which resources satisfy the RPS requirement. To date, more than thirty states and the District of Columbia have renewable electricity standards as they encourage the use of renewable and alternative energy resources, create new jobs, and support new technologies.[53] Collectively, the state policies currently apply to roughly 50 percent of the

U.S. electricity load. Most recently, federal legislation[54] has been introduced (but not passed) to achieve 20 percent renewable energy by 2020, and it has been projected that an RPS of 20 percent by 2020 can have the effect of increasing total renewable energy capacity sufficient to power 47 million homes.[55]

RECs. RPS programs may also involve a system of renewable energy credits (RECs), which are tradable certificates. The REC represents a property right in the environmental and social benefits (such as non-carbon-emitting properties) of renewable resources.[56] RECs can be traded either voluntarily by firms or can be traded in markets established by regulation. Depending on the regulatory design, a utility can purchase RECs from a qualifying renewable or alternative energy provider or on the market to demonstrate compliance with a state regulation in satisfaction of their RPS obligation. In order to ensure that retailers are motivated to meet state goals, penalties are also imposed that are often in excess of the cost of the renewable energy credit.

Today, REC programs are only administered at the state level, which in turn means there is no national market for these tradable certificates. As a consequence of state-only RPS requirements, these markets do not run as smoothly as they should. Each state RPS program is differently constructed. Consequently, the nature, and value, of the REC certificate varies. Even though RPS compliance has been rated at 96 percent to 97 percent effectiveness, a national REC market can improve performance and advance the growth of renewable resources and energy efficiency.[57]

Feed-In Tariffs. Another approach to expand the use of renewable energy for electricity production is known as the feed-in tariff (FIT).[58] The FIT is distinct from, but may be designed to be complementary to, RPS requirements and has met with significant success in increasing the use of renewable energy in Europe. Whereas the RPS requirement sets a mandatory quantity of renewable energy that a utility must purchase at an indeterminate price, a FIT provides a utility with a revenue stream by setting the price, including a reasonable profit, for the renewable energy that a utility purchases from a particular provider or project. In this way, the project developer has a reliable income, and the utility knows what rates it must charge. Further, RPS requirements are generally met through competitive bid solicitation, which both imposes transaction costs on project developers and pricing uncertainty on utilities. By contrast, a FIT is a negotiated contract that can reduce transaction costs as well as pricing uncertainty, thus increasing the investment climate for renewable energy projects. FIT programs are beginning to be used in the United States. In most states, FIT requirements are set on a utility-by-utility basis with the exception of California[59] and Vermont, which have statewide programs.[60]

A FIT program can complement RPS requirements by allowing a utility to enter into long-term power purchase contracts with renewable energy providers. There are several benefits to this arrangement. First, both sides have reliable price information, which reduces financial and business risks. Second, transmission investments also look more secure because the renewable energy projects themselves are more secure and predictable. Third, consumers also receive more accurate price signals regarding the costs of their electricity purchases. Fourth, long-term contracting provides a more secure environment for technological innovation.[61]

Disclosure. State disclosure requirements obligate the iUtility to provide information to its customers and investors about its fuel sources and about its carbon and other GHG emissions profile.[62] Disclosure requirements are intended to set uniform standards to allow consumers to price and compare the resource mix and the energy characteristics of their electricity purchases. Today, more than half of the electricity producers in the United States are subject to disclosure requirements. Through disclosure, customers are provided with information, have increased product choice, and can improve their energy efficiency.

In addition to state PUC rules, private firms are voluntarily providing or are required to provide information about their carbon emissions to their shareholders.[63] Recently, the Securities and Exchange Commission has issued guidelines regarding disclosure requirements for public companies regarding climate change. As examples, the guidelines suggest that a company should make disclosures regarding: (1) the impact of state or federal climate change legislation or regulations if they materially affect the company; (2) the impact of international accords; or (3) if other legal, technological, political, or scientific trends create new opportunities or risks.[64]

Surcharges. In addition to including expenses and capital investments in the rate formula, PUCs will consider adding an additional charge, an energy surcharge, onto customer bills to enable iUtilities to recover expenditures in energy efficiency programs or to comply with other clean energy requirements.[65] To the extent that a utility can assist with the installation of energy efficiency products or energy savings from complying with building codes, the iUtility can recapture those costs through the surcharge.

Another form of surcharge is known as the lost-revenue adjustment. To the extent that utilities are required to implement either energy efficiency programs or obtain a certain percentage of their power from renewable resources, it is possible that a utility will lose profits because of lost sales.

To compensate a utility for participating in such programs, a regulator can award a lost-revenue adjustment. The purpose of the adjustment is to cover a utility's fixed costs, which are otherwise lost due to these regulatory programs. The advantage to such an adjustment is that the utility is indifferent to its investment between traditional electricity and either efficiency or renewable resources. The downside to such adjustments, however, is that they are notoriously difficult to calculate, can overcompensate the utility, and can reward underperforming programs.[66]

iUtility Products and Services

Federal and state regulators are adopting this variety of obligations as part of the transformation of energy policy. In order to respond and satisfy the obligations, the utility of the future will transform itself by reconfiguring its portfolio of energy products and services, thus redesigning the business of the traditional IOU into the iUtility.[67] The iUtility will offer electricity from traditional energy sources and from green sources, and the iUtility will offer to sell energy efficiency as a product. The key insight is that the iUtility is in the energy business, not the electricity business. A sample of products follows. An iUtility can sell any or all of such products and services. The regulatory structure, then, should facilitate the sales and investments in products and services that contribute to greater efficiency and increased carbon reductions.

Green Electricity. Green power is a product that utilities offer customers the option of buying usually at a premium. Consumers can choose to purchase an amount of power that would be generated from a renewable energy technology for which they will pay a premium or a flat fee. In this way, green pricing programs can create markets for clean energy technologies. Another version of green pricing comes when the utility can offer an opportunity for customers to make contributions to support the development of alternative or renewable energy projects. About 25 percent of the nation's utilities have been offering, and more than half of the country's electricity consumers have been purchasing, this product.[68] Recently, the DOE's National Renewable Energy Laboratory reports that in 2006 green power sales exceeded 3.5 billion kilowatt hours.[69]

Energy Efficiency. Energy efficiency is the most economically advantageous method for saving energy, reducing energy bills overall, reducing demand for fossil fuels, and stabilizing the energy system. Energy efficient gains in buildings, appliances, and cars are available to be made, and in this regard, energy efficiency can be treated as a resource. A recent report states that the results from existing energy savings programs, if extrapolated throughout the country, could yield annual energy savings of $20 billion and net social benefits of more than $250 billion over the next ten to fifteen years. Additionally, such programs could defer the need for 20,000 megawatts of electricity or forty new 500-megawatt power plants while reducing U.S. emissions by more than 200 million tons of CO_2.[70] The report also notes that these goals are not being achieved due to barriers including traditional utility regulation.[71]

Energy efficiency is not a new concept. Indeed, when prices rise, the market signals the need for energy efficiency, and regulators have looked to utilities to invest in efficiency with cost savings, in part, inuring to the benefit of customers. There are three notable problems with energy efficiency programs. First, unless electricity prices are stable and predictable, there is little incentive to invest in energy efficiency devices or renewable power when the return on investment is uncertain. Second, as a business, energy efficiency programs must be scalable, and scalability has not been rapidly forthcoming in this arena.[72] Third, regulations, especially rate regulation, must be designed not only to provide sufficient revenue to the utility but to encourage the adoption of appropriate technologies. Demand response programs, for example, require smart meters for consumers and require an advanced metering infrastructure for the utility.[73] The challenge, then, is to devise the appropriate pricing formula to encourage investments in the appropriate technologies.[74]

PED – Personal Energy Device. The iUtility can resemble its namesake, the iPhone or iPad, in one particular way. Imagine a personal energy device, a PED, which provides personal energy information. The device tells you the gas mileage on your car, the amount of energy lost in your home, the current prices of gasoline and electricity, alternative energy suppliers and products, energy efficiency tips, and any other environmental and energy information you desire. Further, such a device, once installed in a home or building, could save energy by monitoring and regulating appliances and temperatures and by reducing or eliminating voodoo electricity; that is, the electricity wasted when we do not shut off our computers and other technologies. The iUtility could provide the information, sell the device or the application, service the plan, and retrieve, compile, and synthesize customer information for its own business planning and for a better informed consumer base. Better coordination of demand and supply information facilitates purchasing and planning and will also increase the efficient use of energy and the resources used to produce it.[75]

The iEfficiency Utility. The nature of an electric utility can be dramatically reconceived. Efficiency Vermont is a unique "utility" and is the first of its kind in the United States. It is a public utility charged with helping state residents save energy and protecting the environment through energy efficiency gains. It was created by the Vermont Public Service Board and is operated by a nonprofit service organization called Vermont Energy Investment Corporation.

The efficiency utility is funded by an energy efficiency charge (EEC) on electric bills. Early reports indicate that the EEC has caused little or no increase in monthly electricity bills for most customers. Essentially, Efficiency Vermont provides technical assistance as well as financial incentives to customers to help reduce energy costs through energy efficient equipment and lighting as well as energy efficient approaches to construction and renovation. Efficiency Vermont reports that during 2006 there was a $6 million reduction in retail energy costs with nearly half of those costs coming from more than 685 businesses. The report also notes that a net lifetime economic value for activities in 2006 could be in excess of $47 million with total costs at roughly $28 million for a net benefit to the economy of $19 million.[76] In its most recent annual report, Efficiency Vermont projects the total benefit to exceed $342 million during the 2009–2011 period.[77]

Energy Hedging. The iUtility is an integrated energy provider. To survive, the iUtility must have a comprehensive understanding of its energy portfolio including the most efficient mix of energy resources, including negawatts,[78] energy efficiency, energy futures, and carbon reduction strategies.[79] This information will enable the iUtility to: (1) produce the electricity and other energy products and services that it will sell; (2) at the lowest cost; and (3) with the highest return. Further, the iUtility will guide its investments in energy products as a key segment of its investment portfolio. Part of this investment strategy can include hedging with energy and other energy-sensitive commodities, other securities, or with investment-grade paper and interest swaps. The iUtility will, then, become an energy trader and investment manager and will have developed a valuable intellectual property and an important service to be sold in the market to all energy users.[80] The iUtility, then, can offer this service to customers to help them plan their future energy investments hedged against other financial investments.

Energy Audits. The iUtility that best manages its diversified energy portfolio and its energy investments can also provide energy advice as an iUtility service. The iUtility can perform energy audits for itself and its customers and then advise businesses, governments, and consumers about how best to realize energy savings, what energy mix is most valuable, how buildings can be constructed with the highest degree of energy efficiency, and which products are most efficient. Additionally, the iUtility can advise those same customers about the range of options for putting together their energy portfolios and planning for future energy use.

Smart Rate Designs

Rate design is the method through which the iUtility recovers its revenue to cover the costs of new regulatory obligations and to recoup its investments in new products and services. Today, iUtility regulators can design policies to promote energy efficiency and clean energy, discourage dirty electricity, and stimulate technological investments.[81] As utilities are required to expand their energy services and products, they are also going to be asked to reduce their rate of capital growth and reduce their investment in traditional electricity production, which will reduce sales made under the traditional formula. New rate designs, therefore, must provide enough revenue to attract capital so that the iUtility can satisfy its service obligation, meet the social objectives set by the regulators, and encourage the traditional utility to reformulate its business plan by transforming itself into the iUtility. Any rate design must balance several interests, including rate stability for customers and the utilities, proper incentives for smart energy, and accurate price signals for optimum efficiency and investment.[82] No single rate design is likely to accomplish all of these goals. Nevertheless, the traditional rate design has outlived its useful life and must be replaced with some type of dynamic pricing, and several alternatives are available.[83]

There are two common themes with smart rate designs. First, from the consumer side, price signals should be set as accurately as possible so that consumers can make efficient demand decisions. Second, and from the producer side, the rate structure must move away from a cost-plus revenue scheme to an incentive scheme that encourages utilities to make efficiency investments and investments in renewable resources. There are four basic incentive structures. A Public Utility Commission can: (1) simply mark up rates as a percentage of spending; (2) award a bonus for achieving energy or capacity savings; (3) split savings between producers and consumers; or (4) increase the rate of return for either efficiency investments or for investments in renewable resources.[84] The following rate designs adopt these principles.

Marginal Cost Pricing. We can no longer afford promotional rates (i.e., rates designed to encourage consumption) and instead must approach the true cost of electricity as closely as possible. Economic theory supports the idea that the true price of electricity, known as marginal cost, should be charged to customers. Historically, consumers have been charged the average cost of electricity, which distorts price signals sent to customers because average costs do not represent the actual cost of production. Under average cost pricing, as a firm realizes economies of scale and as production costs decline, the customer is overcharged. As production costs increase, as is the case in most energy markets today, if a customer is charged the historic average cost, then they are underpaying for the electricity and demand is distorted. If, however, customers pay the cost of producing the next unit of electricity, in other words if they pay its marginal cost, then they would receive accurate price signals and can change their demand accordingly.[85] In other words, as the cost of electricity rises, consumers can consume less.[86] The argument about marginal costs is that the pain of higher costs must be absorbed to get the electricity market working more efficiently for both consumers and producers in the mid- to long-terms.

Inverted Block Rates. A simple fix to the problem of the declining block rate, which promoted consumption, is to invert the rates and make electricity more expensive as consumption increases. Inverted block rates are relatively easy to construct and understand.[87] With the initial blocks set below the anticipated marginal cost, they can protect low-income users who are price sensitive to energy costs while passing more fixed costs on to larger consumers. Yet, whereas inverted rates achieve conservation, there is no built-in incentive for the iUtility to either invest in energy efficiency programming or new technologies. Additionally, this rate structure will reduce consumption, but it does not necessarily reduce consumption during peak hours when the electricity load is most expensive.[88]

The inverted block rate design can be constructed to be revenue neutral while still sending signals to consumers that they should reduce demand. The simplest design would be a two-tiered rate structure in which the first tier falls below the revenue baseline and the second tier above. The utility receives its revenue requirement, and consumers are urged to conserve electricity. This design can advance demand-side management programs favored by regulators but is unlikely to advance more ambitious clean energy goals.[89]

Decoupling. Traditional rate design is known as a two-part rate and is intended to cover fixed costs and variable costs. Curiously, though, these two-part rates were built on the misnomer that all of a utility's fixed and variable costs were clearly separated into their respective components, when in reality some of a utility's fixed costs were allocated, then recovered, in the variable component. A dramatic break from the traditional rate formula is generically known as decoupling.[90] The central idea is to decouple a utility's revenue from its electricity sales.[91]

The idea behind decoupling is to remove the sales incentive while encouraging investments in energy efficiency, renewable resources, and distributed generation as well as smart technologies. The purpose of the design is to allow the iUtility to recover its fixed costs, thus not affecting its net income. In addition, properly designed, decoupled rate designs can reward the iUtility for its investment in smart energy programs.

One form of decoupling is known as a straight fixed variable rate design (SFV).[92] The SFV realigns the costs into their respective and proper fixed and variable boxes. Through a proper alignment, then, the utility will recover all of its fixed costs, such as the cost of capital, in the fixed component and all of its variable costs, such as the cost of energy, in its variable component. In this way, utility income is decoupled from electricity sales. Further, the utility should be indifferent to how much electricity it sells because it will recover all of its fixed and variable costs, and its financial risk due to variance in sales is reduced. This form of rate design is also known as dynamic pricing and is projected to result in significant savings because more accurate signals will reduce peak demand, thus reducing the sales of high-cost electricity.[93]

As regulators impose efficiency and renewable resource burdens on utilities, decoupling provides a mechanism for the utility to earn revenue in the face of decreased electricity sales. At its simplest, the utility submits its revenue requirement to the regulator, and the regulator sets the rates. If the utility earns more than enough to meet its revenue requirement, then the excess is returned to customers; and if it earns less, the customers are charged more. The business case for energy efficiency requirements can be made for consumers and shareholders especially when decoupling rules are clear and can be relied on by the utility.[94] This is a radical departure from the old way of doing utility business. The traditional IOU made its money from greater electricity sales made possible by greater investment in capital plant and associated facilities such as transmission and distribution wires.

Decoupled rates are not unproblematic. First, in designing a decoupled rate, the regulator must make a choice between allowing the recovery of revenue per customer or setting a net revenue requirement and apportioning it among all customers.[95] Under a revenue-per-customer design, the utility will lose income as it loses customers. A net revenue design, however, will raise rates to remaining customers as others depart the system. In addition, decoupling requires accurate forecasts and, to the extent that the forecasts are unreliable and require frequent adjustments, rate stability suffers.

Straight Fixed Variable Rate and Feebate. The SVF rate design alone, however, may not provide the most accurate price signals for two reasons.

First, this rate design relies on short-term marginal costs rather than on the more economically reliable long-term incremental, or marginal, costs.[96] Second, the design concentrates on that part of a customer's bill involving fixed costs. To improve the price signal, another charge referred to as a revenue-neutral energy efficiency feebate (REEF) can be made.[97] The core idea behind REEF is that a baseline electricity charge will be set. Those customers who conserve electricity by using it off-peak will receive a rebate, and those customers who use more costly electricity will pay a higher fee. The fees and rebates offset one another and can induce certain consumer behavior including energy efficiency and conservation, which can reduce generation and should reduce grid congestion and other grid stress.[98]

The iUtility will see no financial effect because the charges and rebates will equal one another. Consumers, however, will see their bills increase or decrease depending on their consumption of electricity. Through such a rate scheme, all of the iUtility's fixed costs are recovered while its variable costs, most importantly its energy costs, will vary according to the demand made by customers. The iUtility will then earn revenues on which it can rely, and customers can use electricity more efficiently. Without too much difficulty, then, the SFV rate structure can maintain the iUtility's revenue requirement and, with the appropriate REEF adjustment, can facilitate efficient consumer choices as long as the baseline is regularly monitored and adjusted.[99] The combined SFV and REEF rate design thus relies on marginal costs, is revenue neutral, and can sharpen price signals to consumers. Efficiency investments, costs of carbon emissions, technology investments, and the like can be accounted for and structured in ways that were not possible under the traditional rate formula.[100]

In any rate structure, a return on investment is needed to attract capital, and this is also the case for the iUtility. The problem, of course, is that we do not want to encourage capital investment in dirty energy. Instead, the desire is to reward capital investment in clean and renewable energy sources. Consequently, any return on investment should be structured to enhance both efficiency and new renewable technologies. One mechanism for achieving those ends would be to reward the iUtility with a higher return on clean investments, a lower return on dirty ones, or both. With the SFV rate, the utility should always recover its prudently incurred fixed costs; therefore, its financial risks are lessened. In exchange for the reduced risk, the iUtility should receive a lower return on equity.

Regardless of which items will be included in the rate formula, a choice must be made as to whether or not the item should be carried as an expense or included in the rate base. If the iUtility expenses are prudently incurred, both consumers and the utility should be indifferent as to which items are carried as expenses. The utility receives a dollar-for-dollar return, and the customer receives equally valued services. In addition, the utility receives immediate recovery, and the accounting is straight forward. A utility and its shareholders would prefer to have items carried in the rate base so that they can earn a return on that investment. To the extent that the iUtility is moving away from traditional supply-side investments to smart supply-side investments, then rate base treatment makes sense as an incentive. However, regulators must not penalize customers and overcompensate the utility by including too many expenditures in the rate base.[101]

These new rate designs have the potential of increasing consumer cost for several reasons. First, marginal cost pricing will likely raise costs. Second, to the extent that customers are subject to renewable portfolio standards, their costs are likely to rise. Additionally, smart meter requirements and smart grid expenditures can raise rates. Finally, energy efficiency expenditures for appliances, or for utility efficiencies, can also raise rates. We can safely assume that rates will rise in the near term. Yet, the hope is that through these expenditures savings will eventually be realized. The realization of savings, however, will only come about to the extent that demand is responsive to price increases as consumers either invest in energy savings devices and appliances, use less electricity, or both.

A hidden assumption in designing incentive rates to capture efficiencies and promote green power is that consumers have accurate information and that they have responsive demand elasticities.[102] In other words, a solid understanding of price elasticity of demand is the necessary variable for rate designs that will impose additional costs at least in the short term. Consumers will respond to rising prices. What is more important, however, is the rate of that response. The rate of response is known as the price elasticity of demand. Perfect, or unitary, elasticity results in a 1 percent decrease in demand for every 1 percent rise in price. Historically, electricity has not been very price elastic. Instead, consumers continue to consume electricity as price rises. However, recent studies indicate that consumers are beginning to show increasing elasticity.[103] We must be careful here not to claim too much in terms of elasticity. Different classes of customers will have different demand elasticities. Large industrial consumers, for example, are able to respond more quickly and flexibly to price increases than residential consumers because large industrial consumers have more bargaining power and they are often capable of switching to cheaper fuels. Still, accurate price signals for the cost of electricity, including the cost of carbon, is a necessary component for a reformulated energy policy including a modern electric grid.

Investing in the Smart Grid

The electricity grid is the infrastructure of the industry in both real and in symbolic ways, and significant upgrades and investments are necessary for the electricity future described here.[104] As we move into that future we should anticipate: a greater reliance on alternative and renewable forms of energy; a greater independence from imported oil; and a reduction of carbon emissions.[105] To achieve all of those gains, a modernized, or smart, grid is necessary.[106] The smart grid will be more efficient and reliable, will help reduce carbon emissions, and will promote security among other benefits and will be able to deploy smart technologies and incorporate renewable resources.[107] Grid investment will be aimed at achieving technological advances and serving new sources of energy. The country is just beginning to make the necessary investments in grid modernization, which promises to be cost-effective.[108]

There are three reasons for improving the existing electric grid. First, although the growth of the electricity industry has slowed, the demand for electricity will continue to rise into the future and the existing grid will need expansion and upgrades. Over the last sixty years, the growth in demand for electricity has slowed appreciably. Post-World War II, the annual increase in electricity production of approximately 7 percent has declined as the infrastructure has been constructed and as the country has realized gains in efficiency. Since 2000, annual growth has fallen to 1.1 percent, with the projection falling lower to approximately 1 percent. The Electric Power Research Institute further estimates that through energy efficiency programs, electricity growth from 2008 through 2030 can be reduced to between .83 percent and .68 percent.[109] Still even at those reduced levels, from 2007 to 2030, electricity demand is expected to increase 26 percent.[110]

The base case for increased demand is that by 2030 the United States will need an additional 214 GW of electricity at a cost of $697 billion. That demand, with its attendant costs, could be reduced by between 38 percent and 48 percent by using energy efficiency and demand response programs.[111] To satisfy increased demand, traditional energy sources such as coal and nuclear power continue to be attractive because coal is abundant and nuclear power appears clean; both are already online with billions of dollars of sunk costs; and both are already connected to the grid. However, both are connected to an aging grid in need of modernization. Most recently, the North American Electric Reliability Corporation estimated that over the next ten years, the United States will need 1,700 more circuit miles of transmission lines to maintain reliability and to integrate new resources.[112] To maximize gains in efficiency and integrate renewable resources, the projected costs for investment in needed transmission and distribution range between $1.5 trillion and $2.0 trillion.[113]

The second reason for investing in the electricity grid is efficiency. The smart grid can be broken down into two major components – smart transmission and smart distribution. Both components promise an increase in energy and economic efficiency. The smart transmission segment of the grid is comprised of a superhighway that will deliver wholesale power across 765 kV extra high voltage (EHV) transmission lines.[114] These lines increase energy efficiency as one EHV line can transmit as much power as six existing 345 kV lines and can reduce the transmission line footprint by a factor of greater than 4 to 1.[115] In other words, the EHV occupies one-quarter of the land occupied by current transmission lines. Further, the cost of transmitting 1 billion watts one mile over a 756 kV line is one-fourth the cost of a 230 kV line.[116] Smart grid investments will not only increase energy efficiency, they will also improve reliability as well as reduce congestion.[117]

The other, and equally important component of the smart grid, involves smart distribution of electricity to end users. Today, distribution is a one-way street with electricity moving from the local utility to the customer and with the utility reading meters for the sole purpose of billing its customers for their consumption. Today's electric distribution system is hardly different from Edison's constructed at the end of the nineteenth century because both are one-way systems. Smart distribution will be a two-way system with information flowing both ways. Smart distribution will provide better information about the price and use of electricity to both parties. Consumers can then use electricity at the lowest costs to them, and producers can acquire information about stress on their load and system. In short, a smarter grid will facilitate demand response programming, more accurate price signals, and real-time pricing that, in turn, will enable producers and consumers to capture more surplus, thus increasing economic efficiency.

The smart grid will require investment in both segments and will require the development of communications technologies throughout the electricity system from producers to end users. Communications technologies are necessary to coordinate regional transmission operations, send supply and demand signals between and among consumers and producers, indicate stresses on the grid, provide information about weather patterns for variable sources such as wind and solar power, and generally fine-tune price signals to improve the electricity market as a whole. This portion of the smart grid has been referred to as "transactive," meaning that the grid network is the platform connecting producers and consumers for the purpose of not only conveying information and improving reliability but also facilitating purchase and sale transactions at lower cost.[118]

Third, the grid can play an important role in reducing carbon emissions by expanding its connections to alternative and renewable resources particularly as improvements in forecasting increase the reliability of intermittent resources such as solar and wind.[119] An integral part of this segment of the grid must incorporate feeder lines to resources such as solar and wind, which are generally not located near the existing transmission corridors[120] and must be connected to a modernized grid.[121] The DOE, for example, reports that the nation can achieve 20 percent wind energy by 2030 only if the transmission grid improved.[122] Additionally, it is estimated that there are more than 4000 MW of large solar power plants scheduled for construction over the next five years that will also need access to the grid.[123]

The development of the smart grid is not taking place in a vacuum. The last few years have witnessed a noticeable uptick in utility investment in transmission and distribution; most recently, federal modernization efforts are underway, and those efforts will need to be coordinated both regionally and locally.[124] Pursuant to the Energy Independence and Security Act of 2007,[125] the DOE was given the authority to engage in smart grid planning. On March 3, 2009, DOE issued an Intent to Issue Funding Opportunities for smart grid demonstration projects.[126] This notice was part of the American Recovery and Reinvestment Act,[127] which provides at least $11 billion for smart grid investments.[128] Additionally, proposed energy and climate change legislation addresses climate change and provides support for the smart grid through smart grid advancement and transmission planning.[129]

For grid modernization to be successful, several things must occur. First, a clear statement of purpose from Congress about energy policy in general and carbon emissions reductions in particular would play an important role in establishing national and uniform goals for grid investment.[130] Next, FERC must have clear authority and must exercise it,[131] especially regarding interconnection rules, transmission line siting,[132] and cost allocation, so that the benefits as well as the costs are fairly distributed across the system[133] and so that incentives exist for utility investment.[134] Also, state regulators will play an indispensable role in setting rates on which utilities can rely for their smart grid investments. In essence, clear lines of authority and clear rules are the necessary components of workable and competitive markets, which themselves are necessary for a smart energy future domestically and internationally.[135]

The Federal Energy Regulatory Commission has been charged with the responsibility to improve the grid[136] including the responsibility to establish incentive rate treatment for new grid investments, including new and renewable sources, reduction of transmission congestion, and improved reliability.[137] FERC has adopted a policy statement addressing its jurisdiction over wholesale transmission; addressing the need to adopt "interoperability standards" to ensure the proper connectivity throughout the system; and addressing its rate-setting authority in anticipation of adding variable power resources and electric vehicles to the grid.[138] Pursuant to that charge, FERC adopted rules enabling it to approve incentive-based rates including a higher return on equity as well as favorable rate base treatment for other expenditures.[139] Additionally, FERC has begun approving rate treatment for transmission investments in the smart grid,[140] and rate treatment will be a contentious issue.[141]

The smart grid will require remaking the physical and corporate structure of the existing grid as well as its regulation. Currently, the grid has more than 520 owners largely constituted by local monopolies that plug into the three major grid systems in the United States.[142] The traditional federal/state division of regulatory authority has enabled IOUs to rely on regulation to support and reward retail distribution. Consequently, private ownership in the transmission and distribution segment has created a set of economic discontinuities, which neither follow the physical laws of electricity nor do they follow the smart grid needs of the future. If the country is to realize gains to be made from investment in the smart grid, then the federal role must increase. That role will include regulating independent transmission system operators, greater siting authority, and uniform transmission and reliability rules, as well as investment incentives and other financial supports.[143]

Improving the grid to improve efficiency and promote clean energy is complicated for many reasons but mostly for balancing the burdens of cost allocation among customers.[144] In August 2009, the United States Court of Appeals[145] rejected a FERC cost allocation scheme and remanded the case back to the agency. There are two significant issues involved with cost allocation. The first question is whether or not FERC has jurisdiction to promote energy efficiency and renewable resources. Recent legislation should grant them that authority. The second problem is more difficult: how to allocate the costs and benefits of transmission investments among utilities and their ratepayers. Prior to the court's rulings, FERC began inquires on cost allocation in anticipation of a formal rule-making proceeding, and the clarification of cost allocation authority is essential for grid investments to proceed at their necessary levels.[146]

Conclusion

The smart grid is central or, according to the DOE, transformational,[147] to the electricity future not only for the physical role it plays in transmission and distribution but also for its symbolic role demonstrating the necessity of technological innovation and the need for a regulatory culture that supports investment in innovation. In part, that regulatory culture should be one in which federal regulators play the leading role in encouraging smart grid investment, set rules for cost allocation and interoperability, and establish reliability standards.[148] The electricity future will reduce carbon emissions, increase economic efficiency, and provide communications networks to expand choices among producers and consumers. In order to achieve these gains, all segments of the industry must change their ways of doing business.

From the perspective of the iUtility, the electricity future requires the traditional utility to broaden its business model from concentrating on electricity to engaging in the energy business more broadly. It also means that for the iUtility, Bigger may not be Better and that smaller-scaled, decentralized dispatch is profitable. From the perspective of the regulator, the old regulatory compact must now be dramatically renegotiated.[149] Whereas the old compact encouraged the development of vertically integrated utilities selling electricity in a guaranteed service territory, the new compact must support innovation, investment in new technologies, a reduced dependence on the volume of sales in favor of customer service, and the recognition that the utility business from wholesale through retail must be more competitive in the long run. Moreover, from the perspective of the consumer, consumers must anticipate more choices and must become more conscious, not only about their sources of power, but about the costs of their consumption habits.

Venture Regulation

[T]he Nation can achieve the necessary and timely transformation of its energy system only if it embarks on an accelerated and sustained level of technology development, demonstration, and deployment along several parallel paths between now and 2020.

The National Academies[1]

Introduction

The previous chapters demonstrated that the century-old U.S. energy policy has entrenched not only private, for-profit firms that provide and distribute energy; traditional energy policy has also entrenched public-sector regulators and bureaucrats that, ostensibly, oversee private actors "in the public interest." Today, policymakers are faced with several demands for a new energy path, and a growing consensus has established the contours of that policy. In addition to climate change, a new energy policy is necessary to respond to threats to national security and economic security, which arise from a dependence on foreign oil; to respond to a desire for environmentally sensitive energy resources; and to respond to the desire for economic growth through the more efficient distribution and use of energy.[2] The problem, then, should be clear: How can a transition to a new, smart energy policy occur in the presence of embedded private and public-sector actors?

Entrenchment occurs with any institution whether it is public or private or whether it is profit or nonprofit. The belief in one's own press, the willingness to maintain a market niche by producing tomorrow what one produced yesterday, the desire to replicate past successes and ways of doing business, and the ambition to exist in perpetuity can render any firm, bureaucracy, or philanthropy ineffective.[3] Business as usual is the death of any organization largely because it prevents an organization from evaluating its status and progress, reassessing its mission, and moving forward strategically.

In the area of government regulation, there are additional distorting effects, not the least of which is caused by spending other people's money. An administrative agency is not likely to go out of business – the sun rarely sets on a bureaucracy funded with taxpayer dollars. Sunsetting provisions in legislation are rare and unwelcome. Such provisions serve neither the bureaucrats who depend on the agency for employment nor the clients they serve regardless of the public interest. Nevertheless, acknowledging that resources are limited and that an institution may have a limited lifetime can have a salutary effect on an organization by sharpening its focus on mission and by sharpening its commitment to successfully completing its assigned tasks by delivering recognized outcomes. Government agencies must change the way they do business if they are to effectively respond to complex, systemic problems such as the relationship between energy and the environment.

Over the last two decades, forward-thinking philanthropies have changed dramatically, and the change is instructive for any organization including government regulators. Historically, foundations engaged in doing good by writing checks either in response to grant requests or as investments in projects or programs undertaken in satisfaction of the foundation's mission. Since the 1990s, though, philanthropy has been changing notably. Change began when private and community foundations began to ask: What is the return on our investment? Has our money made an impact? Have we improved arts and culture? The environment? Education? Or any other field of interest? Too often the answer to these questions was: We do not know. Sometimes, the answers were more disheartening: No.[4]

Such acknowledgments of failure generated a deep rethinking about the nature of philanthropy that, in turn, necessitated new forms of behavior and organizational culture. Instead of foundations behaving as either *grand dames*, respected for their charitable generosity, or as the funder and controlling owner of specific programs, foundations began to transform themselves into organizations with the explicit intent of having a discernible and measurable impact on their communities and on their fields of interest. Instead of acting as the *paterfamilias*, foundations used their financial and human capital to effectuate change through leadership and leverage by using a variety of strategies and activities intended to improve their communities.[5]

Some foundations have taken the bold and risky initiative of putting their capital at risk by adopting aggressive spending policies even to the point of recognizing that in the midterm they may well exhaust their principle and cease to exist. No risk, no reward. The most dramatic example of a philanthropist putting his money into use rather than preserving it in perpetuity is Warren Buffett's gift of $37 billion to the Bill and Melinda Gates Foundation that came with the express stipulation that the money was to be completely spent over a set time period. Another example of the new giving is Sir Richard Branson's $3 billion pledge to the William J. Clinton Foundation for climate change innovations, again with the stipulation that the money be put to use.[6]

Foundations that engage in strategic thinking; consider their grant making and programming as investments; employ continuous evaluation of programs and assessment of their missions; develop a theory of organizational change;[7] and, perhaps most importantly, demonstrate a willingness to discontinue unpromising programs, as well as admit failures, are referred to as venture philanthropies or social entrepreneurs.[8] The hallmark of such organizations is a commitment to social change and continuous improvement with an expectation of a return on their investments. That return can come in the form of demonstrated improvements in a particular program or area or, indeed, in a financial return.

In Chapter 6, the business model of the iPad was used to discuss a new form of utility business – the iUtility. Similarly, the developing business model of venture philanthropy provides a portal for redesigning government regulation into venture regulation. Venture regulators can and should adopt a similar strategy to that of venture philanthropy, including a willingness to sunset themselves or to discontinue any specific program or office. Imagine, for example, an Office within the Department of Energy called the Innovative Energy Technology Office that has (1) a set budget of $x billion; (2) a mission to deploy energy technologies to reduce dependence on fossil fuels; and (3) a charge to either spend or leverage its budget by either a certain date or at a specific spending rate. This charge does not preclude the Office from earning a profit or from attracting outside contributions to supplement its budget. Rather, the charge is intended to create a culture in which performance and results are prized over institutional longevity. More specifically, the charge is to replace a traditional energy policy with one that maximizes the use of energy efficiency and renewable resources to achieve the consensus goals discussed earlier. More importantly, the charge necessarily requires a serious commitment to technological innovation throughout the energy sector.[9]

In order to achieve a breakthrough in energy policy, it is as necessary to redesign the regulatory apparatus as it is to reformulate the economic and policy assumptions that undergird that policy. Historically, the country's scientific and technological community has relied on R&D to provide solutions to technical and scientific problems. Even if we expand the concept of R&D to include demonstration and deployment, we need to go further and craft a system-based innovation policy to bring new technologies into the market and to attain widespread application and diffusion of knowledge.[10] Historically, R&D solved specific problems. A new energy innovation policy must drill deeper and solve the systemic failure of the traditional policy that completely ignored its environmental impacts. Ultimately, the key to achieving a new energy policy is to stimulate market breakthroughs. The key to new markets, including energy markets,[11] has always been, and will continue to be, based on innovation.[12] Innovation, as a market stimulant, is as important to the public sector, especially given the energy challenges the country now faces, as it is to a private-sector firm wishing to either enter a market or to improve its existing market position.

Technological innovation has been said to account for 85 percent of the measured economic growth in the United States.[13] To compete in today's global economy, it is necessary that the United States create high-quality jobs in science and technology, and a high proportion of those jobs will come from the clean energy sector. Two of the most important drivers for job creation are better education and a national commitment to innovation. To facilitate innovation in energy technology, there must be a transformation in traditional R&D policy that has developed as a result of bureaucratic and client incumbency. The new approach will incorporate enhanced and creative intellectual property rules; financial incentives; favorable tax treatment for innovation such as favorable capital gains treatment; and an improved infrastructure such as the smart grid and broadband access.[14]

This chapter examines a new form of regulation for the energy sector by explaining core principles for redesigning that portion of the energy bureaucracy responsible for innovative energy technologies. Energy innovation is not likely the single silver bullet needed to achieve the multiple goals that have been discussed throughout this book. Consumer behavior must change, and international cooperation for wholesale change is also necessary, as examples of further necessary actions. However, without energy technology innovation it will be business as usual; consequently, the country will remain dependent on foreign oil; will continue to foul the commons; will not likely achieve projected energy efficiency goals;[15] will not likely enjoy sustainable economic growth and prosperity; and will incur risks to national security.[16] A sound innovative technology policy is necessary, if not sufficient, and it must be based on the economic and policy assumptions for following the smart energy path and must be responsive to the political dynamics and strategies necessary to successfully implement those policy choices.[17]

Agency Entrenchment

The modern administrative agency developed parallel to and mirrored the structure of the private firms they were monitoring. The history of energy regulation shows that as energy firms transformed from local, competitive enterprises to statewide to regional and then to national firms connected to a countrywide infrastructure, government regulators facilitated and supported that development under the, mostly correct, assumption that economic growth would follow. Energy regulation was not coordinated; instead, it facilitated the independent operation of multiple firms and maintained a certain level of competition among a select group of industries. Energy regulators, nevertheless, demonstrated both a policy bias and, then, a path dependency as they facilitated and maintained the services and products of incumbent fossil fuel firms and industries.

The institutional design of these regulatory agencies followed a set pattern. Initially, a market failure, such as excess production or monopoly pricing, was identified, then Congress would pass legislation directing an administrative agency to address and resolve the identified market failure. The agency, then, would develop the expertise needed to address the issue, and its focus would be narrowed accordingly to fix the identified problem in satisfaction of its legislative mandate.

Other consequences followed. In addition to having narrowed its focus, the agency would develop a constituency and would both draw its staff from and supply professionals to the industry and firms being regulated.[18] This model of institutional design reaches back into the late-nineteenth century. The expert agency pays attention to the tasks assigned to it and is too often prevented by legislative mandate, bureaucratic inertia, or political pressures from exercising independent thought and creativity. A case in point is the former Federal Power Commission (now FERC). The FPC, like FERC today, had jurisdiction over natural gas and electricity wholesale sales. Within the commission, the electric side and natural gas side of operations were treated separately. Indeed, interstate natural gas pipelines are treated differently than interstate electricity transmission, even though the issues surrounding access to both of these network infrastructures, and the need to open access to them, are nearly identical. Quite simply, the FPC narrowed its vision and narrowed its exercise of jurisdiction in the service of the industries it was to watchdog. In fact, the FPC has often been criticized by courts and scholars for its refusal to fully exercise the jurisdiction given to it.[19] By way of counterexample, today FERC finds itself handcuffed in its efforts to modernize the electric grid and design regional electricity transmission markets as a result of enabling legislation, the case law encrusted around it, and the current judicial climate.[20]

To be sure, political pressures from either Capitol Hill or the White House affect agency action and behavior. Nevertheless, institutional design contributes to a bureaucratic culture, which tends to be risk-averse, must cater to Congress for its budget, and can grow comfortable, or ossify,[21] doing business as usual. This is true of the Department of Energy, which when it was created in 1977 was intended to oversee a comprehensive national energy policy and foster energy independence[22] – an independence we have yet to achieve. Instead, a brief look at the organizational chart for the DOE reveals how the agency is segmented with separate offices responsible for overseeing individual industries and projects.[23] Assuming for the moment that the classic design of independent treatment of energy resources was both effective and efficient for the traditional energy path, this lack of coordination retards progress toward the smart path and hampers innovation in energy technologies.

In support of the proposition that bureaucratic design is suboptimal, the DOE has also been criticized for its lopsided approach to R&D[24] and its under-funding of energy innovation.[25] This under-funding has several negative consequences. First, variable funding undermines reliability and the concomitant ability of the private sector to invest. Second, the private sector's $3 billion investment in energy R&D in an industry with annual revenues of more than $1 trillion amounts to less than one-quarter of 1 percent of their annual budgets as compared with the biotech, health care, and information technology industries, which invest 5 percent to 15 percent in R&D. Third, venture capital funding has slowed due to the economic recession of 2008–2010. Further, venture capitalists are also slow to place money into innovative energy projects unless they can see a path to commercialization. Finally, and most problematic, the curve of reduced funding is matched by a corresponding reduction in new patents for innovative energy technologies.[26]

In addition to an insufficient and variable budget, DOE strategies for smart path energy innovations are lacking. Most of the DOE budget is for defense, not energy, R&D. Its R&D programs follow a linear pattern to serve either a specific client or resolve a specific problem instead of widespread innovation or systemic change. There is little coordination of activities across energy boundaries and disciplines. Further, there is weak coordination between fundamental and applied research. Too often, the DOE lacks the capacity or will to either terminate unpromising programs or discontinue funding for clients only because they have received monies in the past. Moreover, unsurprisingly, political influence contributes to a fractured R&D environment.[27] DOE R&D mainly focuses on managing and cleaning up nuclear weapons instead of developing commercial-ready energy technologies. This narrow focus is the result of DOE being formed from previous weapons-related departments and its use of national labs dedicated more to basic research than commercial deployment.[28] For an energy transition to occur, DOE funding must be radically changed.

The Need for Regulatory Redesign

We can assume that it is either inevitable or useful that regulatory agencies mimic the industries and firms that they regulate. Indeed, in order for regulation to succeed and remedy a perceived market failure, the regulatory fix must match the market imperfection. Consequently, the key to regulatory redesign is to take a close look at the problem under investigation, the market structure that created it, and then the new market structure needed to fix it. In the case of future energy policy, certain parameters are apparent. First, the new energy policy must account for new entrants and must break away from traditional fossil fuel favoritism. Second, the new energy policy must be based on innovation and competitive markets. Third, that policy must also overcome the inherent market failures and regulatory impediments that plague the country's inability to achieve either energy independence or adopt an environmentally responsive energy path. Additionally, the new energy policy must incorporate and integrate the four primary energy policy variables – energy, environment, economy, and security.[29]

These variables will not always be consonant. Clean energy is likely, in the near term, to be more costly that the dirty energy we now consume.[30] However, it is also the case that the variables can work together. By way of example, adding solar and wind power to the electricity grid can: (1) promote *environmentally* clean energy; (2) at an *economically* lower cost in the midterm; (3) while promoting national *security* by reducing foreign oil dependence; and (4) while opening *new energy* technologies and markets and while adding jobs to the economy. Hybrid and plug-in vehicles, distributed generation, and efficiency standards, among other innovations known and to be discovered, also contribute to achieving these four goals.

Barriers to Innovation

In designing a new regulatory approach to innovation in energy technologies, four barriers must be overcome. Two of these barriers are associated with classical market imperfections, and two others are the result of an institutional design and structure that has been in place for many decades. Both sets of barriers create entrenched and favored incumbents. Incumbency is not the road to entrepreneurial innovation and breakthrough technologies.[31] The two market imperfections involve the positive and negative dimensions of externalities. The negative externalities of pollution are apparent throughout the fuel cycle as the price of energy fails to account for the full environmental costs of exploration, distribution, and consumption. Nor does our fossil fuel policy fully account for the military costs attributable to maintaining oil imports. Carbon emissions throughout the fossil fuel sector are largely responsible for bringing us to the point at which we question the traditional energy path we have taken. As consumers, we do not pay the full price for either the electricity or the gasoline we consume. We have become too reliant on cheap energy. Consequently, energy markets do not function properly. They operate inefficiently, which is another way of saying that consumer and producer choices are suboptimal.

Innovation is also susceptible to positive externalities. In brief, a positive externality involves the situation in which a producer is unable to capture all of the benefits that emanate from a particular product or service. The field of intellectual property is rife with the problem of positive externalities and attempts to address it through granting various legal protections for copyrights, trademarks, and patents. Still, once an idea is out in the world, consumers and competitors can benefit from it. Consumers may not pay the full price of the product, and competitors can improve on a core idea or sell a product more cheaply, thus eroding the originator's market share.

Clearly, the inability to capture all of the benefits generated by an idea acts as a disincentive to provide information for free. Therefore, because of positive externalities, the market does not optimally supply new ideas and innovations, especially if intellectual property protections are weak. Another way of stating the problem is to recognize that the private sector under-funds innovation at least to the extent that they cannot recoup a desired return on investment either as quickly or at the rate of return they would like.[32]

Market barriers for energy innovation are real and cannot be underestimated. No private firm is likely to engage in pollution reduction if it is acting alone and without government support in a competitive industry.

The direct costs for a private firm to invest in long-term, capital-intensive energy innovation projects, from basic research through commercialization, particularly in the absence of reliable market signals and reliable estimates of future energy and material costs, can be prohibitive.[33] In addition to direct costs, private producers must have ample infrastructure available; must capture the costs of past capital investment; must have an adequate market organization for the distribution of new products; must be able to sufficiently distribute information about those products; and must have sufficient financing options.[34] Likewise on the demand side, consumers are unlikely to purchase "energy-saving devices" unless they realize a timely return on their purchases as well. The failure of consumers to participate in new and efficient energy markets is, in part, an information problem. Mainly, however, the expansion of those markets is simply a cost problem from discovery to marketing.

Not only do private-sector investments for new energy technologies lag behind what is necessary for full deployment, past investments in capital-intensive energy industries and firms further inhibit innovation investment until those costs and returns are recouped. Additionally, along the innovation frontier, the private sector may lack sufficient industrial capacity as well as skilled personnel to move forward vigorously.[35] Innovation, then, is better seen as a form of public good.[36] As is true with all public goods, the private sector will not maximize its production, and therefore the public sector plays a linchpin role. Nevertheless, before widespread innovation occurs, institutional barriers must also be overcome.

Institutional barriers result from having developed a regulatory structure committed to the traditional energy path. Traditional energy policy did not coordinate the production and distribution of particular resources; instead, select individual firms and industries were left to compete in the marketplace as separate and independent entities. Similarly, energy R&D was intended to "facilitate the production of effective hardware to produce or conserve energy," and the "primary idea" was "to support the private sector in its development of energy technologies."[37] In short, traditional R&D narrowly viewed the energy innovation space as a mechanism to advance problem-specific technologies through the use of targeted clients. In other words, hard path industries enjoyed government financing because these industries were perceived as contributing to economic growth.

Such a client-centered role for innovation policy, which looks to individual actors and which relies on the market for social ordering, appears consonant with the core values of democratic capitalism. However, there are two flaws with this model. First, energy markets have not been as free as theory might posit. Instead, fossil fuel energy markets have been greatly favored by government to the extent that direct and indirect financial support and incentives tilt energy markets toward dirty fuels, and R&D dollars have followed this pattern to the detriment of new entrants and new technologies. Indeed, the DOE $20 billion R&D budget greatly favors defense over non-defense energy projects 5:1 as traditional energy sources are favored over alternatives. Second, energy markets have changed dramatically, requiring more integration with environmental consequences, thus requiring broader government coordination than energy industries have enjoyed in the past. A further consequence of the allegiance to a mythical market model has been that innovation investments have been notably influenced by the "tendency of political actors to focus on short-term goals and consequences and ... their reluctance to threaten the existing business models of powerful incumbent actors."[38]

United States' innovation policy has had remarkable successes as evidenced by the Manhattan Project and Project Apollo, the creation of national labs, and a wide range of R&D projects. R&D policy has also experienced lock-in, which has narrowed opportunities for broadscale and systemic innovations. Lock-in can be defined as a system committed to an embedded infrastructure, regulations, and technological regime.[39] The multimillion-dollar investments in synfuels in the late 1970s serve as an example of R&D lock-in to the traditional path. The model of investing in large-scale demonstration projects to replace oil and natural gas with alternative fossil fuels from tar sands, coal, and coal shale failed. To be sure, the failure was driven in large part by declining oil prices, which made the price of synfuels uncompetitive. Nevertheless, the commitment to a large-scale model of innovation in the energy sector by replicating fossil fuels was shortsighted, and the DOE structure was too rigid to broaden its approach to energy innovation. The synfuels experiment constituted a mismatch between our energy needs and our past R&D behavior.[40]

Innovation systems, like other organizations, are greatly affected by multiple forces, which can lead to stasis. Complex sectors, such as energy, develop their own infrastructures. Those infrastructures, in turn, are influenced by established technologies, prevailing economics, public policies and expectations, as well as by trends in science and technical expertise. Once the infrastructure is established, then, "emphasis shifts away from innovation in the overall system towards component innovation in technologies that can be launched on existing platforms."[41] Commercial nuclear power is an example of such an entrenched technology. Not only do we continue to build plants with old technologies, the nuclear industry has become a significant beneficiary of government largesse built on old assumptions.

Similarly, and equally unsettling, R&D dollars in the energy sector have fluctuated wildly, thus serving as a disincentive for private-sector involvement on a long-term, reliable, and sustainable basis.[42] From 1978 until 2009, for example, a study of six DOE-funded fossil fuel and energy efficiency projects, together with a study of five renewable energy projects from 1992 through 2008, revealed that during those periods the average standard deviation for funding these programs was 27 percent. In other words, in an average year there was a one in three chance that a program would receive a significant funding change, either up or down, thus contributing to funding insecurity and an inability to budget effectively.[43]

Considering the approach the U.S. government has taken in the development of the atom bomb and space exploration, the general R&D model is that innovation policy is linear. A problem or issue is targeted; research, exploration, and engineering address it; and, then, demonstration and deployment follow aimed at supplying problem-specific solutions and technologies. This approach is consistent with how energy markets developed and how they were regulated. Again, the regulatory approach to innovation mimicked the narrow market orientation of traditional energy policy. Instead of perceiving the production and distribution of energy as interdisciplinary and crosscutting, past R&D practices, such as reliance on cost reimbursement instead of other financial incentives, were fragmented and were often developed to serve a particular problem or client.[44]

The production and distribution of energy is, of course, primarily a private-sector function. Nevertheless, government regulation has been ubiquitous in this sector. Going forward, government regulators must develop practices that are clearly articulated, sufficiently long-term, and open to industry collaboration along a wide variety of activities. Instead of attempting to solve discrete, identified problems or to promote specific technologies, innovation policy must address the energy system as a whole; be technology-neutral; intend to deploy innovations at scale and in private markets rather than own them; and break down the barriers sustaining fossil fuel industries.[45] Further, innovation policies must better align existing innovation institutions with new ones and must better connect funding with demonstration and deployment for commercialization and marketability at scale.[46] More so than in the past, energy innovation processes and practices must exhibit greater coordination, integration, and performance.[47] Similarly, government regulators must be insulated from shifting political winds. All of these variables combine with the intent of providing the private sector with reliable market signals for their involvement and their investments.

Together, these market imperfections and institutional constraints have inhibited energy innovation, yet they also provide signals for how to restructure and redesign a regulatory regime promotive of innovation in energy technologies.[48] To the extent that the old forms of R&D track traditional energy policy, then, a new regulatory regime must adopt the structure and characteristics of the desirable energy future. That future must be based on: (1) the explicit intent to reduce our dependence on oil and coal; (2) the concomitant intent to reduce carbon emissions; (3) the complementary intent to ramp up production and consumption of renewable fuels and energy efficiency; and (4) the opening of diverse and competitive energy markets. A broad and diverse energy portfolio requires a broad and diverse innovation policy. The public sector will play a necessary role in the design and implementation of a new energy policy, and the regulatory structure must reflect those values.

Institutional Principles and Organizational Behavior

Our energy future depends on a radical break with past policy in general and with the transformation of technology innovation processes in particular. Innovation policy can no longer rely on linear thinking and single-issue problem solving. Instead, innovation policy must acknowledge that future energy technologies will be the result of complex and nonlinear processes and will require a systems approach for a full complement of solutions that, hopefully, will go viral reminiscent of the digital revolution in the 1990s.[49] Energy innovation, then, will likely occur as a result of "multiple dynamic feedbacks between the stages of the [innovation] process,"[50] and it is likely to emanate from a variety of institutions including universities, established firms, start-up companies, and governments at every level.[51]

Innovation is often a complex process and no more so than for energy technologies. Without a price on carbon or a price on national security, market signals are murky. Without reliable price signals, the private sector will necessarily be reluctant to heavily invest in large-scale, long-term technologies. Private-sector investment is also at risk because there is a multiplicity of energy technologies and, because the direction of energy policy is changing, those new technologies must compete with incumbents who already enjoy market position. Market realities, including the complexities of innovation processes, must be managed by a public-sector regulatory regime that is designed to match the challenges and the multiple goals of going forward with a transformative energy policy.

Any number of reasons can be given for why an effective energy innovation strategy will not succeed, including the long-term, dynamic, and uncertain nature of markets and the complexity of innovation systems. To these perhaps ever-present challenges in the face of broad social change must be added one of the deepest conflicts: Design an energy innovation environment that is open and free and that is also required to produce commercially viable products and services. The conflict between open research and marketability is real. The private sector, naturally, expects a return on investment and can secure those returns through ownership of their innovations and inventions. The public sector, however, must treat energy innovation as a public good. The public sector must mediate its role as provider of public goods with its role as an actor in the commercialization of energy technologies. Through the effective design of an energy innovation environment, innovation must be treated as a public good that is not necessarily "owned" by individual R&D clients; in which the problem of positive externalities is minimized through transparency and open source information;[52] and with a willingness to spend money on risky ventures while maintaining accountability through appropriate benchmarks and protocols. Still, at some point along the innovation frontier, commercialization and private ownership must come into play. The public and private sectors must perceive innovation policy as a joint venture. In short, the success of a smart energy future is dependent on a multiplicity of innovation institutions built on a sound and clear mission.

The Belfer Principles Plus

In several recent papers,[53] Harvard's Belfer Center for Science and International Affairs has articulated a set of five principles for the effective management of energy innovation institutions. Although these principles provide a solid foundation for redesigning institutions and processes, supplementation is required. Institutional redesign is intended to match the challenges presented by our energy future with the regulatory tools that government has at its disposal. Regulators can no longer continue to focus exclusively on large-scale demonstration projects regardless of whether they expand our use of fossil fuel resources or develop large-scale alternatives to them. This is not to say that carbon capture and sequestration and advanced nuclear plant designs should not be addressed. Rather, it is to say that the energy innovation horizon must be broadened. New innovations must range in scope from basic science such as fusion research to the broader deployment and commercialization of advanced batteries and electric vehicles. Similarly, innovation must range in scale from CCS facilities to energy-saving light bulbs and consumer electronics.

The Belfer principles are drawn from lessons learned from successful innovation initiatives at both private-sector research institutions and national laboratories. The principles include: mission, leadership, culture, structure, management, and funding.[54] An institution must keep these elements in balance to ensure a robust and vigorous innovation culture. In addition to these principles, external political influences and the dynamic nature of innovation in general will also have impacts on effectiveness; therefore, outside intervention must be managed watchfully as well as aggressively. Energy innovation policy and its regulation involves more coordination and integration in the design and development of a future energy policy than we have experienced in the past.[55]

Mission. Although it may appear obvious that a well-defined mission is central to a successful organization, the reality is that too often organizations spend too little time focusing on mission and, even more often, too little time revisiting and reevaluating their missions. This failure to focus on mission affects the public and private sectors alike. The experience of the U.S. automobile industry and the market threats to such major corporations as IBM and Xerox over the last decades are examples of firms that were slow to adjust their missions in light of changing technologies and markets. Energy regulators have no better record.

A clearly articulated mission defines an organization's purpose and core values, helps attract talent, serves as measures of progress and success, and will shape internal design and management. Additionally, a clear mission will serve as the basis for evaluating and coordinating projects and structuring a budget. An energy innovation organization must have, as part of its mission, the intent to provide innovation as a public good, the creation of new energy markets, the promotion of a diversity of energy portfolios and products, and the encouragement of new entrants across the innovation spectrum. Due to the heterogeneity and dynamic nature of the developing energy economy and the innovations needed to respond to it, an innovation organization must be flexible and adaptive while simultaneously applying rigorous standards and benchmarks against which funded projects are measured.

Leadership. Similarly, sound leadership may appear to be an obvious and necessary component of any organization. However, finding the appropriate leader for a public-sector organization can too often become embroiled in politics. Aside from political constraints, the job description for the leader of an innovation organization will be challenging. A CEO must bring with her scientific and managerial excellence, an understanding and sympathy with the organization's mission, and the ability to coordinate and integrate both internal and external activities in both the public and private sectors. Such a leader must have one eye on the public goods mission of innovation policy and another on for-profit marketability. The innovation CEO must create an environment that encourages people to develop their skills and that is adaptable to changes in markets, and the CEO must have the vision to move an organization forward while regularly reevaluating its articulated mission.

Culture. At bottom, the energy innovation organization must create an entrepreneurial culture that stresses commitment and excellence from top to bottom. Staff must recognize that personal initiative and creativity are prized, that openness is key, and that problem solving and commercialization are the core objectives. The culture must also value collaboration and interaction as well as experimentation. Such a culture should prove to be an attractive employment opportunity and career path as well as a continuous learning environment. Whereas an innovation culture invites the exploration of multiple approaches to problem solving, it must remain accountable to mission and to articulated processes and protocols.

Management. The most important element of the management structure for the new energy organization is to break down the linear approach to innovation. Walls that separate basic and applied research into separate disciplines must be dismantled. Project directors and managers must be given a degree of independence so that they can react to new information as well as scientific, technological, economic, and political developments. A critical mass of researchers for each project must be attained so that sufficient expertise and diversity of viewpoints are directed at projects. Senior directors must have regular access to their chief officers so that budgets and directions can be adjusted accordingly.

Consistent with mission, leadership, and culture, management must have as a principal task the development of mid-level managers who can nurture scientists, inventors, and problem solvers. They must be able to carry out meaningful performance reviews of personnel and programs and assess both against the overall mission of the organization. The intent is to create a pipeline of professional staff capable of both carrying out programs as well as managing and evaluating them.

Funding. Whereas the Obama administration has announced its commitment to increase R&D in the energy sector, increased funding must be substantial. According to a National Science Foundation study, for example, federal non-defense energy R&D declined in real terms from about $7 billion in 1980 to about $1 billion in 2006.[56] Current budget proposals increase non-defense DOE R&D to $2 billion. Further, the budget increases funding for efficiency and renewable resources and reduces funding for fossil fuels. Active and proposed innovation organizations within the DOE are discussed later in this chapter.

Stable, sufficient, and reliable long-term funding is a necessary prerequisite for successful innovation. More specifically, there must be sufficient and sustained support for early-stage and exploratory development.[57] Similarly, senior managers and directors must be afforded a degree of discretion and flexibility over their budgets and the ability, within limits, to set funding priorities. The challenge of balancing flexibility with reliability in funding requires commitment to all of these principles. More importantly, however, variability in funding should be driven by the success or failure of projects according to internal measures, performance reviews, and by external market forces rather than driven by external policies or politics.

The relationship of the private sector to public-sector innovation funding must occur across the continuum. To the extent that basic science and early-stage research is involved, the active involvement of the private sector may be minimal. Government should, however, involve the private sector in policy development and in the identification of promising innovation paths. As innovation moves toward commercialization, the private-sector role will increase and grow in importance. Innovation policy, then, becomes a dynamic public-private partnership that will require clear protocols and lines of authority.

The Belfer principles are sound and fit any organization. They do not go far enough. An energy innovation organization operates in a more dynamic space than other nonprofit, or most for-profit, organizations. Energy innovation is more like the communications and digital markets of the last two decades. Rapid technological change, increasing economies of scale, network effects, very large potential profits, increased competition, new markets, and intellectual property challenges are all elements of an energy innovation economy that innovation policy and design must anticipate and manage. To the Belfer principles must be added two elements that are intended to impose more organizational rigor and accountability.

Sell Disciplines. Successful professional money managers operate with two core ideas. First, they have a market niche and have a plan or model for investing in that niche. To remain successful, they regularly and systematically test and refine that model. Second, an integral part of the model is a sell discipline. Managers cannot afford to lose their objectivity regarding their investments regardless of how attractive and profitable a stock or other investment has been in the past. A crucial dimension of that objectivity is to develop a sell discipline, which is a set of protocols for divesting themselves from a particular stock or investment when the objective indicators show that the investment no longer fits their model.

The energy innovation organization must similarly adopt a sell discipline for project investments. The organization must establish benchmarks and protocols against which to measure the development, and likely success, of a particular budgeted project. When the project fails to meet those benchmarks and protocols, then funding must end. New innovation policy must look more like venture capital investments than as a source of long-term government financing. If the program does not produce, then "second round funding" will not be made.

Benchmarks, Assessment, and A Theory of Change. These last elements for a successful energy innovation organization are all aimed at internal organizational measurements. The organization must continually monitor and measure itself against its mission and against its articulated objectives. Although energy innovation metrics are neither tightly defined nor well-tested, they must be a necessary part of any innovation organization.[58] There are limited quantitative metrics that can be usefully measured. The most obvious input metric involves the amount and flow of funding over time. Additionally, human resources (e.g., the number of scientists and technicians) can be measured quantitatively as can the number of projects, consortia, and the like.

Measuring outcomes through benchmarks may appear to be a quantitative task; but, too often, the work of establishing and refining benchmarks falls somewhere between art and alchemy rather than between math and science. Nevertheless, quantitative markers have some value as long as the organization continues to rethink its assessment tools and its benchmarks against its mission and against the organization's theory of change.

The more important output and outcome metrics are less quantifiable albeit no less significant. The number of scientific and scholarly papers, patents, and marketable technologies can be counted.[59] However, it is more difficult to calculate the market penetration of successful strategies to reduce oil dependence, lower carbon emissions, or increase energy efficiency. The core problem of designing adequate metrics is the fact that the quantifiable costs of energy innovation are not commensurable with the non-quantifiable benefits of a healthier environment.

Benchmarks are quantifiable measures of success. Although they are imperfect and often cannot measure the overall purpose of funding, they can provide useful data. Public-sector and private-sector investments in education, for example, are intended to instill a love of learning in students so that they can become valuable and contributing citizens with a lifelong commitment to continuous learning. Although it is impossible to measure a love of, or commitment to, learning, other quantitative benchmarks such as attendance, graduation, college completion, and the like provide useful data.

Successful investments in energy innovation should involve changes in complex energy systems across disciplines and, ultimately, in patterns of production and consumption. Quantitatively testing whether or not systems have been changed by any specific innovation project may prove difficult if not impossible. Nevertheless, quantitative data such as the number of projects brought to demonstration or to commercialization will be available and valuable. Herein lies one of the primary and necessary elements for a successful across-the-board innovation policy and program: It is imperative for both the public and private sectors to develop a set of common metrics to measure and assess the success of their investment dollars. Ideally, a metric should be designed that indicates how much energy is saved per dollar invested in a specific project. Similarly, a sound energy metric should also indicate how many carbon emissions (or emissions reductions) result per dollar spent as well. These metrics, then, will help shape the necessary benchmarks.

Assessment is a different process than benchmarking and is aimed at measuring the overall success of individual and collective projects. A project, for example, may meet or exceed its benchmarks. However, after assessment, the project may not serve the organization's mission to a sufficient degree. Innovations in battery technology, for example, may meet a benchmark of increasing battery life, yet battery size may not significantly decrease or may be too costly to affect consumption at scale. Benchmarks may be satisfied, but the mission may not be advanced.

An organizational theory of change is more long-term than either benchmarks or assessment.[60] A theory of change is a strategic planning device that requires an organization to be clear about its assumptions, establish long-term goals, articulate its measurement tools, and recognize the interconnections among all of these elements. The idea behind the theory is to define and measure outcomes throughout every step of the innovation process as a way to assess the feasibility of the organization's behavior, programs, and activities. It is not unusual for the theory of change to be initially vague.

The usefulness of the theory, however, comes as it is refined over time and as benchmarks and assessments contribute to measuring an organization's effectiveness in reaching its long-term goals.

The long-term goal of an energy innovation organization is to move policy away from fossil fuels to one that stresses energy efficiency and renewable resources for the purpose of energy security and independence as well as economic and environmental well-being. The organization, then, must test itself against those goals and ask whether or not, as an organization, it is making a significant contribution to that change. Developing a theory of change loops back to mission, ties into benchmarks and assessment, and incorporates organizational planning and underlying assumptions while continually focusing on long-term outcomes and goals.

Venture Regulation

Ultimately, energy innovation policy and implementing organizations must be committed to creating an innovation process that moves public-sector research and development into private-sector demonstration and deployment.[61] The public-sector energy innovation organization must take a leadership role, but in order to successfully accomplish the creation of new energy policies and technologies, it must leverage resources in the private sector as well.[62] Just as private markets cannot exist without a framework of government rules, government energy programs are not valuable or meaningful without competitive markets.[63] Energy innovation, then, is market driven rather than technology driven. To the extent that innovations are adopted and deployed in private markets, public investment in innovative technologies will create valuable assets both in terms of facilities and products as well as intellectual property and markets. Further, there is no one-size-fits-all public energy innovation organization, and they range from the specific, such as a national lab dedicated to renewable resources,[64] to a Green Bank intended to loosen credit in a tight market for clean projects.[65]

The energy innovation organization, then, will be a different sort of regulatory agency. Agencies of the past were specifically dedicated to resolving identified problems and monitoring specific industries. The new energy agency, however, will be less narrowly focused and more dynamic. The new energy agency will not act as a top-down rule maker or regulatory cop. It will not perceive its role only as the originator or owner of specific projects; instead, it will act as facilitator, convener, collaborator, and partner. A primary function will be to bring together a diversity of public and private actors to address both general policy and specific strategies for moving the country into a smart energy future. The new system will incorporate interactive information links so that data is shared in a timely way. Information will also be open so that multiple organizations can enhance their productive behavior. A new information structure will change its social role and physical presence and "[i]t may develop new laws, new organizations, new technologies, people with new skills, new kinds of machines or buildings."[66] The energy innovation organization will undertake a role of strategic leadership; and, whereas it will play a significant funding role, it must exercise that role to maximize leverage in terms of finances as well as talent, to promote ideas as well as goods, and to construct consortia as well as markets.

Venture Regulators

The most recent DOE budget shows a significant shift to a smart energy portfolio. Even though in 2010 the president indicated a budget freeze for non-mandatory spending items, the Department of Energy budget was slated for a $2 billion increase in fiscal year 2011.[67] The total DOE R&D budget is scheduled to increase more than $500 million to $10.7 billion. Energy-related R&D would total $2.2 billion, which is a 5 percent increase over the previous budget and is an increase over the years from $1.0 billion to $1.5 billion.[68] Significantly, investments in renewables such as solar and wind power show increases of 82.9 percent to $320 million and 36.4 percent to $75 million, respectively, while biomass R&D increases 8.3 percent to $235 million among other renewable projects.[69]

Concomitantly, fossil fuel R&D is reduced 4.5 percent to $469 million exclusive of carbon capture and storage. Coal R&D drops 41.7 percent, which includes the cancellation of the Clean Coal Power Initiative. Further, DOE proposes to eliminate funding for gas and oil technology R&D as well as cancel $50 billion in mandatory funding for deepwater oil and gas exploration R&D programs. Additionally, for fiscal year 2011, the Obama administration is proposing to reduce tax breaks for the fossil fuel industry in the amount of $39 billion over the next decade. Thus, the Obama administration's R&D budget demonstrates a shift away from fossil fuels to smart energy.[70]

The shift to a smart energy portfolio is directly connected to technological innovation. The Office of Management and Budget (OMB) already adopts the goals of the consensus energy policy. OMB recognizes that scientific discovery and technological innovation are central and serve as "major engines of increasing productivity and are indispensable for promoting economic growth, safeguarding the environment, improving the health of the population and safeguarding our national security and the technology-driven 21st century."[71] In its directive to the heads of executive branch departments and agencies, the OMB set out administrative priorities that included applying science and technology strategies for economic growth; promoting innovative energy technologies to reduce foreign energy dependence and mitigate climate impacts while creating jobs; and employing these technologies for national security.[72] These objectives are to be achieved by increasing the productivity of research centers and institutes as well as those of public and private laboratories. Agencies also were directed to develop common goals for their science and technology activities and to establish procedures and timelines for evaluating performance in targeted investments in promising and high-performing programs. Further, agencies were directed to develop a "science of science policy." This policy would require agencies to develop tools for the management of their research and development portfolios so that they can better assess the impact of their science and technology investments.[73]

The Obama administration is cognizant of the need to transform traditional energy R&D policies and procedures to those that resemble the concepts of venture regulation discussed here.[74] The federal government, especially the Department of Energy, has existing institutions, as well as proposals for new ones, that will facilitate this energy transition by restructuring its R&D programming and by transforming old-style R&D into forward-looking innovation policy.

The administration is currently developing innovation organizations within the Department of Energy that will interact with the private sector as well as provide reliable funding over a set period of years. Additionally, the DOE will have an opportunity to record information and create an institutional learning mechanism to assess gains and evaluate processes of innovation as well as assess the overall impact on energy policy. Ideally, this information loop will help sharpen the selection and execution of specific projects.

The new form of energy innovation will require the DOE to assess and adjust its core energy R&D programming away from the traditional path. If we consider the Manhattan Project and Project Apollo, both are examples of a "pipeline" method of innovation that relied heavily on "pushing" technology into the marketplace. The second predominant model approaches innovation from the opposite end of the spectrum. Government would enact a regulation such as a "best available technology" requirement, and the market would respond by providing it and "pulling" those new technologies into the market. Both of these techniques are linear. A problem is identified and resources are aimed at solving it from either the technology push or the market pull ends of the pipeline.

Today's energy needs, however, must reflect the complexity, magnitude, and multi-disciplinary dimensions of the energy problem. Relative to innovation, then, instead of the simple technology push or market pull models,[75] energy innovation must follow an organizational model that incorporates both. At the heart of the organizational model is an array of institutions and mechanisms designed to facilitate the evolution of new technologies in response to both forces.[76] Further, the end focus must be on the "point of market launch" and on attempts to transcend the "valley-of-death" segment of the innovation stream that prevents an idea from reaching commercial reality.[77]

Innovation policy will cross a range of activities from experimental technologies such as hydrogen fuel cells and fusion to disruptive technologies such as wind and solar and must keep reconsidering existing technologies for upgrades and advanced innovation. The regulatory tools to be used by the regulators will include front-end financial support through R&D and demonstration investments; back-end financial incentives through tax credits, loan guarantees, and the like; and sustaining regulations such as mandates and standards. Additionally, government energy innovation organizations must maintain neutrality regarding which technologies are to be developed and must foster the free exchange of information between and among collaborators and other participants.

Energy Frontier Research Centers. In 2009, the Obama administration through the DOE funded forty-six energy frontier research centers (EFRCs) for a total commitment of $377 million for fiscal year 2009. EFRCs are based on a 2001 report of the Basic Energy Sciences Advisory Committee (BESAC),[78] which conducted a study to assess the scope of fundamental scientific research necessary to achieve energy efficiency, greater use of renewable resources, improved use of fossil fuels, safe nuclear energy, future energy sources, and reduced environmental impacts. The BESAC study brought together a wide group of scientists, which further refined the most critical issues for our energy economy. A further report identified the most pressing challenges to achieving that future, and the response to those challenges was the creation of the EFRCs. The administration also announced that over the next five years, $777 million would be dedicated to this program. These centers will engage largely in basic energy science research in a broad array of fields.[79] Supported in part by the American Recovery and Reinvestment Act, the EFRCs are intended to accelerate scientific breakthroughs and will be housed at various universities, national laboratories, nonprofit organizations, and private firms.

The forty-six projects under this initiative are funded at between $2 million and $5 million per year for a planned initial five-year allocation. All the awards fall into four basic categories including: (1) renewable and carbon neutral energy such as solar and advanced nuclear power; (2) energy efficiency in the areas of clean and efficient combustion as well as superconductivity; (3) energy storage and nanotechnologies; and (4) crosscutting science including materials science.[80] For fiscal year 2011, the administration includes a request for $140 million, which is an increase of $40 million over the previous year's appropriation. The EFRCs, thus, are directed to harness basic science and advanced discovery research in order to establish a sound scientific foundation for a fundamentally new U.S. energy economy.

DOE Energy Innovation Hubs. The core idea behind energy innovation hubs is to engage long-term funding for the integration of basic and applied research from engineering through commercialization. The innovation hubs will involve large and integrated teams directed to solve priority technology challenges. National labs have been criticized recently for not creating technological breakthroughs, and the creation of these hubs is an attempt to construct an innovation program modeled after the early successes of national labs as well as those private-sector labs such as Bell Labs, Xerox, and IBM Research.

Funding is to be based on five-year renewable contracts commissioned for promoting scientific excellence and for developing a cadre of professional staff dedicated to such research. Funding has been proposed at $135 million with initial year funding at $22 million with up to $10 million for infrastructure start-up costs, equipment, and instrumentation. It is anticipated that each hub will be funded at up to $25 million a year for the remaining four years of the project.[81]

The DOE has launched three hubs that are to expected to pursue transformative breakthroughs in technology to meet our energy needs. The hubs are intended to engage energy science and engineering from the early stages of research to the point at which technology can be then handed off to the private sector. The hubs are to involve cross-disciplinary collaborations between science and technology and will be directed to address three challenges: (1) using solar power in an efficient and economical way; (2) design, construct, and retrofit commercial and residential buildings to enhance energy efficiency; and (3) employ modeling and simulation technologies to significantly improve nuclear reactor design and engineering. These efforts are directed to producing technologies that will lead to a clean and secure energy future. The hubs will organize research teams from universities, private industry, nonprofits, and government laboratories with the intent to become a world-leading R&D center in each of these three topical areas.[82]

Waxman-Markey Energy Innovation Hubs (ACES-E). In 2009, the U.S. House passed the American Clean Energy and Security Act of 2009 (ACES).[83] This bill also contained provisions for energy innovation hubs that would be awarded to local consortia of large research universities, energy-related government offices, and NGOs. These hubs would be given the authority to make smaller awards to entities that would perform a range of innovation functions.

One estimate under the act is that $240 million would be available by 2012 and that over time these funds would increase. Each hub would receive a minimum of 10 percent of the annual allocation with a maximum of 30 percent of the allowances dedicated to the hubs, which would mean initially receiving anywhere from $24 million a year to $72 million a year.

Advanced Research Projects Agency – Energy (ARPA-E). ARPA-E was established under the DOE as part of the America Competes Act of 2007[84] and was modeled on the successful Defense Advanced Research Projects Act (DARPA), which is responsible for stealth technologies as well as the Internet. ARPA-E was initially funded at $400 million in 2009 under the American Recovery and Reinvestment Act. Its mission is to fund projects to develop transformational energy technologies and accomplish the goals of a smart energy policy. The agency is specifically directed to increase our energy independence, reduce greenhouse gas emissions, improve efficiency, and ensure that the United States remains a technological and economic leader in deploying advanced energy technologies.

It is also understood that ARPA-E will focus exclusively on high-risk, high-payoff concepts to dramatically change the way we generate, store, and utilize energy.[85] The agency has initially funded projects involving building efficiency, carbon capture, direct solar power, biomass, energy storage, and others.[86] The funding came in such esoteric areas as fuel-secreting bacteria, liquid batteries, and creating solar energy by mimicking photosynthesis. Currently, ARPA-E projects are funded for three years; and, given the high-risk nature of these investments, in the future the projects may need to be of longer duration and, in any event, will need to be protected from political intrusion should the risks appear to overtake the rewards.

One criticism of ARPA-E is precisely that it was modeled after DARPA, which was a technology-driven model of innovation.[87] The Defense

Department was assigned the task to develop certain technologies without clear evidence of market demand. The needs in the energy sector, however, are intentionally market driven. Without a direct and constant focus on commercialization, most energy innovations have little value or utility. Of course risks will be taken, and failures will occur. Nevertheless, at the outset, most innovation initiatives will look to scale and marketability as well as commercial feasibility as measures of success if not for the initiation of projects.

Energy Innovation Council and Energy Technology Corporation. The agencies described previously are either in existence or appear in proposed legislation. Legislators and technology analysts have other suggestions as well.[88] Legislation, for example, has been introduced to create a comprehensive approach to the deployment of innovative technologies by creating an administrative agency dedicated to fostering clean energy programs.[89] An NGO report suggests the creation of an Energy Innovation Council (EIC), which would be responsible for developing a national R&D strategy. The Council would develop a plan to integrate federal energy R&D programs over a multi-year period and would examine the use of both direct-spending cheers for technology support and indirect financial incentives intended to promote demonstration.[90] Under the proposal, the EIC would be based in the White House and would be comprised of representatives from the key federal agencies involved with energy, the environment, and national security.[91]

The national innovation strategy would set program priorities, schedules, and resource requirements. The strategy would be based on modeling and simulation tools as well as relevant engineering and cost data. The national innovation strategy would identify alternative technological pathways as well as assess the inevitable trade-offs. The EIC would be assisted by a national advisory group from public-sector and private-sector institutions, and the national strategy would be submitted to Congress for its review and endorsement. The strategy would then serve as the basis for a five-year authorization and appropriation for energy innovation programming.

In addition to developing an energy innovation strategy of national scope, energy R&D would be integrated to the ends of discovering new ideas for energy supply and efficient use; accumulating scientific and engineering data as the basis of deployment; constructing needed R&D facilities; and establishing mechanisms for interaction between technical experts and market savvy entrepreneurs. Once developed, these ideas and programs would be brought across the valley of death and to market under the auspices of a semi-public Energy Technology Corporation specifically focusing on demonstration.[92]

National Institutes of Energy. Modeled after the National Institutes of Health (NIH), a National Institutes of Energy (NIE) has been proposed to create a network of regionally based energy innovation organizations.[93] The NIE would be dedicated to applying new energy technologies and coordinating a variety of labs, research universities, and private-sector innovators. The NIE model is comprised of twenty decentralized institutes and seven multi-disciplinary research centers coordinated by a central federal office. The NIE is intentionally result oriented and maximizes information sharing with the intent of maximizing discoveries.

Each center would have its own director, thus offering a degree of independence and flexibility. Funding for two dozen institutes would range from $50 million to $300 million per institute per year, with additional funding available for several larger projects. Each institute would have a directed mission such as solar energy, CCS, biofuels, electricity storage, or grid modernization. The singular mission of the NIE would be to develop the commercial and affordable clean energy technologies of the future while working closely with the private sector.[94] An overarching goal of the NIE would be to raise public awareness and education of the need for innovative energy technologies as well as generate political support for a new clean energy policy and enabling legislation.

Energy Discovery Innovation Institutes (e-DIIs). Another suggestion for stimulating innovation, as well as its commercialization, is the creation of a dozen Energy Discovery Innovation Institutes (e-DIIs), which would be comprised of a national network of regionally based research centers.[95] The proposal calls for an interagency process to competitively award up to $200 million per year for each institute operated by either an individual university, a national laboratory, or a consortia. The funding would be augmented by industry, investors, universities, and governments with the goal of reaching $6 billion per year in innovation spending.

The goal of the institutes would be to foster partnerships for cutting-edge and application-oriented research among a diversity of participants and disciplines. Additionally, the institutes would be designed to transfer innovation technologies and information about innovation processes rapidly and broadly.[96] A midterm goal would include building a knowledge base and the human capital necessary to sustain innovation efforts. Further, the institutes would help encourage regional economic development by creating a number of start-up firms, private research organizations, suppliers, and other aligned groups and businesses.[97]

Conclusion

Innovations in energy technology are a necessary dimension of our energy future. More significantly, as the country continues to wrestle with climate change legislation and as that legislation appears to recede from sight, the necessity of formulating an energy policy that is environmentally sound, contributes to national security, and is economically promising becomes a national imperative. Technology and innovation have always been the fundamental building blocks for any economy because private-sector competition and wealth creation thrive on them. Unfortunately, private-sector participation is under-supplied as a result of a number of market defects and entrenched institutional arrangements.

Energy innovation falls victim to multiple market problems. Private firms will under-invest because they cannot capture all of the benefits that innovative ideas generate. Private firms will under-perform because the energy sector is not linear; instead, it is an interdependent network of firms and industries that are susceptible to not being able to generate the sorts of system-wide ideas and solutions that will have an impact on our energy profile. Further, agency and information problems plague this arena. As private-sector or public-sector actors, as business men and women, as the politicians responsible for guiding the nation's energy future, and even as ordinary citizens, it is both difficult and daunting to try to understand and then act on a multi-dimensional, long-term problem such as energy in light of the innumerable social, technical, scientific, and economic uncertainties embedded within it. Innovation may well not be the answer; it is, undeniably, a necessary part of it.

Smart Energy Politics

For decades, we have known the days of cheap and easily accessible oil were numbered. For decades, we've talked and talked about the need to end America's century-long addiction to fossil fuels. And for decades, we have failed to act with the sense of urgency that this challenge requires. Time and again, the path forward has been blocked – not only by oil industry lobbyists, but also by a lack of political courage and candor.

President Barack Obama[1]

There is much to be said about President Obama's remarks. We have talked for decades and decades about oil dependence and about the need to "break our addiction." Government regulation has been held captive by oil lobbyists. We do lack the political will to move away from the traditional energy path. And, tragically, the country may now have an opportunity to focus on an energy transition because of the ecological disaster in the Gulf of Mexico. Later in his remarks, the president also called the transition to a clean energy future a national mission to be furthered by technological research and innovation and through a new energy economy. This chapter explores the political impediments that have prevented the transition and argues, more hopefully, that a new energy politics is emerging. The new politics is based on a non-partisan consensus energy policy and is supported by a broad array of public and private actors employing a wide set of strategies to make a transformation to a clean energy future possible.

The current wisdom regarding the contemporary politics of energy is that it appears that we lack the political will to effectuate change. This chapter contests that perceived wisdom on two counts. First, we must think more clearly about an energy transformation and must distinguish it from climate change. Although both are intimately related and largely complimentary, the fossil fuel industry has done what it can to obfuscate the connection.

By arguing, among other things, that climate change is a myth or is based on bad science or that better science exists and that global warming is simply the natural progression of things, the industry has taken the focus off energy policy. Second, we must reconceive energy politics. Although it is true that Washington plays a crucial role in policy change, it is not true that only Washington can change the politics of energy. Washington's diminished role is all the more apparent by Congress's failure to enact climate legislation and the Republican success in the 2010 midterm elections. Still, over the last decade or two, cities, states, and regions have initiated changes in energy policy that have been more forward-looking and progressive than anything to come out of Washington. More significantly though, energy politics is changing from the ground up as new actors enter the political marketplace. To be sure, many of the new actors are most interested in opening and profiting from new economic markets, and that is how it should be. A bottom-up change will be more effective, more sustained, and more successful that the central government could bring about.

The Failure of Traditional Energy Politics

As much as we might wish for Washington to be a national and international leader in transforming patterns of energy production and use leading to a low-carbon energy economy, Washington is, unfortunately, not likely to do so in the near term. Adding a smart energy policy to the current list of political failures is consistent with the stasis in our nation's capital. Further, history has revealed the power of incumbency in the economic marketplace as well as in the political arena as capital markets, vested interests, and a friendly, or complacent, bureaucracy weaken or destroy progressive reform.[2] Fossil fuel dominance is apparent as government policymakers and regulators continue to support these fuels and as economic markets continue to favor large-scale energy. Traditional energy policy is a powerful and persistent drag on new energy markets largely because as a society we have simply misidentified the correct public good to support.

Traditional energy policy was based on two failures. First, traditional energy policy misdiagnosed the market failure at the heart of our energy economy in two complimentary ways. To emphasize the point that the social costs of carbon have not been accounted for in the price of energy, especially fossil fuel energy,[3] global warming has been called the "greatest market failure that the world has seen."[4] Next, as a matter of public policy, cheap and abundant energy has been treated as a public good; a good that by definition the market will over supply. Cheap energy is not a public good.

Instead, whereas energy is a necessary input for all economic activity, fossil fuels can be adequately produced in private markets without the helping hand of government. Further, according to microeconomic theory, fossil fuels should be priced competitively and should not be subsidized; otherwise it will be underpriced and oversupplied. Treating cheap energy as a public good leads directly to an overproduction of carbon emissions. It is simply foolish to assume that cheap and abundant fossil fuel energy continues to be in the national interest.

Regardless of whether or not the assumption that cheap energy contributed to a vibrant economy in the past, and was not simply a cover argument for industry interests, it is no longer true today as fossil fuel energy has weakened our economy and our standing in the world. Nevertheless, there is a sense in which energy is a public good. However, instead of cheap and abundant energy, the correct public good is clean or low-carbon energy that, now, must be the object of government support and private investment.

The second failure of traditional energy policy is regulatory failure. The bureaucracy that was created to sustain fossil fuel industries and promote their growth and expansion has failed us. As David Orr notes, our energy/environmental failures include: ignoring warning signs and failing to anticipate climate trends; little to no public education about the risks we face; ignorance, willful or otherwise, of the security consequences of global warming; no effort to "recalibrate the economy" for a low-carbon energy future through the serious and sustained promotion of energy efficiency and renewable resources; and the contribution to economic and fiscal policies that led to the recession we are continuing to experience.[5] As he goes on to note: "Policy failure at this scale certainly reflects the stranglehold of coal and oil money on public policy."[6] The Deepwater Horizon and Upper Big Branch Mine case studies confirm his charge as they expose the regulatory weakness of government enforcement; the lopsided support for fossil fuels; and the general lack of federal government enthusiasm for alternative and renewable resources. We can characterize the relationship between industry and government as either capture or simply another example of politics as usual; still, the outcome is the same – a constricted and narrow energy policy that must be transformed.

The argument for a new energy politics is based on two ideas. First, as noted, Washington cannot be relied on to lead the transformation. In his Inaugural Address, President Obama committed himself to rolling "back the threat of a warming planet,"[7] and in his speech to the nation about the BP disaster, he emphasized the need for an alternative. Yet, even though he has appointed a green team to his administration, climate legislation has stalled. Similarly, even though there are senators and representatives who would support a new energy policy, draft legislation has been modest. Proposals for climate legislation that necessarily have the effect of changing energy policy, such as the Waxman-Markey and Kerry-Lieberman bills, are steps in the right direction but do not go far enough. Similarly, the Obama administration's support for offshore drilling and nuclear power more than hint of continued commitment to the traditional energy path, even if interpreted as pragmatic imperatives. Time will tell if the BP debacle changes much.

Reliance on Washington to lead the country to a smart energy future may be misplaced, which, however, is not to admit defeat. The transformation of energy politics, then, is based on a second idea – a transition will occur as a result of bottom-up, not top-down, policy making. Washington will be a necessary participant but as facilitator, support system, and partner, not as sole leader and certainly not as the controlling owner of policies susceptible to further regulatory capture. Similarly, the transition will be successful as a result of a horizontal politics encompassing a broad array of participants instead of a vertical politics moving downward from Washington and along narrow partisan paths and moving within established regulatory silos.

Indeed, the new politics of energy is progressing on many fronts and in many guises, and the new politics is led by a variety of new actors including bipartisan coalitions, trade associations for new entrants, think tanks and research centers, NGEOs, and private-sector investors. Further, and in no small part, private-sector investors, particularly venture capitalists, who are looking for financial returns in the multitrillion-dollar energy economy, have the potential to effectively address climate change, energy efficiency, competition, and energy independence.[8] In short, if the country is to realize a transformed energy policy, that transformation will occur as new political actors literally educate our elected politicians and policymakers about the need for a smart policy; and it will occur as a widespread array of activities take hold and occur organically and spread virally.[9]

An energy policy transformation will encompass certain basic elements. The transformation will be led by new public and private actors and new for-profit and nonprofit organizations operating in new markets with new technologies. The Great Recession of 2008 lingers, and investments in new markets lag behind expectations. Still, there is something of a silver lining in this troubled market, although the lessons have been learned painfully. Robust markets do not exist without reliable and transparent rules, and no good markets exist without government vigilance. In addition, widespread public education about the virtues of a new energy economy must take place. In part, public education must embody a new language that marries concerns about energy with concerns about the environment and that no longer treats energy and the environment as separate disciplines. Such an education should enable consumers to make smarter energy choices in more competitive energy markets. The final building block is a transformed regulatory structure that breaks down fossil fuel industry and regulatory silos; engages in systems change; uses science more intelligently and cost-benefit analysis less politically; and perceives a regulatory structure that behaves as a venture regulator bringing an array of smart energy technologies to commercial scale.

New Actors

Traditional energy policy successfully linked government and markets in a mutually supportive relationship to the great financial benefit of fossil fuel industries. The most effective conduit between industry and government has been industry trade associations such as the National Petroleum Council, the American Gas Association, the American Petroleum Institute, the National Mining Association, the U.S. Chamber of Commerce, and the like. These organizations helped consolidate industry interests, concentrate lobbying efforts in Washington and in state capitals, and served to funnel substantial campaign contributions to both sides of the aisle. As a result of these efforts, the fossil fuel industry has enjoyed special access not only in congressional corridors but throughout the bureaucracy and right into the White House.

As noted throughout this book, the fossil fuel industry has successfully lobbied for financial perquisites, has had a special seat at the table during the Bush administration for national energy planning, and has effectively neutralized regulatory enforcement of health, safety, and environmental regulations that the industry deemed as too expensive and unnecessary. These efforts reveal industry's Janus-like approach to regulation. Under the banner of free-market ideology, industry fights against social regulations aimed at protecting the health and safety of workers and the environment. However, industry curries government economic regulations that reduce its cost of doing business, reduce its taxes, and otherwise increase its profitability.

As climate change continues to garner headlines, the fossil fuel industry has not been idle. Instead, industry's approach to climate change has been to increase lobbying against it, financing junk science as a public relations strategy,[10] promoting the misuse of cost-benefit analyses,[11] and continuing the drumbeat of economic growth through dirty energy. The industry approach to matters of climate change and the need for a smart energy future is to ramp up their political activities rather than improve or change their business model to accommodate a change in the energy climate. Sound policy, then, becomes subject to politicalization. Such is life in a pluralistic democracy, however, and advocates for an alternative energy future must play in the same political arena. Fortunately, there is growing political activity in favor of a low-carbon energy economy as an array of public and private actors attempt to counter traditional energy politics.

Clean and Green Venture Capitalists

Traditional energy interests have sunk costs of trillions of dollars. That simple fact alone indicates the reality that fossil fuel firms will not walk away from their investments. It also underscores the reality that if a low-carbon energy economy is to replace them, then they must win in the market and they must win big. One strategy for increasing market penetration of new energy technologies is through the use of venture capital (VC). There are multiple clean and green VC firms and intermediary companies promoting these markets.[12] In the first quarter of 2010, VC investment was estimated to be $733.3 million, up 68 percent from a year ago of $435.5 million but lagging behind pre-recession 2008 at $844.7 million.[13] Further, it should not be at all surprising, then, that venture capital firms are not only entering the market they are entering with political clout.

The California venture capital firm of Kleiner Perkins,[14] as an example, named former vice president Al Gore as a partner working in their Greentech division. Gore has been an outspoken and Nobel prize-winning advocate of efforts to combat climate change.[15] Greentech has an investment portfolio intended to promote clean power, clean water, and clean transportation as the world's population continues to urbanize. Kleiner Perkins invests in early-stage breakthrough ventures that promise to create new markets. To date, they have invested in nearly two dozen companies ranging from geothermal development of biofuels to solar power through renewable fuel cells powered by oxygen and hydrogen.[16]

The touchstone of success for any venture capital firm is commercialization. Innovative technologies must have the capability of widespread adoption. Kleiner Perkins investments are intended to achieve that goal. Its investment in the software company Verdiem, for example, is aimed at allowing consumers to monitor power use over their personal computers (a form of personal energy device) and by using special power settings to realize cost savings.[17] In June 2010, it was announced that the Verdiem software was to be used by several California state departments and agencies serving nearly 70,000 personal computers with a goal of reducing energy consumption by up to 60 percent.[18] Accessing new energy markets through PCs opens those markets to hundreds of millions of consumers. Similarly, its investment in Bloom Energy is an attempt to expand the use of energy saving technologies, such as high-efficiency servers and distributed generation. Bloom Energy builds on-site generation systems powered by fuel cells that produce clean, reliable, and reasonably priced power. Distributed generation reduces the cost of power by eliminating the need for large-scale transmission and distribution networks and other fixed costs. Bloom Energy's products promise to be cost-effective with an estimated three-to five-year payback time.[19]

One Kleiner Perkins alumnus, Vinod Khosla, began his own venture capital firm, Khosla Ventures, and named former British Prime Minister Tony Blair as senior advisor. The firm has dedicated more than $1 billion for clean and information technologies. Khosla Ventures focuses on building sustainable companies through leveraging relationships and building teams to assist new firms in becoming billion-dollar businesses.[20] The firm's portfolio runs from battery development and building materials to utility-scale generation and cellulosic alcohol. It has invested in nearly four dozen clean tech companies such as Altarock Energy, which develops engineered geothermal systems. The firm has also invested in Calera, a company engaged in CCS. Calera has developed a process of capturing carbon dioxide and other emissions including mercury and converting those pollutants into sustainable building materials and water.

The approach of these two VC firms is instructive. Both are invested in a range of clean and green technologies. The investments are intended to bring new technologies to scale as well as build sustainable businesses. Further, both firms approach the greening of our energy economy by realizing the necessity of not only partnering with other firms but with having the public and private sectors linked to expand these markets. It is no accident that people such as Al Gore, Colin Powell, and Tony Blair have become attractive to VC firms. These former national and international leaders have joined these firms to open doors, help build coalitions, and otherwise broaden the base for new energy initiatives as well as respond to climate change challenges. More importantly, the linkage between venture capital and political clout underscores the pervasive geopolitics of energy. The history of fossil fuels is a history of government support. To effectively counter that history, our energy future must well understand the old energy politics just as it must create a new one.

Non-Partisan Coalitions

In the early years of the twenty-first century, two reports stand out as catalysts for widening the national conversation regarding the necessity for a new energy policy. Most importantly, these reports emanated from new energy coalitions intent on transforming our energy economy with the express intent of ridding the country of its past fossil fuel dependency. Since these two reports were published, innumerable other reports and studies have been issued, many of which are cited throughout this book. The two, though, are notable for taking up a discussion that was begun more than thirty years ago; by relying on small groups of prestigious thinkers, politicians, and business persons to bring the discussion more into the mainstream; and putting their policy proposals into action.

The first report, *Challenge and Opportunity*, was published in 2003 by the Energy Future Coalition.[21] The coalition was formed in 2001 specifically to address the inadequacies in U.S. energy policy. The coalition was comprised of more than 150 national leaders from business, labor, government, the academy, and the NGO community. The coalition created six working groups to present recommendations in the areas of transportation, bioenergy and agriculture, the future of coal, energy efficiency, the smart grid, and energy financing. In the beginning, coalition discussions focused on the need for oil independence, risks to our economy, the need for national security, and the threat of climate change. In writing the report, the coalition sought to identify policy changes that would expand innovation in energy technologies. Additionally, the coalition realized that political lines needed to be crossed and that partisan gridlock needed to be broken.

The steering committee for the coalition includes former Senators Timothy Wirth and Tom Daschle, entrepreneurs such as Richard Branson, Ted Turner, and Vinod Khosla, and representatives from environmental groups, philanthropies, and various business organizations and NGOs. The coalition approaches its work for a new energy future by searching for alternatives to an outdated past energy policy that has, for too long, proceeded along narrow and partisan lines. In its place, the coalition seeks practical solutions to address national concerns about energy independence and the environment by bringing together a diverse and inclusive group of actors from all sectors of society.

Its report focuses on energy innovations that expand consumer choice, open clean energy markets, move the country away from oil and coal, and change the base of the transportation sector. The coalition's policy initiatives recognize the need for a transition to sustainable energy but also recognize the need for immediate action. The coalition also aims at reducing risks to the global economy by concentrated oil states and looks to the United States to take a leadership role in developing a domestic agenda as well as take an international leadership role.

The second report, *Ending the Energy Stalemate*, was published in 2004 by the National Commission on Energy Policy.[22] The commission was formed in 2002 and consisted of a bipartisan group of twenty national leaders in the energy fields drawn from industry, government, labor, the academy, as well as from consumer and environmental protection organizations. Several members of the commission have held significant senior positions in government and business. Co-chair William K. Reilly, as an example, has been a senior adviser to industry and was former administrator for the U.S. Environmental Protection Agency. Similarly, co-chair Susan Tierney is a principal of the consulting firm, the Analysis Group, and a former Assistant Secretary of Energy. Other commissioners include financial advisors, environmentalists, utility executives, and venture capitalists.

Ending the Energy Stalemate was written over a two-year period and was based on more than thirty-five original research studies. The commission continued to develop its energy analysis by publishing a series of legislative recommendations, including one for raising fuel economy standards, which were ultimately adopted. The commission sees its role as overcoming the political barriers to energy policy reform by facilitating outreach to government, business, and the NGO communities. Its current agenda includes a focus on issues of oil security, climate change, and the development of an adequate and well-located energy infrastructure.

The National Commission on Energy Policy is part of the nonprofit Bipartisan Policy Center established by four former Senate Majority Leaders – Howard Baker, Tom Daschle, Bob Dole, and George Mitchell. The center's mission is to attract public support to achieve progress on a number of fronts including health care, national security, financial services, and transportation as well as energy.[23] The center has published a variety of papers and reports on its policy areas and crosscutting reports on such topics as the economy and the proper role of science in regulation.

The approach of the National Commission and the Bipartisan Policy Center is similar to that of a venture capitalist. First, policy initiatives must be based on sound data, must be non-partisan, and must incorporate public and private actors. Second, the approach is nonlinear. There is no single solution to economic reform nor is there a single solution to reforming energy policy. Instead, a panoply of technologies and innovations must be supported; political and legal barriers to progress must be examined and reformed; and the overall goal must be to bring new technologies to commercial markets at affordable prices by a diversity of sustainable busi-nesses, not by a small handful of bureaucracy-capturing mega-firms. The approach of both of these organizations has been bipartisan. More impor-tantly, however, the approach has been an integrative one linking the need for new energy policies within the need for environmental protection as well as economic growth and national security.

Think Tanks and Research Centers

There is no shortage of university-based think tanks and research centers and institutes directed to climate change, energy policy, environmental pro-tection, or some combination. Major public and private universities have become key players in shaping the energy future, and their research activi-ties are becoming increasingly interdisciplinary and multi-disciplinary. At Duke, for example, science, law, policy, and mathematics are at the heart of its Climate Change Partnership, which combines the research efforts of the university's Center on Global Change, the Nicholas Institute for Environmental Policy Studies, and the Nicholas School of the Environment and Earth Sciences together with industry partners such as Duke Energy and ConocoPhillips.[24] Yale's School of Forestry and Environmental Studies has been actively publishing its research on energy and the environment[25] as has the Energy and Resources Group at University of California at Berkeley.[26] Additionally for many years, Harvard's Electric Policy Group has been a center of excellence for its analyses and publications on all aspects of the electric industry.[27]

While wide-ranging, these research centers and institutes share certain characteristics. The Princeton Environmental Institute, for example, is struc-tured as a classic academic interdisciplinary research center. Housed within the institute, however, is the Carbon Mitigation Initiative[28] (CMI), which has brought together a group of scholars, as well as individuals with signif-icant experience in the public and private sectors, together with NGEOs. Additionally, the initiative has engaged research partners from other uni-versities in the United States and abroad in efforts to develop a sustainable response to carbon buildup and climate change.

The initiative draws on the academic strength of its research components and partners as well as on the business strength of its private firm partners in an attempt to fuse fundamental science, technological development, and business practices in an effort to accelerate technological development from initial discovery through scalable application. The primary goal is to design methods of CCS that effectively capture carbon in a safe and reliable man-ner, cost-effectively, with little negative environmental impact, and without dramatic disruptions of energy consumption. The initiative's research program has four components starting with the science group, which collects raw data directed to understanding climate change phenomena. Research is conducted, separately, on storage and on capture, and all of these components are synthesized by the integration research group aimed at developing mitigation strategies and communicating them to industry, government, NGOs, and the general public.

One of the most notable products of the initiative has been its study on stabilization wedges. The core idea is the conservative assumption that carbon emissions from fossil fuels are expected to double in the next fifty years. That doubling is unacceptable, and a strategy must be designed to at least keep the rate of increasing emissions flat, which still means reaching a CO_2 concentration of 500 ppm. To keep that rate flat requires reducing projected carbon output by seven billion tons per year until 2054 or, in CMI terms, in seven wedges of one billion tons each. The CMI wedges include: efficiency and conservation; CCS; alternative fuels; nuclear power; and increasing natural sinks.[29] In effect, the wedges serve as measuring devices, and each wedge is susceptible to several strategies. Can, for example, we achieve one billion tons of carbon reductions through greater energy efficiency in one wedge and another one billion tons through the increased use of alternative fuels? CMI starts with the premise that we already have the fundamental scientific, technical, and industrial capacity, as well as a portfolio of technologies, to achieve these gains. After ten years, CMI reports having produced significant results in original research and reports that it has significantly expanded its partnerships and research activities along a broad front.[30]

The Belfer Center for Science and International Affairs at the Harvard Kennedy School is another excellent example of an academic institution creating an interdisciplinary research community of national and international scholars and researchers and practitioners to develop policy solutions to pressing social issues. The center's mission is to develop leadership and policies particularly in the areas of integrating science, technology, and environmental policy while also preparing future leaders. Further, the center sees public education as a key element in promoting public policies intended to further science and technology as well as improve the economy.

Although not limited to matters of the environment and energy, the center has produced an impressive set of research papers directly addressing the development of innovative energy technologies that respond to climate challenges. Most specifically, the center's work is policy specific and is intended to have an impact on how government operates and how the public understands issues of science and technology as well as energy and the environment. The center's work on energy technology, for example, has helped shape energy policy in China, India, and the United States. It has also played a significant and influential role in international climate negotiations. The center also believes that information technologies and nanotechnologies will play an important role in the future and must be part of any emerging energy policy. The center's work in energy addresses biofuels, CCS, alternative transportation, reduction in oil consumption, greenhouse gas emissions, and the like.

The center examines U.S. and global budgets and programs for technological development that affect access to energy, energy security, mitigation, and environmental harms and risks. The center has analyzed DOE research and development budgets over the last few years and has published papers with legislative analyses and proposals, many of which have been cited throughout this book. Central to developing a new energy policy, the Belfer Center focuses on technology as the means for developing public policy options that recognize the interconnections between energy, the environment, and the economy.

Academic research centers and institutes, then, are moving beyond their traditional boundaries. Whereas they do take on the traditional role of engaging in their own research, they move beyond it by establishing and setting policy agendas, building coalitions, and promoting public education outside the academy. They are also broadening their research by partnering with other departments, other universities, private-sector actors, and experts from numerous disciplines to address the complexities of energy policy transformation and climate change.

New Energy Organizations

The United States Climate Action Partnership (USCAP) is an example of a new form of organization that recognizes the need for cooperation and collaboration to achieve gains in the formation of a new energy policy.[31] USCAP brings together industry and traditional advocacy groups to seek common solutions. Formed in January 2007, USCAP is a coalition of businesses and leading environmental organizations such as Chrysler, Dow Chemical, DuPont, General Electric, and Duke Energy together with a Natural Resources Defense Council, the Nature Conservancy, and the Environmental Defense Fund, among other organizations and private firms.

USCAP's primary purpose is to support government efforts to achieve significant reductions in greenhouse gas emissions. In its report, *A Blueprint for Legislative Action,*[32] USCAP attempted to craft an integrated and consensus approach to climate legislation and energy policy that addresses concerns of private companies as well as NGO environmental and energy groups. Central to the success of such a policy will be the development of new and emerging technologies for carbon emission reductions that can be deployed and implemented by the private sector. Private-sector involvement will, accordingly, help secure economic growth, provide jobs, and attune American businesses, as well as the public, to a new energy future.

The path to achieving these goals should, by now, be familiar. Energy efficiency is central and must be increased. Domestic supplies of fossil fuels must be responsibly developed, deployed, and consumed. New transportation technologies and fuels are a necessity for a healthy future. Additionally, the electricity sector must switch to low-carbon resources, coal burning with CCS, and other clean coal technologies. Given the mix of industry and environmental organizations, carbon reduction strategies are paramount for USCAP. Consequently, they call for a mandatory climate program to reduce greenhouse gas emissions from the major sectors of the economy including large stationary sources such as utilities, heavy manufacturing firms, and transportation, as well as from commercial and residential buildings. Given the magnitude of the problem and the uncertainties involved, USCAP has no single approach to carbon emission reductions and instead promotes a flexible approach to establishing a carbon price that includes a number of policy tools such as market-based incentives, performance standards, cap-and-trade regulations, tax reforms, and R&D incentives.

Another hybrid organization is the American Council on Renewable Energy (ACORE). From one perspective, ACORE looks like a traditional trade association promoting its members' interests in renewable energy products and services. ACORE does conduct vendor trade fairs such as its annual Renewable Energy Technology Conference – RETECH. However, ACORE is a more ambitious organization founded in 2001 bringing together renewable energy advocates, innovators, and businesses to increase the presence of renewable energy in the American economy and to affect policy change and public education.

ACORE has a membership of approximately 500 companies including utilities, professional service firms, and financial and educational institutions, as well as nonprofit groups and government agencies. The organization has hosted a number of annual meetings focusing on expanding the use of innovation in the renewable energy sector. The underlying principle of ACORE is that renewable energy cannot become a vibrant part of the new energy economy without adequate financing. Consequently, it is cognizant of the need to bring Wall Street to annual renewable energy conferences and to bring together project developers and financiers in this expanding market. Additionally, ACORE has entered the policy arena by hosting conferences to promote an understanding of renewable energy research, science, and technologies with the aim of scaling this sector to the level of commercialization.

ACORE differs from the traditional trade organization through its publications. In addition to publishing summaries of annual conferences,[33] ACORE initiates studies such as a paper written with the Electric Power Research Institute addressing strategies for using renewable energy to improve security, create jobs, and reduce emissions[34] and periodic outlooks for renewable energy.[35] Other publications include a compendium of best practices by state and local governments that promote energy efficiency and renewable energy projects; how to leverage stimulus funds; and reports on individual projects.[36]

ACORE attempts to look at renewable energy along a broad front and has established committees in the areas of climate change, electric utilities, and biomass while also building collaborations among corporate CEOs, the American Bar Association, and renewable energy credit working groups. In addition to serving as a traditional membership trade association, ACORE, then, perceives its role as convener through major conferences; educator through policy fora; publisher for reports and news updates; and proselytizer for renewable energy more generally.

The New NGEO Politics

The several examples just listed of new actors, coalitions, investors, think tanks, and organizations are representative of the new entities dedicated to transforming energy policy. These new entities can be grouped under the label – non-governmental energy organizations or NGEOs. They are distinguished from traditional trade associations, which tend to be single-issue advocates for an industry or a segment of an industry. In this regard, then, there are traditionally structured trade associations such as the American Wind Energy Association or the Solar Electric Power Association that are similar in design and purpose to traditional fossil fuel trade associations such as the National Mining Association or the American Petroleum Institute. These single-issue energy organizations are essentially lobbyists for their members. Their interest in policy is narrow; they reduce lobbying costs through membership fees; and they seek a narrow range of legislative benefits.

The new NGEOs, however, have a broad interest in policy, perceive coalition building as a necessary component of their work, and combine public-sector and private-sector actors. Similarly, as we have seen, university-based research centers and institutes have also broadened their agendas, and instead of focusing on a particular area or topic, they have brought together a variety of partners to look at energy policy and climate change in multiple dimensions. They have also changed their approach to their work. In many instances, they adopt a policy advocacy stance instead of undertaking neutral or non-specific academic research.

NGEOs are also distinguished from traditional advocacy groups such as the NRDC and the Environmental Defense Fund. These environmental organizations began their lives as litigants but are now turning toward policy making. They are also distinguishable from think tanks such as the Brookings Institute and the American Enterprise Institute, which are overtly partisan. The NGEOs are political, yet they are intentionally nonpartisan. Further, NGEOs do not behave like such political organizations as the Center for American Progress or the Cato Institute, although these bodies have made significant contributions to the energy/environmental conversation. The NGEOs are intentionally less ideological; and, to the extent that they are policy advocates, their policy advocacy seeks a broadbased common good. The NGEOs do look to both sides of the aisle to effectuate change.

NGEOs behave differently than traditional NGOs both in their membership and in their approach to political participation. These new energy organizations have a diverse and diffuse membership often combining former adversaries, such as utilities and environmentalists, in common cause. Membership is also diverse by combining public and private, profit and nonprofit, and academic and business interests under one umbrella. An NGEO can be a significant coalition-building organization, such as the Climate Action Network, which has a membership of more than 450 NGOs committed to bringing climate change to ecologically sustainable levels.[37] The coalitions formed by these NGEOs have the benefit of spreading costs among a larger membership; expanding the range of policy options; broadening the potential for benefits; and, hopefully, increasing the possibility of political success.

NGEOs also engage in a wider range of activities than traditional NGOs. As we have seen, these new energy organizations are policy oriented, see public education as a necessary element, engage in research and publication, and attempt to influence politics along a broader front with the national interest, rather than sectarian interests, in mind. At a general level, NGEOs share certain commonalities. They embrace the multiple policy goals of the consensus energy policy. They perceive government as a facilitator playing a supporting role to the development of new energy markets and technologies. Additionally, their interests are long-term rather than focused on quarterly profits or biennial elections.

NGEOs seek to go beyond the sorts of economic and social regulation that characterized the administrative state from the New Deal through the Great Society. Instead, they seek a new form of government regulation that, in Lord Anthony Giddens' analysis, attempts to achieve a political and economic convergence of interests.[38] Energy and the environment are naturally linked, and there is no reason that both cannot be promoted in economically beneficial ways.[39] Further, there is no reason that interest in energy and in the environment cannot become part of a national dialogue of the sort begun and sustained by such nonprofits as Focus the Nation[40] and 350.org[41] that use civic engagement and social networking to empower students, young leaders, and citizens, more generally, to participate in efforts to create a national conversation about a clean energy future.

NGEOs approach energy policy and climate change with different levels of ideological intensity. Some, for example, are more protective of the environment and advocate for strict carbon reduction standards. Others are more tolerant of market mechanisms as well as more tolerant of transitional roles for fossil fuels and nuclear power. Nevertheless, these new energy organizations share a commitment to sound policy, reliable science, and dependable economic analyses. Further, the NGEOs contribute an important element to public discourse – non-partisanship. Their commitment to providing new information rather than sound bites and their direct participation in policy formation have an impact on legislation and implementing regulations.

This broad-based coalition-building approach to energy policy and climate change is a necessary approach. Neither the design of a new energy policy nor a prescription for addressing climate change is a single-issue matter. Nor can a single-issue approach effectively engage the intergenerational, multi-disciplinary, multi-faceted, and multi-lateral issues involved. Instead, NGEOs with research and policy-making capabilities are the necessary actors in the coming energy transformation. They are the aggregators of knowledge and information, and they are the coordinators of diverse groups for effective political participation and public education.[42] The NGEOs are emblematic of the types of public/private profit/nonprofit coalitions that must be built for the new energy politics. Still, the private sector must also be part of the developing new energy economy, and there are signs that the private sector is paying attention.

The Corporate Role in the New Energy Economy

Private actors, commercial and investment banks, venture capitalists and general investors, established corporations and start-ups are all necessary and desirable participants in the new energy economy. Indeed, we can have no new energy economy without them, nor can we effectively address climate change. New technologies and new markets are indispensable. We have encountered several of these actors already; yet, we must be careful not to assume too much. BP has not traveled too far Beyond Petroleum. Duke Energy, although a vocal player in climate change discussions, has not traveled too far beyond coal-fired and nuclear power generation. Moreover, commercial banks, although marginal participants in lending to a low-carbon economy, understandably look to reliable returns from large clients. Many actors talk green but walk along the dominant energy path.[43] Nevertheless, all of these actors will have necessary parts to play in a transformed energy world.

The role of business relative to energy policy transformation and climate change, at first, may appear incongruous. After all, for decades, businesses and environmentalists were at odds. Today, however, they are finding common ground for several reasons. Energy and economic security top the list of overarching goals. Yet, to achieve those goals, businesses have a need for regulatory certainty so that they can plan and invest. Similarly, businesses are aware of the power of innovation in creating markets and jobs. Further, businesses also seek leadership roles in society and can do so through exercising corporate social responsibility.[44]

Corporate Behavior

Although the economy may have slowed green investment, and although the corporate sector may behave opportunistically, that behavior is moving in a low-carbon direction. Private-sector corporations, including utilities, will ultimately adapt to the new energy economy. Forward-thinking firms are already moving in that direction. Over the last few years, for example, corporations are paying more attention to sustainability even to the point of appointing Chief Sustainability Officers as members of the corporate senior management team. These officers are tasked with the job of looking for ways that their companies can become more sustainable, increase energy efficiency, and reduce their carbon footprints.[45]

Over the last few years, much has been made of corporate leaders, such as Lee Scott of Wal-Mart and Jim Rogers of Duke Energy, who have examined the role of their firms in the emerging green economy. Corporations have multiple paths to demonstrate their leadership in a sustainable world. Internally, corporations can do what they can to reduce their carbon footprint such as greening their vehicle fleets, improving building efficiency, and operating transparently and accountably regarding their energy consumption and use.[46] Further, forward-thinking leaders can also build and sustain a corporate culture that will focus less on the short-term than on the corporation's responsibility for the protection of future generations, for the values inherent in the institutions in which they work, and which reconceive the way business is conducted.[47]

Externally, corporate leaders have become more amenable to sitting down with competitors and former antagonists to consider the long-term energy future. Both climate change and the need for a new energy policy are problems of complexity, size, duration, and uncertainty, the dimensions of which we have not seen before. No single entity, government or corporate, can effectively address the challenges before us. Consequently, private firms must collaborate and partner more so than in the past, and they can no longer "go it alone." Instead, they will participate in public/private and profit/nonprofit networks in efforts to stimulate the advance of new technologies and the creation of new markets. Nowhere can this commitment be better demonstrated than through private investments.

Investments

In 2007, several major banks including Bank of America, Citigroup, and J. P. Morgan Chase, met with major electric power companies and environmental groups, such as Environmental Defense and NRDC, for the purpose of designing a framework to identify the risks and opportunities regarding a future carbon policy. As a result of this meeting, a non-binding set of Carbon Principles was adopted.[48] The principles were intended to establish a set of best practices for investment in low-carbon projects and to establish a baseline understanding for investing in traditional coal-fired utilities. Commercial banks, understandably, were concerned about the growing number of federal and state regulatory actions that would have the effect of significantly raising the cost of coal-fired electricity or that might stop projects altogether.

The Carbon Principles were an attempt to bring some order and stability to the climate change environment, particularly in light of the continuing uncertainty surrounding federal climate legislation. The principles were never intended to fix rules or set compliance standards. Instead, they were intended to promote discussion about alternatives including energy efficiency and renewables, increase understanding of carbon risks, and generally broaden discussion about energy options. In the end, the principles did not establish specific performance criteria nor were they intended to establish an agreement as to the specific types of transactions that banks must avoid. Instead, the principles established a process for investigating and analyzing carbon emission risks while financing the electric power industry.

The Carbon Principles project has not met with great success. Relatively few institutions became signatories, and the process for becoming a member has been suspended. Still, the idea behind the project is a valuable one. In the first instance, commercial lenders gather information from their clients as well as from environmentalists. Second, the discussion about best practices can become part of a process for balancing risk and expanding options. No doubt, the economic collapse of 2008–2010 contributed to the suspension of this program. However, ideas such as the Carbon Principles, the American Carbon Registry,[49] the Carbon Disclosure Project,[50] and SEC carbon disclosure requirements[51] are public/private efforts to bring more transparency to private financial transactions that contribute to a dirty energy economy.

More positively, the last several years have seen an expanding investment base in clean energy initiatives. REN21 is a global policy network dedicated to enhancing renewable energy. REN21 facilitates discussions among governments, international institutions, NGEOs, businesses, and other partners in furthering a clean energy economy. The organization also tracks green investments, and it has reported that annual investment in renewable energy grew to $120 billion in 2008. It further reports that between 2004 in 2008 solar photovoltaic capacity increased six-fold, wind power capacity increased two and a half times, and overall renewable capacity increased 75 percent with all sectors gaining at increasing rates.[52]

Green and clean investments are continuing to increase. The Pegasus Sustainable Century, a private equity firm and a member of the Clinton Global Initiative,[53] for example, has announced that it seeks $2 billion in growth capital over a five-year period to bring financial, operating, and policy expertise to businesses to improve sustainability and resource efficiency.[54] This effort is an attempt to bring emerging energy efficient technologies to scale. Similarly, the Cleantech Group, an organization of international financial advisors and investors, reports that during the first quarter of 2010, global clean investments topped $2 billion.[55]

Investment in green technologies can contribute to a positive economic feedback loop. As emerging technologies are brought to scale, markets and jobs are created. Navigant Consulting, for example, reports that by adopting a national renewable portfolio standard of 25 percent by 2025, 274,000 jobs can be created.[56] Additionally, increased energy efficiency, as opposed to increased energy production and consumption, can be economically beneficial, thus breaking the energy economy paradigm that has dominated energy thinking. McKinsey & Company, for example, reports that projected global energy demand growth can be cut in half by capturing opportunities to increase energy productivity. The company further reports that these gains can be achieved by using existing technologies that pay for themselves.[57] McKinsey estimated annual energy savings of up to $900 billion by 2020 based on an average internal rate of return of 17 percent.[58] Key areas of investment include setting energy efficiency standards for appliances and equipment; financing energy efficiency upgrades in new and remodeled buildings; raising corporate standards for energy efficiency; and investing in energy intermediaries to publicize the savings available to all consumers.[59] More significantly, even during this period of economic downturn, growth in energy demand worldwide is projected to continue significantly and carbon emissions will grow rapidly.[60]

Green-sector investments are projected to be significant in the coming years in spite of the economic recession. The Global Climate Change Initiative of the Pew Center, for example, notes that in general investments decreased 53 percent in the first quarter of 2009 relative to the previous year due to the global recession. However, even in a bad economy, investment in clean energy markets fell only 6.6 percent to $262 billion in 2009. Further, Pew anticipates that investment in global clean energy markets will grow significantly in the coming decades and can have a positive impact on the competitiveness of U.S. firms. The center projects that annual investments can reach between $106 billion and $230 billion a year by 2020 and as much as $424 billion a year in 2030, reaching a decade-long total of $2.2 trillion. Current trends in clean energy investments are positive. Between 2004 and 2009 investments in renewables, efficiency, biofuels, CCS, nuclear power, and other low-carbon technologies achieved an annual compound rate of 39 percent. Further, wind and solar markets have sustained annual growth rates above 30 percent for the last decade.[61]

Pew's positive projections for clean energy markets are echoed by Deutsche Bank. The bank's analysis is that global warming is accelerating and that greenhouse gas emissions are increasing at the risk of severe economic and environmental consequences. In response to those risks, and in furtherance of energy security, investment in low-carbon growth becomes a significant priority, especially for economic recovery. In undertaking its analyses, Deutsche Bank looks to the necessity of pricing carbon through government policies as well as the necessity of government policies to facilitate the development of new energy technology markets. As the world population continues to increase, demands for energy, food, and water increase and so does the demand for clean energy products and services. Chief among the low-carbon technologies will be new sources of power generation, energy storage, and cleaner buildings and transportation.

Whereas public-sector investments are a necessity, private-sector investments will drive the transition. Institutional investors, such as pension funds that are sensitive to socially responsible and environmentally sound investments, will help shape emerging markets. Nevertheless, as markets emerge investors of all types will be attracted to them. Consequently, Deutsche Bank argues that clean energy investments should be part of any asset allocation and a necessary component of a sound investment portfolio. Its own analysis indicates that 6 percent of its investments should be allocated to climate change sectors.[62]

Conclusion

Energy is again headline news. In the middle of the BP oil catastrophe in the Gulf, there were news reports that Abdel Baset Al-Megrahi had been released from prison in August 2009. Mr. Megrahi was the only person convicted in the 1988 bombing of Pan Am Flight 103 over Lockerbie Scotland that killed 270 people. Ostensibly, he was released under Scottish guidelines for "compassionate release." These guidelines were to apply to persons who were expected to live another three months. A year later, Mr. Megrahi was still alive and the subject of speculation that he was released in a trade between Britain and Libya for a multibillion-dollar oil deal that would largely benefit BP.[63] Similarly, when a vice president of one of America's largest oil companies was asked whether the invasion of Iraq was about oil, the answer was "Absolutely, yes."[64]

These stories are recounted only to highlight the scope and the scale of fossil fuel energy and its politics. To be sure, oil politics is global in nature and involves billions of dollars of investment and opportunities. The same can be said for big coal and big electricity. If energy production and distribution were our only concerns, then we would encounter problems enough.

The energy picture, however, is complicated by another global problem – climate change. Climate change itself makes regular headline news as an "ice island" four times the size of Manhattan has recently broken off from Greenland.[65]

The premise of this book is that regardless of one's position on climate change, domestic energy policy must be transformed.[66] Domestically, transforming energy policy and paying attention to climate change are complementary activities. There is no necessary conflict between them. The global politics of energy and climate change, however, differ and require international cooperation and agreement.[67] As the world's largest energy consumer and largest polluter, the United States plays a world role regardless of its desires. Yet, American Triumphalism aside, the United States should play an international role, as should the European Union, China, India, other world powers, and, indeed, all nations. Copenhagen 2009, however, demonstrated the difficulties in bringing together all nations in an effort to craft a climate change response. Consequently, the prelude to the United States taking an international leadership role is that domestically, it must take seriously the need to put in place a low-carbon energy economy.

The next chapter will conclude with a discussion of political strategies designed to advance the consensus energy policy and lead to a low-carbon future. Suffice it to say here that national energy policy is not being written on a blank slate. Instead, for more than three decades policy thinkers and analysts have provided ample data with which to construct a consensus energy policy. The most recent policy studies have bipartisan support and lay out directions for a new energy economy. Local and state politicians have acted to conserve energy, reduce carbon emissions, and otherwise generate new ways of energy production and consumption. Equally notably, venture and commercial capital is being invested in a variety of new energy technologies and applications. In short, there is ample evidence that public and private sectors are actively participating in its design. We are beginning to see a convergence of political and economic interests in a green economy at the ground level. These several efforts need national coordination and leadership and need comprehensive planning and funding.[68] Hopefully, activities on the ground will serve to educate not only the public but will educate and embolden Washington to act and to lead as well.

Conclusion – Strategies for the Energy Future

I want to make somewhat of a startling assertion that, at present, we have no politics of climate change. In other words, we do not have a developed analysis of the political innovations that have to be made if our aspirations to limit global warming are to become real.

Lord Anthony Giddens[1]

Today, we can echo Lord Giddens and say that the United States has no politics of energy transition, at least, no comprehensive politics most certainly at the federal level. Successfully achieving a transformative energy policy is as much about paradigm and systems change as it is about crafting new substantive provisions. Moving away from fossil fuels to a low-carbon energy strategy is quite simple to articulate. It is more difficult, however, to manage the change in a multibillion-dollar, century-old economic and political energy system that continues to influence the way we think about our energy needs. The government has not banned offshore drilling in deep waters. Consumers have not stopped buying SUVs. There is no federal plan for financing the smart grid. And, environmentalists have not stopped opposing wind and solar projects. All serve as examples of the reluctance to abandon old fuel habits to the benefit of fossil fuel incumbents.

Systems change can occur if we accurately analyze the challenge and then apply the proper regulatory and market responses. Three themes emerge if we are to succeed. First, the endpoint is low-carbon energy, and it is not necessarily dependent on accepting either the science or the rhetoric of climate change. Energy policy must be transformed regardless of how one perceives the warming of the earth. Second, future energy policy is about integration, not separation. Historically, energy and the environment were separate ways of viewing the world. Similarly, fossil fuel industries are treated independently of each other, yet curry governmental favor and receive its largesse when it serves their best self-interest. A smart energy future recognizes the interrelatedness of energy and environmental strategies. We can treat our oil addiction, for example, best by electrifying the transportation sector with clean energy, thus reducing oil dependence and opening new economies. Third, we must create future energy policy as a transformative moment, not simply as an incremental transition moving slowly away from fossil fuels to new resources. Delay in transformation is not in the best interests of national security, the environment, or the economy.

Throughout the book, a wide variety of policy proposals have been put forward. In Chapter 4, a consensus energy policy was described. This chapter will describe the political strategies necessary for accomplishing an energy transformation. Currently, no one is sanguine that the federal government will move forward with an aggressive energy agenda. Instead, it is the thesis of this book that federal leadership must be prodded from below by the private sector as well as by state and local governments. Nevertheless, a desirable energy policy entails federal participation as well as participation from every sector of society.

By way of illustration, the following principles and policy suggestions are central to a transformation to a clean energy future. At the *federal level*: (1) carbon must be priced either through a carbon tax but more likely through cap-and-trade regime; and (2) federal subsidies must move completely away from fossil fuels and nuclear power to efficiency and renewable energy resources. These two proposals are, at once, the most important and least likely to occur in the short-term.

Nevertheless, federal involvement can occur along a number of fronts including: (3) developing FERC rules on grid modernization and cost allocation; (4) the development of a federal innovation policy in which R&D funding is at least doubled from $20 billion to $40 billion, most of which is targeted for efficiency and renewable energy; (5) federal development of energy efficiency standards; (6) federal adoption of national renewable energy portfolio standards and renewable energy credits; and (7) general public education about the benefits of a new energy economy.

Innovation expenditures should be designed with several principles in mind: (1) expenditures should be technology neutral; (2) funding should be along a continuum from basic science to commercialization, with commercialization and scalability receiving a substantial portion of funding; (3) carbon capture and sequestration must be funded; (4) the transportation sector must wean itself from fossil fuels and rely on electricity and non-petroleum liquids; (5) significant investment in battery technology is also key; and (6) innovation policy must foster broadly designed public/private and for profit/nonprofit partnerships and consortia.

At the *state level*, utility regulators play a crucial role in: (1) promoting grid connections for renewable resources; and (2) for designing rate treatment that encourages the use of renewable resources and energy efficiency. States can also participate in: (3) regional carbon emission reduction plans; (4) setting efficiency standards for appliances, buildings, and vehicle fleets; (5) redesigning land use policies to promote green and efficient development; and (6) ramping up renewable portfolio standards.

Finally, the *private sector* must: (1) continue and increase its investment in energy innovations; and (2) continue commercial lending for clean energy initiatives. Further, the private sector, (3) through public education, can promote a better understanding of climate science and energy transformation. Within the private sector, (4) research institutes and centers can be supported with both funding and private participation; and (5) public utilities and other energy providers must redesign their business models with a view to leading the energy transition in a more competitive environment.

These two dozen policy proposals are mainstream. Now, political will power and political strategies must be galvanized to make the transition happen.

Traditional energy policy is dead and should be pronounced so. The bureaucracy sustaining it should be taken off life support. A reconceived energy policy will come about because of the proliferation of new actors and new markets stimulated, but not controlled, by government financial incentives and supporting regulations. Future energy politics, then, will be progressive as to new ideas; non-partisan as to their implementation; and pragmatic as to its solutions.

The elements of a future energy policy have coalesced substantially over the last generation. The contours and variables of that new policy are quite clear, the justifications for transition are equally so, so too are the regulatory and economic next steps to be taken. Still, serious and substantial strategic political work needs to be done. First, the national conversation about energy and the environment, and their nexus to the economy, must be clarified so that policy discussion is intelligible and understandable. Second, political failures must be cured so that the necessary legislation can be passed. Third, bureaucratic failures must also be cured so that effective regulations can be implemented.

The National Conversation

Energy and climate change, like so much else in our society, have become highly politicized, thus making reasonable conversation difficult.[2] Public opinion polls show a confused division among the American public between their beliefs about future energy policy and their beliefs about the environment. Polls demonstrate that more Americans believe that America's and the world's environments are worsening. Yet, going behind the numbers, there is a deep split along political party lines; 52 percent of Democrats are of the opinion that climate change is occurring and 51 percent believe it is caused by human behavior. Republican opinion, however, indicates that only 15 percent believe climate change is occurring and only 21 percent believe it is caused by human behavior.[3] The 2010 midterm elections have made conversation even more difficult. Half of the incoming House Republicans, for example, deny the science behind global warming as an article of political faith and plan to hold multiple hearings challenging the science behind EPA's actions.[4]

When the conversation turns to energy, though, there is a firm consensus about the need for a new energy policy. In a June 2010 New York Times poll, for example, 89 percent said that fundamental changes are needed in U.S. energy policy and 61 percent said that alternatives to oil should be developed with in twenty-five years.[5]

Aside from the partisan divide, public opinion remains unsettled and conflicted about the relationship between energy and the environment. In a 2008 poll, for example, more than 90 percent of Americans believe that the United States should act to reduce global warming even if those actions have economic costs.[6] Yet, in 2010 polls conducted by the same organizations, opinion had shifted. In November 2008, 71 percent polled believe that global warming was occurring, and in June 2010 only 61 percent believed it to be the case. Similarly, there has been a decline in public opinion regarding the belief that global warming is occurring as a result of human action; fewer people are worried about global warming; and more people believe, against the evidence, that there is disagreement among scientists regarding global warming.[7] Americans see global warming as a less significant priority in 2010 than they did in 2008.

Public opinion, though, is increasing that clean energy should be a high priority on the American agenda. Polls indicate notable support for funding renewable energy research, for tax rebates for efficient cars and solar panels, for regulation of carbon dioxide, and for increasing building efficiency. A public consensus exists that energy policy should promote energy efficiency, create jobs and new energy sources, and protect national security.[8] At the same time, Americans support offshore drilling and nuclear power.[9] The tension between the hard and soft paths continues.

This conflict between climate change skepticism and growing support for clean energy is both instructive and historic. The conflict indicates that whereas public opinion is more comfortable with the language of clean energy, there remains a need for a broad political and economic conversation with a new language and a new vocabulary about a future energy policy that incorporates environmental concerns. Still, energy advocates and environmental advocates continue to talk past one another and that disconnection is continuing as the polling data discussed previously indicate.[10] The tension between energy and the environment persists.

Historically, energy advocates spoke a language of markets and production while environmentalists spoke a language of conservation and protection. One attempt to bridge the gap between them was the tool of cost-benefit analysis so that energy production and environmental protection could be compared with one another. This effort failed because cost-benefit analysis suffered two crippling flaws.[11] As a practical matter, cost-benefit analysis was politicized and, thus, unreliable. As a theoretical matter, cost-benefit analysis of energy and the environment compared apples to oranges and pitted two sets of values against one another.

At a deeper conceptual level, energy types clothed their actions in the cloak of economic growth just to be opposed by environmentalists who feared it. Both sides of the debate are wrong. Economic market advocates must acknowledge the limits to growth inherent in a fossil fuel economy. Moreover, environmental advocates must acknowledge that "limits to growth" does not translate into "no growth." Indeed, a successful energy future depends on the development of markets and their economies through technological innovation and increased efficiencies.

At the heart of the conflict between energy and environmental proponents was the reluctance to acknowledge the need for trade-offs between the two as well as to acknowledge competing sets of normative values. It cannot be doubted that energy is a vital input to the economy. Nor can it be doubted that unchecked fossil fuel energy production and use impacts not only the environment, it threatens the economy. A low-carbon energy economy can bridge these two disciplines. Nevertheless, we must be clear about the inevitable trade-offs. Do we delude ourselves, for example, if we are too quick to embrace the continuing use of fossil fuels as "transitional" fuels? Can we continue to use coal without demanding CCS? Can we continue to attempt to expand the domestic production of oil in fragile ecosystems or continue to search for synthetic fuels? Is natural gas truly a promising transitional fuel? Each of these questions entails continued carbon emissions.

Regarding alternatives, though, can the economy afford clean energy? Can the environment afford not to have it? Perhaps most problematically, what should our nuclear future look like?

These questions are posed, in part, to suggest that a half-hearted energy transition may entail too many environmental and economic risks. In other part, the questions are posed to underscore the need to develop a common metric to talk about energy production and environmental protection. The common metric would entail measuring the returns on investments in energy production and in environmental protection. For debate and conversation to move forward, we should be able to calculate the carbon emission consequences for every dollar invested in any energy producing activity. How much carbon is emitted per dollar invested in coal with and without CCS? How much carbon is emitted by oil, synfuels, and natural gas? What are the carbon consequences of solar, wind, and nuclear power? What is the cost of carbon-free energy? On the other side of the ledger, how much energy is produced, or saved, per dollar invested in any existing or emerging technology? What costs are associated with energy efficiency initiatives?

Clearly, the national conversation about energy and the environment continues to reflect the conflict between them. Public education, then, is a necessary and valuable tool to achieve an energy transition.[12] Further, public education must be directed to establishing a common understanding, if not about climate change, then about low-carbon energy. To do so, a common vocabulary based on sound science and economics can help promote clear and consistent policy signals. A common metric and language can help us assess economic trade-offs between how much energy is produced at what levels of pollution. Moreover, a common metric and language can help us understand the investments that need to be made in the portfolio of energy technologies required for a successful transformation.[13]

This book opened with the statement that the country lacks the political will to seriously engage the challenge of creating a low-carbon energy economy.[14] In no small part, the political intransigence on energy is the result of more than a century of fossil fuel policy and the billions of dollars of investment from which it is difficult to walk away. The government/corporate partnership that designed the policy and created its sustaining institutions makes change all the more difficult. Today, we can identify a range of political and regulatory failures that constitute barriers to reform. More hopefully, however, there are cures available. These political and regulatory failures can be overcome through a series of strategies designed to create a new energy policy and advance it through a broad political coalition.[15]

Political Failure and Its Cure

Every generation since the Founding experiences a different relationship between citizens and their government. In alternating cycles, there are periods of relatively activist government followed or preceded by periods of relatively minimalist government. Over the last decade, we have been coming off a cycle of "free-market" enthusiasm and now look to government to actively involve itself in programs for the betterment of society, such as reforming health care and financial markets or cleaning the Gulf of Mexico and its shorelines.

The previous period can be labeled as either the Reagan Revolution, or deregulation, or even neo-liberalism. At the bottom of all of these labels is a belief that markets can cure all ills and that the role of government should be limited to correcting gross market failures and should do so with minimal intrusion. There were two dimensions to this free-market ideology. The first dimension was positive: Markets appeared to work well over the last generation as investment portfolios increased, as digital technologies brought greater convenience and choice, and as economic growth appeared to move continuously upward. Appearances, though, are deceiving. We know, now, that the mistakes of the last generation of government regulation have led to financial, environmental, and other social dislocations.[16] Over the last generation, disparities in the distribution of wealth have increased, public education and health care continue to perplex us, the middle class has lost economic power, and the job creation necessary to lift us out of the recession is very slow in coming.

If all of the problems of the last generation of free-market government were limited to correcting markets, then technical fixes may well be possible. However, the second dimension of free-market ideology was decidedly, and cripplingly, political. It was not enough for market advocates to extol and advocate the virtues of competitive markets. Competitive markets can create wealth and promote innovation. However, competitive markets are difficult to create and more difficult to sustain. At bottom, markets cannot exist without government even if the role of government is limited to enforcing basic legal rules for the creation, transfer, and protection of property rather than through deeper intrusions by the regulatory state. Instead, free-market advocates sought, and received, a government that closed its eyes to market abuses. A political system open to influence by interest groups, by definition and by operation, will distort markets in favor of those interests. We have seen this with fossil fuel industries.

Free-market ideology, then, was not content to advocate for markets, and it went on to demonize government. The "starve the beast" philosophy of the last decades attempted to close down government. These ideologues were relentless in undermining government regulations intended to promote health, safety, and the environment. These social regulations were deemed too costly and, therefore, needless. To bolster the claims of costly regulation, science was undermined, cost-benefit analysis was politicized, and "liberal" became an evil word to be avoided. On the other side of the ledger, economic protections were afforded businesses through limits on liability, the breaking down of barriers to trade in opaque financial instruments, and the exultation of property rights. As part of the assault on government, even the judiciary and the legal profession were not immune from attack. "Judicial activism" became synonymous with liberalism, and liberalism was a code word for redistribution. The irony, of course, is that free-market ideology has caused an historic redistribution of wealth to the top 1 percent in the country.[17] At its core, free-market ideology was more about politics than economics. Politics became determinative of economic policy to the detriment of both.[18]

The correction for the ills of free-market ideology is not to be found in rejecting markets nor is it to be found in substituting more government for other social institutions. Instead, government must be seen as an active participant in shaping social policy, and unnecessary and cumbersome regulations must be reformed or eliminated. It is not government or markets; it is the proper mix of both. Government can create and support markets where necessary, but more importantly it must convene a broad array of actors and must lead and facilitate an active conversation about the common good.[19] If forced to put a label on an emerging, or a reemerging, ideology the label "pragmatism" fits. Pragmatic government should be bipartisan, dedicated to solving problems and preventing catastrophes, and forward thinking. The necessity of a forward-thinking government is starkly presented by the need for a new energy policy.

A pragmatic political solution to energy policy takes capitalism and markets seriously. Transitioning to a low-carbon energy economy does not mean abandoning a desire for economic growth or the improved well-being of the citizenry. It does not embrace a doomsday view of limits to growth. Instead, the new energy politics sees markets as the *deus ex machina* for a successful transformation. Innovative energy technologies are a necessary component for achieving energy and economic security and environmental protection. Further, a clean energy economy can create jobs and raise standards of living.[20]

Although the education reforms hoped for by the No Child Left Behind legislation have not been deemed a success, when passed it was hailed as an example of progressive bipartisanship. Energy reform should also be seen as such. No sector of the economy is unaffected by energy. No economic class is unaffected. Nor is any political party. Both sides of the political aisle have participated in crafting traditional energy policy, and now both sides must take responsibility for its reform. The consensus energy policy recognizes a commonality of interests in energy reform; recognizes the need for pragmatic solutions; and recognizes the potential benefits across the political and economic spectrums. For an effective pragmatic political philosophy to emerge successfully, government regulation must likewise be reformed.

Regulatory Failure and Its Cures

Problems as complex, multi-disciplinary, and inter-generational as an energy transition and climate change deserve a regulatory regime worthy of them. Today that regime is lacking. Over the last 100 years of modern government regulation, we have fallen into a trap. Past problems were perceived and addressed in a similar way; we can look to Los Angeles smog or the Cuyahoga River catching fire as examples. First, a problem was identified as a market failure. Next, a market fix was crafted. Then an agency was designated to fix the problem. Such an approach, however, of identifying market failures and finding remedies for them is understandable. Quite often, problems are discrete and susceptible to relatively simple and short-term solutions, and the costs and benefits of regulating them are fairly easily discernible. Neither energy policy nor climate change can be characterized as discrete, simple to fix, or susceptible to easy cost-benefit analyses or to single-approach solutions.

A larger failure surrounds the current regulatory culture. Regulatory agencies became politicized silos. The DOE, as an example, is more defense and energy production-oriented than it is concerned about the environment.[21] Agencies, particularly those exercising top-down command-and-control authority, became political battlegrounds often resulting in corporate favoritism to the disadvantage of private citizens. In other words, the public interest gave way to private interests. It is a simple matter of political choice theory that a small number of actors can capture regulatory benefits. Consequently, agencies that were dedicated to a particular problem often looked to the industries it regulated for information, expertise, and revolving-door employment. Agencies thus narrowed their vision and were content to monitor their regulatees within their statutory mandates rather than anticipate new problems or develop creative responses. Further, bureaucratic culture has a disincentive to solve problems; once a problem is solved, then the agency is out of business and regulators are out of jobs. Bureaucratic perpetuation is in a regulator's self-interest. Agencies so constructed have no incentive to coordinate their activities with other regulators, engage in systems thinking, or reduce politicization.

A low-carbon energy policy cannot succeed within such a narrow system. Instead, a regulatory structure must be built that avoids silos, coordinates activities among other agencies, moves away from sustaining incumbents, promotes new entrants, is willing to end programs, and looks to changing regulatory systems rather than solving discrete problems. More specifically, a low-carbon energy policy is as much about the environment and the economy as it is about finding new energy producers and distributors. The White House Office of Energy and Climate Change Policy serves as an example of a crosscutting organization that is charged with addressing energy and the environment in economically sound ways.

The new regulatory regime must do what it can to reduce politicization; increase reliance on good science and on good cost-benefit analyses; and design data-based regulations. Rather than looking ex post to solve problems that have already occurred, regulators must act ex ante with a focus on innovation, investment, and entrepreneurship. To the extent that new forms of energy regulations are market-based, those markets must be based on level playing fields, and fossil fuel subsidies must be redirected to low-carbon activities.[22] Further, free riders, such as low-cost dirty energy producers, should be eliminated, and environmental liability should not be limited. Polluters should pay, and carbon should be priced.

Ex ante regulations must look to public/private partnerships; fund at appropriate levels to attain scale; be willing to sunset programs and even agencies; streamline or fast-track appropriate regulatory requirements; and consider agency integration.[23] Indeed, policy and regulatory integration constitutes the new politics. The idea of integration is neither a form of central planning nor a form of industrial policy. Rather it is an integration of themes, strategies, and actors that are aware of and acknowledge one another. More specifically, political and economic issues surrounding energy and the environment must converge and be shaped into effective policies.[24]

Thematically, the values of energy, the economy, the environment, and security must be seen as complementary, not antithetical. Strategic activities must take place in the public and private, as well as the profit and nonprofit, sectors. Additionally, regulators must use as much leverage as possible and try to maximize network effects. Government agencies must understand these themes, must accommodate these several actors, and must operate transparently. Further, regulators must be accountable, and they must be held to verifiable benchmarks. In many instances, benchmarks will be identifiable and understandable. We can measure the creation of markets and the number of new entrants into them. We can measure innovations created and brought to scale. We can measure enforcement actions. Integration of themes, actors, policies, and strategies should operate along a broad front from multiple directions and multiple levels for the purpose of changing the regulatory system responsible for transforming energy policy.

Our energy future no longer resides in fossil fuels; it resides in a substantial ramping up of energy efficiency and renewable resources. On the one hand, the challenges of climate change may span the timeframe of a century. On the other hand, we must act within a generation or the costs may well be beyond our ability to pay them.

Notes

Introduction

1. *See, e.g.,* Charles F. Sabel & William H. Simon, *Minimalism and Experimentalism in the Administrative State* (April 28, 2010) available at http://papers.ssrn.com/sol3/papers.cfm?abstract_id=1600898

2. DAVID W. ORR, DOWN TO THE WIRE: CONFRONTING CLIMATE COLLAPSE 28 (2009).

3. Joseph P. Tomain & Sidney A. Shapiro, *Analyzing Government Regulation*, 49 ADMIN. L. REV. 377 (1997).

4. SIDNEY A. SHAPIRO & JOSEPH P. TOMAIN, REGULATORY LAW AND POLICY (3RD ED. 2003).

5. *See, e.g.,* Steven Ferry, et al., *FIT in the USA: Constitutional Questions About State-Mandated Renewable Tariffs*, 148 PUB. UTIL. FORT. 61 (June 2010).

6. "To work effectively, in other words, markets have always required energetic, flexible, and imaginative government to set the rules, level the playing field, enforce the law, and protect the larger public good over the long term." ORR at 37. *See also* CHARLES LINDBLOM, THE MARKET SYSTEM: WHAT IT IS, HOW IT WORKS, AND WHAT TO MAKE OF IT (2001); JOHN CASSIDY, HOW MARKETS FAIL: THE LOGIC OF ECONOMIC CALAMITIES (2009).

7. CHRISTOPHER BEDDOR, ET AL., SECURING AMERICA'S FUTURE: ENHANCING OUR NATIONAL SECURITY BY REDUCING OIL DEPENDENCE AND ENVIRON-MENTAL DAMAGE 16 (2009) available at http://www.americanprogress.org/issues/2009/08/pdf/energy_security.pdf ("For more than three decades the United States has repeatedly erred on the side of inaction. Policymakers, rather than pursuing long-term sustainable goals, were swayed by the prospect of immediate benefits or political risks. And the public was too focused on falling prices at the pump after successive energy crises to see the bigger picture of ever-escalating oil imports.")

8. *See* Testimony of Sidney A. Shapiro Before the Subcommittee on Administrative Oversight and the Courts of the Senate Committee on the Judiciary, Hearing on Protecting the Public Interest: Understanding the Threat of Agency Capture (August 3, 1 2010) available at http://www.progressivereform.org/articles/Shapiro_AgencyCapture_080310.pdf; *see also* BARBARA PRAETORIUS, ET AL., INNOVATION FOR SUSTAINABLE ELECTRICITY SYSTEMS: EXPLORING THE DYNAMICS OF ENERGY TRANSITIONS (2009), ch. 3.

9. Stuart Eizenstat, *The U.S. Role in Solving Climate Change: Green Growth Policies Can Enable Leadership Despite the Economic Downturn*, 30 ENERGY L.J. 1 (2009).

10. ANTHONY GIDDENS, THE POLITICS OF CLIMATE CHANGE 2 (2009).

11. Jonathon B. Weiner, *Radiative Forcing: Climate Policy to Break the Logjam in Environmental Law*, 17 N.Y.U. ENVT'L L.J. 210 (2008).

12. Union of Concerned Scientists, *Scientific Consensus on Global Warming* available at http://www.ucsusa.org/ssi/climate-change/scientific-consensus-on.html; INTERGOVERMENTAL PANEL ON CLIMATE CHANGE, CLIMATE CHANGE 2007: SYNTHESIS REPORT 36–41 (2007).

13. Anders Enkvist, Tomas Naucler, & Jens Riese, *What Countries Can Do About Cutting Carbon Emissions*, MCKINSEY QUARTERLY (2008) (estimating that changes to address carbon emissions can be achieved at a cost of 0.6% to 1.4% of global GDP by 2030). *See also*, NICHOLAS STERN, THE ECONOMICS OF CLIMATE CHANGE: THE STERN REVIEW (2007).

14. RICHARD A. POSNER, CATASTROPHE: RISK AND RESPONSE 43–45, 253–60 (2004). "Turning to global warming, we again confront a danger that seems to lie comfortably in the future. But in this case a wait-and-see policy would be perilous." *Id.* at 253. *But see* CASS R. SUNSTEIN, WORST-CASE SCENARIOS 214–18 (2007) (criticizing Posner's analysis).

Chapter 1

1. Regulatory capture (a form of public choice) and public interest regulation are the dominant theories of government regulation, and there is substantial evidence and examples to support both theories. For purposes of this chapter, however, I am agnostic about which is the superior theory because the regulatory demands of changing energy policy or responding to global warming challenge both theories. Useful discussions of both theories can be found in George L. Priest, *The Origins of Utility Regulation and the "Theories of Regulation" Debate*, 36 J. LAW & ECON. 289 (1993); STEVEN P. CROLEY, REGULATION AND PUBLIC INTERESTS: THE POSSIBILITY OF GOOD REGULATORY GOVERNMENT (2008); SIDNEY A. SHAPIRO & JOSEPH P. TOMAIN, REGULATORY LAW AND POLICY: CASES AND MATERIALS ch. 3 (3rd ed. 2003).

2. JEFFREY D. SACHS, COMMON WEALTH: ECONOMICS FOR A CROWDED PLANET 5 (2008).

3. Tom Zeller, Jr., *A Battle in Mining Country Pits Coal Against Wind*, N.Y. TIMES (August 14, 2010) available at http://www.nytimes.com/2010/08/15/business/energy-environment/15coal.html.

4. The difference between 20 mbd of petroleum consumption and the 15 mbd of crude oil consumption consists of blends, oil distillates, and liquefied petroleum gases.

5. For a history of the development of world energy, *see* JOHN GARRETSON CLARK, THE POLITICAL ECONOMY OF WORLD ENERGY: A TWENTIETH-CENTURY PERSPECTIVE (1991).

6. *The Standard Oil Co. of New Jersey v. United States*, 221 U.S. 1 (1911).

7. Daniel Yergin, The Prize: The Epic Quest for Oil, Money, and Power 113 (1991).

8. Richard D. Cudahy & William D. Henderson, *From Insull to Enron: Corporate (Re)Regulation After the Rise and Fall of Two Energy Icons*, 26 Energy L.J. 35 (2005).

9. Jerry L. Mashaw, *Recovering American Administrative Law: Federalism Foundations, 1787–1801*, 115 Yale L.J. 1256 (2006); Jerry L. Mashaw, *Administration and "The Democracy": Administrative Law From Jackson to Lincoln, 1829–1861*, 117 Yale L. J. 1568 (2008); Robert L. Rabin, *Federal Regulation in Historical Perspective*, 38 Stan. L. Rev. 1189 (1986).

10. *Munn v. Illinois*, 94 U.S. 113 (1877).

11. *Id. at* 127–30. *See also Fed. Power Comm'n v. Hope Natural Gas Co.*, 320 U.S. 591 (1944); *Missouri ex rel. Southwestern Bell Telephone Co. v. Pub. Serv. Comm'n of Missouri*, 262 U.S. 276 (1923); *Smyth v. Ames*, 169 U.S. 466, modified by *Smyth v. Ames*, 171 U.S. 361 (1898); *Jersey Cent. Power & Light Co. v. Fed. Energy Regulatory Comm'n*, 810 F.2d 1168 (D.C. Cir. 1987).

12. *See, e.g.*, Paul A. Samuelson & William D. Nordhous, Economics 911 (12th ed. 1985); Sanford V. Berg & John Tschirhart, Natural Monopoly Regulation: Principles and Practice (1988); James Cummings Bonbright, Albert L. Danielsen & David R. Kamerschen, Principles of Public Utility Rates 17–24, 33–36 (1988); Michael A. Crew & Paul R. J. Kleindorfer, The Economics of Public Utility Regulation (1986); Stephen G. Breyer, Regulation and Its Reform 15–19, 30–31 (1982); Alfred E. Kahn, The Economics of Regulation: Principles and Institutions Volume I 11–12 (ed. 1991); Alfred E. Kahn, the Economics of Regulation: Principles and Institutions Volume II 113–71 (ed. 1988); Richard A. Posner, *Natural Monopoly and Its Regulation*, 21 Stan. L. Rev. 548 (1969). *Contra see* Harold Demsetz, *Why Regulate Utilities?* 11 J. of Law & Econ. 55 (1968); Peter Z. Grossman & Daniel H. Cole (eds.), The End of a Natural Monopoly: Deregulation and Competition in the Electric Power Industry (2003).

13. *Omega Satellite Prods. Co. v. City of Indianapolis*, 694 F.2d 119, 126 (7th Cir. 1982) (Posner, J.) ("The cost of the cable grid appears to be the biggest cost of a cable television system and to be largely invariant to the number of subscribers the system has.... [O]nce the grid is in place ... the cost of adding another subscriber probably is small. If so, the average cost of cable television would be minimized by having a single company in any given geographical area; for if there is more than one company and therefore more than one grid, the cost of each grid will be spread over a smaller number of subscribers, and the average cost per subscriber, and hence price, will be higher. If the foregoing accurately describes the conditions in Indianapolis ... it describes what economists call a 'natural monopoly,' wherein the benefits, and indeed the very possibility, of competition are limited.")

14. Joseph P. Tomain, *The Dominant Model of United States Energy Policy*, 61 U. Colo. L. Rev. 355 (1990).

15. *Id.* at 363.

16. JOHN GARRETSON CLARK, ENERGY AND THE FEDERAL GOVERNMENT: FOSSIL FUEL POLICIES, 1900–1946, 25–26, 259 (1987).

17. BRUCE ACKERMAN, WE THE PEOPLE: FOUNDATIONS chs. 3–5 (1991); BRUCE ACKERMAN, WE THE PEOPLE: TRANSFORMATIONS chs. 9–10 (1998); G. EDWARD WHITE, THE CONSTITUTION AND THE NEW DEAL (2nd ed. 2002).

18. ALAN BRINKLEY, THE END OF REFORM: NEW DEAL LIBERALISM IN RECESSION AND WAR (1995).

19. PETER H. IRONS, THE NEW DEAL LAWYERS (1982); JOSEPH P. LASH, DEALERS AND DREAMERS: A NEW LOOK AT THE NEW DEAL (1988).

20. *Panama Refining Co. v. Ryan*, 293 U.S. 388 (1935).

21. Public Utility Holding Company Act of 1935, 15 U.S.C. §§ 79a et seq. (Repealed 2005).

22. Richard D. Cudahy & William D. Henderson, *From Insull to Enron: Corporate (Re)Regulation After the Rise and Fall of Two Energy Icons*, 25 ENERGY L.J. 35 (2005).

23. Federal Power Act, 16 U.S.C. §§ 791a et seq.

24. Natural Gas Act of 1938, 15 U.S.C. §§ 717 et seq.

25. Oil pipelines were similarly regulated by the then Interstate Commerce Commission under the 1906 Hepburn Act, 34 Stat. 584.

26. *Phillips Petroleum Co. v. Wisconsin*, 347 U.S. 672 (1954).

27. U.S. Senate Subcomm. Admin. Practice and Procedure, *Report on Regulatory Agencies to the President-Elect* 55 (December 1960) (also know as the Landis Report).

28. Although the economic signals clearly pointed producers to the intrastate markets, natural gas producers who had "dedicated" their gas to the interstate market were constrained from leaving that market. Thus, they tried to simply hold back gas from that market. SHAPIRO & TOMAIN at ch. 5.

29. PAUL W. MACAVOY, THE NATURAL GAS MARKET: SIXTY YEARS OF REGULATION AND DEREGULATION (2000); ARLON R. TUSSING & CONNIE C. BARLOW, THE NATURAL GAS INDUSTRY: EVOLUTION, STRUCTURE, AND ECONOMICS (1984); ELIZABETH M. SANDERS, THE REGULATION OF NATURAL GAS: POLICY AND POLITICS, 1938–1978 (1981); Stephen Breyer & Paul W. MacAvoy, *The Natural Gas Shortage and Regulation of Natural Gas Producers*, 86 HARV. L. REV. 941 (1973).

30. VITO A. STAGLIANO, A POLICY OF DISCONTENT: THE MAKING OF A NATIONAL ENERGY STRATEGY 10 (2001).

31. STAGLIANO at 10–12.

32. Atomic Energy Act of 1964, 42 U.S.C. §§ 2011 et seq.

33. Price-Anderson Act of 1957, 42 U.S.C. §§ 2210 et seq.

34. STAGLIANO at 18.

35. I write "closer" to market levels because in actuality there was no competitive oil market. On the one hand, the Middle East cartel controlled a significant amount of oil production, and on the other hand, oil supplies were negotiated by government officials. "In 1973, the buying and selling of oil was carried out mainly through diplomatic arrangements among government officials. The oil market resembled a geopolitical chess board on which, internationally, U.S. and European agents alternately jockeyed and colluded for political and economic

influence in the Persian Gulf, and where domestically the accommodation of federal and state governmental interests took precedence over the efficient delivery of fuel to consumers." *Id.* at 26.

36. Joseph P. Kalt, The Economics and Politics of Oil Price Regulation: Federal Policy in the Post-Embargo Era (1981); David Glasner, Politics, Prices, and Petroleum: The Political Economy of Energy (1985); Robert Sherrill, The Oil Follies of 1970–1980: How the Petroleum Stole the Show (and Much More Besides) (1983).

37. Stagliano at 19–22.

38. Energy Supply and Environmental Coordination Act of 1974, 15 U.S.C. §§ 791 et seq.

39. Energy Policy and Conservation Act of 1975, 42 U.S.C. §§ 6291 et seq.

40. Energy Conservation and Production Act, Pub. L. 94–385, 90 Stat. 1125.

41. Joseph P. Tomain & Richard D. Cudahy, Energy Law: In a Nutshell 232–33 (2004); Stagliano at 27–30.

42. Joseph P. Tomain, *Institutionalized Conflicts Between Law and Policy*, 22 Houston L. Rev. 661 (1985); Alfred C. Aman, Jr., *Institutionalizing the Energy Crisis: Some Structural and Procedural Lessons*, 65 Cornell L. Rev. 491 (1980); Clark Byse, *The Department of Energy Organization Act: Structure and Procedure*, 30 Admin. L. Rev. 193 (1978).

43. President's Address to the Nation, Pub. Papers 656 (April 18, 1977).

44. National Energy Act of 1978, Pub. L. 95–618.

45. Power Plant and Industrial Fuel Use Act of 1978, Pub. L. No. 95–620, 92 Stat. 3289.

46. Public Utility Regulatory Policies Act of 1978, Pub. L. No. 95–617, 92 Stat. 3117.

47. National Energy Conservation Policy Act of 1978, Pub. L. No. 95–619, 92 Stat. 3206; The Energy Tax Act of 1978, Pub. L. No. 95–618, 92 Stat. 3174.

48. Tomain & Cudahy at 271–74.

49. *FERC v. Mississippi*, 456 U.S. 742 (1982) (upholding the constitutionality of PURPA

50. Richard D. Cudahy, *PURPA: The Intersection of Competition and Regulatory Policy*, 16 Energy L.J. 419 (1995).

51. The full avoided cost requirement was upheld as constitutional in *Am. Paper Inst. v. Am. Elec. Power Serv. Corp.*, 461 U.S. 402 (1983).

52. President's Address to the Nation, Pub. Papers 609 (April 5, 1979).

53. The Crude Oil Windfall Profits Tax, Pub. L. No. 96–223, 94 Stat. 229 (1979).

54. Report of the President's Commission on the Accident at Three Mile Island, the Need for Change: The Legacy of TMI (1979).

55. Nuclear Energy Institute, *New Nuclear Plants* available at http://www.nei.org/keyissues/newnuclearplants/

56. President's Address to the Nation, Pub. Papers 1235 (July 15, 1979).

57. Biomass and Alcohol Fuels Act, Pub. L. No. 96–294, 94 Stat. 683.

58. Renewable Energy Resources Act, Pub. L. No. 96–294, 94 Stat. 715.

59. Solar Energy and Energy Conservation Act of 1980, Pub. L. No. 96–294, 94 Stat. 719.

60. Geothermal Energy Act, Pub. L. No. 96–294, 94 Stat. 763.

61. United States Synthetic Fuels Corporation Act, Pub. L. No. 96–294, 94 Stat. 633.
62. Joseph T. Kelliher & Maria Farinella, *The Changing Landscape of Federal Energy Law*, 61 Admin. L. Rev. 611, 634–50 (2009).
63. NaturalGas.org, Industry and Market Structure, available at http://www.natu-ralgas.org/business/industry.asp#industry
64. FERC promulgated rules that eliminated mandatory minimum charges for customers, FERC Order No. 380, FERC Stats. & Regs. ¶ 30,571 (1984) affirmed in *Wisconsin Gas Co. v. FERC*, 770 F.2d 1144 (D.C. Cir. 1985); allowed producers to sell gas to entities other than the pipeline with which it had a contract (known as the special marketing program, a producer and the pipeline could amend their contract so that a producer could sell to someone else and credit the proceeds of the sale to the pipeline contract. Producers who received income and pipelines were relieved of their contractual obligation. However, consumers paid the higher price under this arrangement, and the courts required FERC to revisit its special marketing program), *Maryland People's Counsel v. FERC*, 761 F.2d 768 (D.C. Cir. 1985); *Maryland People's Counsel v. FERC*, 761 F.2d 780 (D.C. Cir. 1985); *Maryland People's Counsel v. FERC*, 768 F.2d 450 (D.C. Cir. 1985); and separated the merchant and transportation roles of pipeline companies, FERC Order No. 436, FERC Stats. & Regs. ¶ 30,665 (1985); FERC Order No. 436-A, FERC Stats. & Regs. ¶ 30,675 (1985); FERC Order No. 436-b, FERC Stats. & Regs. ¶ 30,688 (1986). This last series of rules was intended to serve as an initial comprehensive package aimed at pipelines and designed to have them transport gas to third parties as well as to their own customers on a non-discriminatory basis and generally open access for producers and consumers. Most of the Order No. 436 series of rules was upheld in *Associated Gas Distribs. v. FERC*, 824 F.2d 981 (D.C. Cir. 1987).
65. FERC Order No. 636, FERC Stats. & Regs. ¶ 30,939 (1992); FERC Order No. 636-A, FERC Stats.& Regs. ¶ 30,950 (1992) (most of this order was upheld in *United Gas Distribution Companies v. FERC*, 88 F.3d 1105 (D.C. Cir. 1996)). *See also Interstate Natural Gas Ass'n of Am. v. FERC*, 285 F.3d 18 (2002) (upholding most of the regulatory modifications to Order No. 636).
66. James H. McGrew, Federal Energy Regulatory Commission ch. 7 (2nd ed. 2009).
67. MacAvoy at ch. 6; Paul W. MacAvoy, The Unsustainable Costs of Partial Deregulation ch. 3 (2007) (arguing that deregulation of the natural gas market has not gone far enough and the failure to do so has imposed significant consumer losses [over $600 million estimated from 1995–2010]).
68. FERC Order No. 888, FERC Stats. & Regs. ¶ 31,036 (1996).
69. FERC Order No. 2000, 89 FERC ¶ 61,285 (1999).
70. Federal Energy Regulatory Commission, White Paper: Wholesale Power Market Platform (2003) available at http://www.nwcouncil.org
71. Walter R. Hall II, et al., *History, Objectives, and Mechanics of Competitive Electricity Markets*, in Capturing the power of Electric Restructuring 1, 13 (Joey Lee Miranda ed., 2009).
72. Jacqueline Lang Weaver, *Can Energy Markets Be Trusted? The Effect of the Rise and Fall of Enron on Energy Markets*, 4 houston Bus. & Tax L.J. 1 (2004); Nancy B. Rapoport, Jeffrey D. Van Niel, & Bala G. Dharan, Enron

AND OTHER CORPORATE FIASCOS: THE CORPORATE SCANDAL READER (2nd ed. 2008).

73. For a general discussion of expired rate caps resulting in increased prices for consumers, *see* GREGORY BASHEDA ET AL., WHY ARE ELECTRICITY PRICES INCREASING? AN INDUSTRY-WIDE PERSPECTIVE 96 (2006) available at http://www.eei.org/ourissues/finance/Documents/Brattle_Report.pdf

74. Energy Policy Act of 1992, P.L. No. 102–486, 100 Stat. 2776.

75. NATIONAL ENERGY POLICY DEVELOPMENT GROUP, NATIONAL ENERGY POLICY: RELIABLE, AFFORDABLE, & ENVIRONMENTALLY SOUND ENERGY FOR AMERICA'S FUTURE (May 2001) available at

76. Energy Policy Act of 2005, Pub. L. 109–58.

77. Energy Independence and Security Act of 2007, Pub. L. 110–40.

78. U.S. House of Representatives Committee on Oversight and Government Reform, *Energy Policy Under the Bush Administration: A Record of Failure*, available at http://www.majorityleader.gov/docUploads/AdministrationEnergyFailures091608.pdf

79. *See, e.g.,* Press Release, Office of the Press Secretary, *President Signs Energy Policy Act* (Aug. 8, 2005) available at http://georgewbush-whitehouse.archives.gov/news/releases/2005/08/20050808-6.html

80. U.S. House of Representatives Committee on Government Reform, *Flash Report: Key Impacts of the Energy Bill-H.R.* 6 (July 2005) available at http://henrywaxman.house.gov/uploadedfiles/rep.pdf

81. Bloomberg.com, *U.S. Energy Industry's Lobbying Pays Off With 411.6 Bln in Aid* (July 27,2005) available at http://www.bloomberg.com/apps/news?pid=10000103&cisd-agbeVim04Ec&refer=us; U.S. PIRG & Friends of the Earth, *Final Energy Tax Package Overwhelmingly Favors Polluting Industries* (July 27, 2005) available at http://newenergyfuture.com/final2005energybilltaxanalysis.pdf

82. Roger Stark, *A Continuing Reign of Incoherence*, PUB. UTIL. FORT. 51 (Dec. 2005) ("U.S. energy stakeholders have for too long been fooled into believing that patchwork reforms are a substitute for coherent policies. The Energy Policy Act of 2005 (EPACT) is the latest, and hopefully the last, example of this tradition.")

83. Cudahy & Henderson at 37–38 ("[D]uring a financial bubble driven by rapid changes in network industries ... regulatory officials will inevitably buckle under political pressure and (a) fail to issue new rules that might interfere with financial 'hijinks' and (b) fail to vigorously enforce laws already on the books.").

84. Joseph P. Tomain, *To a Point*, 52 LOYOLA L. REV. 1201, 1210–16 (2006).

85. CONGRESSIONAL RESEARCH SERVICE, ENERGY INDEPENDENCE AND SECURITY ACT OF 2007: A SUMMARY OF MAJOR PROVISIONS (2007) available at http://energy.senate.gov/public/_files/RL342941.pdf

86. American Recovery and Reinvestment Act of 2009, Pub. L. 111–5.

87. Congressional Research Service, *H.R. 2454: American Clean Energy and Security Act of 2009* available at http://www.govtrack.us/congress/bill.xpd?bill=h111-2454&tab=summary

88. Brad Johnson, *The Wonk Room: A First Look at the Details of the Kerry-Lieberman American Power Act* (May 12, 2010) available at http://wonkroom.thinkprogress.org/2010/05/12/kerry-lieberman/

89. Matthew L. Wald, *In the Heartland, Still Investing in Coal*, N.Y. Times F1 (November 17, 2010); Clifford Krauss, *There Will Be Fuel: New Oil and Gas Sources Abound, but They Come With Costs*, N.Y. Times F1 (November 17, 2010).

Chapter 2

1. Press Release, Office of the Press Secretary, *President Signs Energy Policy Act* (August 8, 2005) available at http://georgewbush-whitehouse.archives.gov/news/releases/2005/08/20050808–6.html
2. Amory B. Lovins, Soft Energy Paths: Toward A Durable Peace (1977), ch. 2.
3. U.S. Energy Information Administration, *Who Are the Major Players Supplying the World Oil Market?* (January 28, 2009) available at http://tonto.eia.doe.gov/energy_in_brief/world_oil_market.cfm
4. For a list of the top private companies in the world, see Forbes.com, *Special Report: The Global 2000* (April 8, 2009) available at http://www.forbes.com/lists/2009/18/global-09_The-Global-2000_Rank.html
5. Nuclear Energy Institute, Status and Outlook for Nuclear Energy in the United States 1 (2009) available at http://www.nei.org/resourcesandstats/documentlibrary/reliableandaffordableenergy/reports/statusreportoutlook/
6. *See, e.g.*, Jeff Goodell, Big Coal: The Dirty Secret Behind America's Energy Future (2006); *see also Caperton v. A. T. Massey Coal Co., Inc.*, 129 S.Ct. 2252 (2009) (discussing the influence of the defendant coal company in the election of a West Virginia Supreme Court justice).
7. Nuclear Energy Institute, The Cost of New Generating Capacity in Perspective 5 (2009) available at http://www.nei.org/financialcenter/industrydata/
8. David Schlissel, Allison Smith & Rachel Wilson, Coal-Fired Power Plant Construction Costs 1 (2008) available at http://www.synapse-energy.com/Downloads/SynapsePaper.2008–07.0.Coal-Plant-Construction-Costs.A0021.pdf
9. *See* The Brattle Group, *Brattle Group Estimates $1.5 Trillion Needed in Utility Infrastructure Investment Through 2030* (April 21, 2008) available at http://tdworld.com/business/brattle-group-infrastructure-investment-findings/
10. *See* Joseph P. Tomain, *Smart Energy Paths: How Willie Nelson Saved the Planet*, 36 Cumberland L. Rev. 417 (2006).
11. *See* Joseph P. Tomain, *"To A Point,"* 52 Loyola L. Rev. 1201 (2006); Herman E. Daly, *Economics in a Full World*, Sci. Am., Sept. 2005, at 100; Lovins, at chs. 1–2; Jeffrey D. Sachs, Common Wealth: Economics for a Crowded Planet 29–30 (2008).
12. *See, e.g.*, James Russell, *Fossil Fuel Production Up Despite Recession* (October 15, 2009) (fossil fuel production up 2.9% in 2008) available at http://vitalsigns.worldwatch.org/vs-trend/fossil-fuel-production-despite-recession
13. As of October 2009, more than thirty nuclear units were in some stage of planning or licensing. For a list of these plants, see Nuclear Energy Institute,

New Nuclear Plant Statistics available at http://www.nei.org/resourcesandstats/ documentlibrary/newplants/graphicsandcharts/newnuclearplantstatus/

14. Matthew L. Wald, *U.S. Rejects Nuclear Plant Over Design of Key Piece*, N.Y. TIMES, October 16, 2009, at A13 (discussing how the U.S. Nuclear Regulatory Commission rejected a new reactor design, which was intended to lower construction costs and serve as a prototype plant).

15. LOVINS at 27 (footnote omitted).

16. NATIONAL PETROLEUM COUNCIL, FACING THE HARD TRUTHS ABOUT ENERGY: A COMPREHENSIVE VIEW TO 2030 OF GLOBAL OIL AND NATURAL GAS 5 (2007) available at http://www.npchardtruthsreport.org/

17. *See, e.g.*, DAVID W. ORR, DOWN TO THE WIRE: CONFRONTING CLIMATE COLLAPSE 7 (2009). ("[D]ecades of such governmental and political failure have brought us uncomfortably close to the brink of global collapse.")

18. PETER W. HUBER & MARK P. MILLS, THE BOTTOMLESS WELL: THE TWILIGHT OF FUEL, THE VIRTUE OF WASTE, AND WHY WE WILL NEVER RUN OUT OF ENERGY xii, xxvi (2005). *See also* chapter 7 in which the authors argue in favor of a direct and positive correlation between energy consumption and GDP: "[I] t's always obvious: supply the worker with more, better-ordered power, and he produces more; if the new machines are different enough, both the factory's and the worker's productivity improve beyond recognition," at 129.

19. COUNCIL ON ENVIRONMENTAL QUALITY & U.S. DEPARTMENT OF STATE, THE GLOBAL 2000 REPORT TO THE PRESIDENT: ENTERING THE 21ST CENTURY (1980).

20. JULIAN L. SIMON & HERMAN KAHN (EDS.), THE RESOURCEFUL EARTH: A RESPONSE TO GLOBAL 2000 25 (1984).

21. S. FRED SINGER (ED.), FREE MARKET ENERGY: THE WAY TO BENEFIT CONSUMERS 43 (1984).

22. *See, e.g.*, JACOB WEISBERG, IN DEFENSE OF GOVERNMENT: THE FALL AND RISE OF PUBLIC TRUST (1996); STEVEN P. CROLEY, REGULATION AND PUBLIC INTERESTS: THE POSSIBILITY OF GOOD REGULATORY GOVERNMENT (2008); MIKE FEINTUCK, THE PUBLIC INTEREST IN REGULATION (2004).

23. U.S. Energy Information Administration, Weekly Retail Gasoline and Diesel Prices available at http://tonto.eia.doe.gov/dnav/pet/hist/LeafHandler.ashx?n= PET&s=MG_TT_US&f=w

24. JOSEPH P. TOMAIN, NUCLEAR POWER TRANSFORMATION (1987).

25. PAUL W. MACAVOY, THE UNSUSTAINABLE COSTS OF PARTIAL DEREGULATION ch. 3 (2007) (arguing that price reductions occurred as a result of deregulation but that deregulated prices did not reach competitive levels and could not be sustained, as firms did not generate enough revenue to expand or innovate). *See also* Jay Apt, *Competition Has Not Lowered U.S. Industrial Electricity Prices*, 18 ELECTRICITY J. 52 (March 2005); Price C. Watts, *Heresy? The Case Against Deregulation of Electricity Generation*, 14 ELECTRICITY J. 19 (May 2001); Severin Borenstein, *The Trouble with Electricity Markets (and Some Solutions)* (Working paper for the Program on Workable Energy Regulation, 2001) available at http://www.ucei.berkeley.edu/PDF/pwp081.pdf

26. JUSTIN FOX, THE MYTH OF THE RATIONAL MARKET: A HISTORY OF RISK, REWARD, AND DELUSION ON WALL STREET (2009); JOHN CASSIDY, HOW

Markets Fail: The Logic of Economic Calamities (2009); Richard A. Posner, A Failure of Capitalism: The Crisis of '08 and the Descent into Depression (2009); Joseph E. Stiglitz, Freefall: America, Free Markets, and the Sinking of the World Economy (2010).

27. *See, e.g.,* Christopher H. Schroeder & Rena Steinzor (eds.), A New Progressive Agenda for Public Health and the Environment: A Project of the Center for Progressive Regulation (2005).

28. *See, e.g.,* Jedediah Purdy, *The Politics of Nature: Climate Change, Environmental Law, and Democracy,* 119 Yale L.J. 1122 (2010); The National Parks: America's Best Idea (PBS 2009). A description of this PBS film is available at http://www.pbs.org/nationalparks/

29. *See Tennessee Valley Auth. v. Hill,* 437 U.S. 153 (1978). The *TVA v. Hill* case was an early victory for the environmentalists in the United States Supreme Court. The Court indeed interpreted the Endangered Species Act to require work to stop on the nearly completed federal Tellico Dam intended to be used for electric power production. After the court decision, however, Tennessee Senators Howard Baker and Jim Sasser successfully included an amendment in a Senate appropriations bill to override the decision and have the dam completed.

30. *See* E. F. Schumacher, Small is Beautiful: Economics as if People Mattered (1973).

31. Sachs at 83 and ch. 5.

32. Donella Meadows et als., The Limits to Growth: A Report for the Club of Rome's Project on the Predicament of Mankind ch. IV (1974).

33. *See, generally,* Huber & Mills; Simon & Kahn; and Singer.

34. *See* The World Commission on Environment and Development, Our Common Future (1987).

35. *See, e.g.,* Daly at 100; Lovins, at chs. 1–2.

36. It is necessary to distinguish this curve from the Environmental Kuznets Curve (EKC), which posits that continued economic growth will lead to a reduction in environmental harms. The hypothesis of the EKC is that continued economic growth in more-developed societies poses no threat to the environment because those societies will move from heavy manufacturing to information technologies and such societies will consciously protect against environmental degradation. The hypothesis is hopeful. However, two significant barriers exist. First, the empirical data is not conclusive. Second, the political landscape described here and elsewhere indicates that if energy policy is left alone, our society will continue down the hard path of fossil fuel use with its attendant environmental degradation. *See, generally,* David I. Stern, Michael S. Common & Edward B. Barber, *Economic Growth and Environmental Degradation: The Environmental Kuznets Curve and Sustainable Development,* 24 World Development 1151 (1996) (empirical evidence does not support the EKC hypothesis); William T. Harbaugh, Arik Levinson & David Molloy Wilson, *Reexamining the Empirical Evidence for an Environmental Kuznets Curve,* 84 Rev. of Econ. and Stat. 541 (2002) (also asserting that empirical evidence does not support EKC hypothesis); Susmita Dasgupta, et al., *Confronting the Environmental Kuznets Curve,* 16 J. of Econ. Perspectives 147 (2002) (showing empirical support for the

EKC); *but see* David I. Stern, *The Rise and Fall of the Environmental Kuznets Curve*, 32 WORLD DEVELOPMENT 1419 (2004) (critiquing Dasgupta, et al.).

37. Other curves are possible. *See* DONELLA MEADOWS, JORGEN RANDERS & DENNIS MEADOWS, LIMITS TO GROWTH: THE 30-YEAR UPDATE 147–67 (2004) (discussing scenarios of economic oscillation and collapse).

38. *See, e.g.,* ROBERT L. BRADLEY, JR., CAPITALISM AT WORK: BUSINESS, GOVERNMENT, AND ENERGY (2009). Bradley, a former Enron employee, argues that Enron failed not because of too much capitalism but because of too little. He further argues that fossil fuels are the key to a healthy global economy.

39. NATIONAL PETROLEUM COUNCIL, FACING THE HARD TRUTHS ABOUT ENERGY: A COMPREHENSIVE VIEW TO 2030 OF GLOBAL OIL AND NATURAL GAS (2007) available at http://www.npchardtruthsreport.org/

40. COMMITTEE ON HEALTH, ENVIRONMENTAL, AND OTHER EXTERNAL COSTS AND BENEFITS OF ENERGY PRODUCTION AND CONSUMPTION, HIDDEN COSTS OF ENERGY: UNPRICED CONSEQUENCES OF ENERGY PRODUCTION AND USE 16 (2009).

41. The major citation in this field is Garrett Hardin, *The Tragedy of the Commons*, 162 SCIENCE 1243 (1968).

42. *See* SCHUMACHER.

43. *See, e.g.,* Rebecca Smith, *Bechtel to Back Small Nuclear Plants*, WALL. ST. J. (July 14, 2010) available at http://online.wsj.com/article/NA_WSJ_PUB:SB100014 24052748703834604575365290043159312.html; World Nuclear Association, *Small Nuclear Reactors* (July 2, 1 2010) available at http://www.world-nuclear.org/info/inf33.html

44. SCHUMACHER; *see also* FRANK ACKERMAN & LISA HEINZERLING, PRICELESS: ON KNOWING THE PRICE OF EVERYTHING AND THE VALUE OF NOTHING (2004).

45. SCHUMACHER at 42.

46. MARK SAGOFF, THE ECONOMY OF THE EARTH: PHILOSOPHY, LAW, AND THE ENVIRONMENT 1 (1988).

47. *See* MARK SAGOFF, PRICE, PRINCIPLE, AND THE ENVIRONMENT (2004).

48. *See, e.g.,* SIDNEY A. SHAPIRO & JOSEPH P. TOMAIN, REGULATORY LAW AND POLICY ch. 2 (3rd ed. 2003).

49. *See, e.g.,* CASS R. SUNSTEIN, AFTER THE RIGHTS REVOLUTION: RECONCEIVING THE REGULATORY STATE ch. 1 (1990).

50. *See, e.g.,* JEFF MADRICK, THE CASE FOR BIG GOVERNMENT (2009). "In fact, enlightened regulation has been imperative for economic growth at least in Jefferson's policies for governing the distribution of land. When done well, regulation keeps competition honest and free.... The open flow of products and services information is critical to a free-market economy. The conditions for healthy competition have simply not been maintained under a free-market ideology of minimal government that professes great faith in competition." *Id.* at 127–28.

51. Joseph T. Kelliher & Maria Farinella, *The Changing Landscape of Federal Energy Law*, 61 ADMIN. L. REV. 611, 621 (2009).

52. *See, e.g.,* WORLDWATCH INSTITUTE, 2009 STATE OF THE WORLD: INTO A WARMING WORLD ch. 4 (2009). *See also infra* chs. 3 and 4.

53. *See* John S. Applegate, *The Taming of the Precautionary Principle*, 27 Wm. & Mary Envt'l L. & Pol. Rev. 13 (2002). For a critical analysis of the precautionary principle, see Cass R. Sunstein, Worst-Case Scenarios ch 3 (2007).

54. *See, e.g.,* United States Climate Action Partnership at http://www.us-cap.org

55. *See, e.g.,* American Council on Renewable Energy at http://www.acore.org/

56. *See, e.g.,* Princeton Environmental Institute at http://www.princeton.edu/pei/; Yale School of Forestry & Environmental Studies at http://environment.yale.edu/; Harvard Electricity Policy Group at http://www.hks.harvard.edu/hepg/; University of California at Energy Institute at http://www.ucei.berkeley.edu/

57. *See, e.g.,* Cutler J. Cleveland & Robert K. Kaufmann, *Oil Supply and Oil Politics: Déjà Vu All Over Again*, 31 Energy Policy 485 (2003) (further development of domestic oil will not significantly increase U.S. production); Center for Security Policy, *"Set America Free": A Blueprint for U.S. Energy Security* available at http://www.setamericafree.org/blueprint.pdf

58. John P. Holdren, *The Energy Innovation Imperative: Addressing Oil Dependence, Climate Change, and Other 21st Century Energy Challenges*, 1 Innovations 3 (Spring 2006).

59. *See, e.g.,* Press Release, *Environmental Defense Fund, Environmental Defense Announces Jon Anda as President of New Environmental Markets Network* (January 25, 2007) available at http://www.edf.org/pressrelease.cfm?contentID=6063

60. *See, e.g.,* Eric Lambin, The Middle Path: Avoiding Environmental Catastrophe (2007); Ted Nordhaus & Michael Shellenberger, Break Through: From the Death of Environmentalism to the Politics of Possibility (2007).

61. Chris Greenwood, et al., Global Trends in Sustainable Energy Investment 2007: Analysis of Trends and Issues in the Financing of Renewable Energy and Energy Efficiency in OECD and Developing Countries 3 (2007) available at http://sefi.unep.org/fileadmin/media/sefi/docs/publications/SEFI_Investment_Report_2007.pdf

62. *See* Worldwatch Institute, 2008 State of the World: Innovations for a Sustainable Economy (2008).

63. *See* William J. Baumol, Robert E. Litan & Carl J. Schramm, Good Capitalism, Bad Capitalism, and the Economics of Growth and Prosperity (2007); William J. Baumol, The Free-Market Innovation Machine: Analyzing the Growth Miracle of Capitalism (2002); Daniel S. Esty & Andrew S. Winston, Green to Gold: How Smart Companies Use Environmental Strategy to Innovate, Create Value, and Build Competitive Advantage (2006); Jason Furman, et al., *An Economic Strategy to Address Climate Change and Promote Energy Security* (October 2007) available at http://www.brookings.edu/papers/2007/10climatechange_furman.aspx; Florian Bressandd, et al., *Wasted Energy: How the U.S. Can Reach its Energy Productivity Potential* (June 2007) available at http://www.mckinsey.com/mgi/publications/wasted_energy/index.asp

64. *See* Donella H. Meadows, et al.; *compare with* H. S. D. Cole, et al. (eds.), Models of Doom: A Critique of The Limits of Growth (1973);

Paul L. Joskow, *Energy Policies and Their Consequences After 25 Years*, 24 ENERGY J. 17 (2003).

65. *See* Amory B. Lovins, *Energy Strategy: The Road Not Taken?*, 55 FOREIGN AFFAIRS 65 (October 1976). Written nearly thirty years ago, Lovins' analysis is remarkably accurate today.

66. *See* the Regional Greenhouse Gas Initiative (RGGI) at http://www.rggi.org/ (a multi-state effort to control carbon emissions).

Chapter 3

1. ROBERT BOYERS (ED.), GEORGE STEINER AT THE NEW YORKER 237 (2009). (George Steiner quoting Lévi-Strauss in his review of the book TRISTES TROPIQUES).

2. WORLD COMMISSION ON ENVIRONMENT AND DEVELOPMENT, OUR COMMON FUTURE (1987).

3. Amory B. Lovins, *Energy Strategy: The Road Not Taken*, 55 FOREIGN AFFAIRS 64 (OCTOBER 1976).

4. AMORY B. LOVINS, SOFT ENERGY PATHS: TOWARD A DURABLE PEACE (1977).

5. *See, e.g.*, AMORY B. LOVINS & JOHN H. PRICE, NON-NUCLEAR FUTURES: THE CASE FOR AN ETHICAL ENERGY STRATEGY (1975); AMORY B. LOVINS & L. HUNTER LOVINS, BRITTLE POWER: ENERGY STRATEGY FOR NATIONAL SECURITY (1982).

6. Timothy E. Wirth, C. Boyden Gray & John D. Podesta, *The Future of Energy Policy*, 82 FOREIGN AFFAIRS 132 (July/August 2003).

7. Daniel Yergin defines "energy security": "The objective of energy security is to assure adequate, reliable supplies of energy at reasonable prices and in ways that do not jeopardize major national values and objectives." Daniel Yergin, *Energy Security in the 1990s*, 67 FOREIGN AFFAIRS 110, 111 (Fall 1988). Yergin's concern involved the negative impact of oil shocks on the economy and interruptions in supply manipulations, leading him to argue for excess of supply and production capacity as a cushion.

8. Wirth, Gray & Podesta at 138.

9. DONELLA MEADOWS ET AL., THE LIMITS TO GROWTH: A REPORT FOR THE CLUB OF ROME'S PROJECT ON THE PREDICAMENT OF MANKIND 151–53 (1974).

10. *Id.* at 24.

11. *Id.* at 34–38.

12. *Id.* at 173–74.

13. Lovins does not say that we must eliminate the grid, only that we can make more efficient energy choices, especially for small-scale consumers. Later, in Chapter 7, I argue that the grid must be improved along smart technology lines, and it must be extended to integrate solar and wind capacities.

14. ROBERT STOBAUGH & DANIEL YERGIN (EDS.), ENERGY FUTURE: REPORT OF THE ENERGY PROJECT AT THE HARVARD BUSINESS SCHOOL (1979).

15. *Id.* at 54. The authors also argued that domestic production should be maximized, including exploration in Alaska under stringent environmental controls.

16. "Worldwide, gas resources are more than sufficient to meet projected demand to 2030, though there are doubts about whether sufficient investment can be mobilised in all regions." International Energy Agency, World Energy Outlook 88 (2009).

17. *See* Rena Steinzor & Michael Patoka, *Comments from The Center for Progressive Reform, Hazardous and Solid Waste Management System: Identification and Listing of Special Waste; Disposal of Coal Combustion Residuals from Electric Utilities,* Docket ID No. EPA-HQ-RCRA-2009–0640 (November 19, 2010) available at http://www.progressivereform.org/articles/Coal_Ash_Comments_Steinzor_111910.pdf

18. Stobaugh & Yergin at 97.

19. *Id.* at 135.

20. *See, e.g.,* MIT, The Future of Nuclear Power: An Interdisciplinary Study (2003). This study recognizes that the cost of electricity generated by a nuclear power plant exceeds that of a coal fire plant. *Id.* at 7. Those costs, however, might be brought more in alignment if a carbon tax is set at a sufficiently high level to make nuclear power comparable with coal. Nevertheless, the MIT analysis recognizes the cost disparity. *See also* University of Chicago, The Economic Future of Nuclear Power: A Study Conducted at the University of Chicago (2004). This study also recognizes the higher cost of nuclear power but argues that under certain conditions, such as standardized plant design as well as reduced construction cost based on the experience of building standardized plants, those costs might be brought more in alignment. *Accord* Paul L. Joskow & John E. Parsons, *The Economic Future of Nuclear Power,* 138 Daedalus 45, 56 (2009):

> To stimulate a true nuclear renaissance that leads to significant investments in new nuclear plants, several changes from the status quo will need to take place: (a) a significant price must be placed on CO_2 emissions, (b) construction and financing costs for nuclear plants must be reduced or at least stabilized, and the credibility of current cost estimates verified with actual construction experience, (c) the licensing and safety regulatory frameworks must demonstrate that they are both effective and efficient, (d) fossil fuel prices need to stabilize at levels in the moderate to high ranges …, and (e) progress must be made on safety and long-term waste disposal to gain sufficient public acceptance to reduce political barriers to new plant investments.

> *See also* Amory B. Lovins & Imran Sheikh, *The Nuclear Illusion* (2008) available at http://www.rmi.org/rmi/Library/E08-01_NuclearIllusion. In this paper, the authors argue that, not only is nuclear power more expensive than other resources, its contribution to carbon emissions reductions is overstated and may worsen climate change. *See also* Joseph P. Tomain, *Nuclear Futures,* 15 Duke Envt'l L. & Pol Forum 221 (2005).

21. Stobaugh & Yergin at 177 (italics removed).

22. Ford Foundation, Energy: The Next Twenty Years (1979) (hereinafter Ford Study).

23. Ford Foundation, A Time To Choose: America's Energy Future (1974); Ford Foundation, Nuclear Power Issues and Choices: Report of the Nuclear Energy Policy Study Group (1977).

24. FORD STUDY at 71.
25. *Id.* ch. 4. The energy disruptions in the 1970s not only resulted in higher energy prices, but they had other economic costs as well. First, energy prices fluctuated wildly, making planning and investment difficult. Second, the price of oil per barrel quadrupled, creating inflation. Third, higher prices and inflation contributed to an economic recession.
26. *Id.* at 116.
27. *Id.* at 6.
28. *Id.* ch. 3.
29. *Id.* chs. 10 & 11.
30. *See id.* at 329–35. The authors were perhaps more cautious than they should have been about climate change: "Scientists generally agree that whatever the precise relationships and ultimate limits may be, the effects of CO_2 will be small and gradual over the next several decades and appear to be reversible if they do not go too far. Projected increases in worldwide use of coal over the rest of this century are not likely to take the world to a point where it is physically or economically impossible to control CO_2 buildup." *Id.* at 34.
31. *Id.* ch. 11.
32. *Id.* at 140–42.
33. *Id.* at 543.
34. SAM H. SCHUR, ET AL., ENERGY IN AMERICA'S FUTURE: THE CHOICES BEFORE US (1979) (hereinafter RFF STUDY).
35. The report looked at projected energy consumption levels from various sources. All of which overestimated what consumption would be at the turn of the millennium. *Id.* at 180.
36. *Id.* at 84.
37. *Id.* at 34.
38. *Id.* ch. 9.
39. *Id.* chs. 12 & 13.
40. *Id.* ch. 7.
41. *Id.* ch. 5.
42. *Id.* chs. 8, 10 & 11.
43. *Id.* at 384.
44. Critics of LIMITS often argued that the book made Cassandra-like claims that the world was facing the end of certain resources. This is not true. "Nowhere in the book was there any mention about running out of *anything* by 2000. Instead, the book's concern was entirely focused on what the world might look like 100 years later. There was not one sentence or even a single word written about an oil shortage, or limit to any specific resource, by the year 2000." Matthew R. Simmons, *Revisiting The Limits to Growth: Could the Club of Rome Have Been Correct, After All?* 11 (October 2000) available at http://www.greatchange.org/ov-simmons,club_of_rome_revisted.pdf
45. In an analysis of *Limits*, Graham Turner emphasizes the importance of this linkage: "By linking the world economy with the environment it was the first integrated global model." Graham Turner, *A Comparison of the Limits to Growth with Thirty Years of Reality* 1 (Commonwealth Scientific and Industrial Research Organisation Working Paper Series, Working Paper No. 1834–5638, 2008) available at http://www.csiro.au/files/files/plje.pdf (citation omitted).

46. *See* Paul L. Joskow, *Energy Policies and Their Consequences After 25 Years*, 24 ENERGY JOURNAL 17 (2003). "Even after 25 years, there is still not widespread agreement about the absolute or relative importance of various energy policy goals." *Id.* at 23. Joskow argues that (1) because it is a public policy issue, the importance of energy changes over time; and (2) because energy policies affect such large segments of the economy, the economic impacts on industries, regions, and different groups of people make consensus difficult to achieve.

47. DEPARTMENT OF ENERGY, ENERGY INFORMATION ADMINISTRATION, ANNUAL ENERGY REVIEW 2008 68 (June 2009).

48. *See* DONELLA MEADOWS, JORGEN RANDERS & DENNIS MEADOWS, LIMITS TO GROWTH: THE 30-YEAR UPDATE, 6–7 (2004) (hereinafter THE 30-YEAR UPDATE).

49. *See* The World Bank, *Measuring Poverty at the Country Level* available at http://web.worldbank.org/WBSITE/EXTERNAL/TOPICS/EXTPOVERTY/0,,contentMDK:20153855~menuPK:373757~pagePK:148956~piPK:216618~theSitePK:336992,00.html. For a discussion of wealth disparity, *see* JEFFREY D. SACHS, COMMON WEALTH: ECONOMICS FOR A CROWDED PLANET 18, 30–31 (2008); The World Bank, *World Development Report 2009: Reshaping Economic Geography* (2008).

50. Press Release, The World Bank, *New Data Show 1.4 Billion Live on Less Than US$1.25 a Day, But Progress Against Poverty Remains Strong* (August 26, 2008) available at http://web.worldbank.org/WBSITE/EXTERNAL/NEWS/0,,contentMDK:21881954~pagePK:34370~piPK:34424~theSitePK:4607,00.html

51. United Nations Development Programme & The World Health Organization, *The Energy Access Situation in Developing Countries: A Review Focusing on the Least Developed Countries and Sub-Saharan Africa* 1 (2009) (finding that "[c]urrently, about 1.5 billion people in developing countries lack access to electricity and about 3 billion people rely on solid fuels for cooking"). UN-Energy, *The Energy Challenge for Achieving the Millennium Development Goals* 2 (2005) (explaining that "[w]orldwide, 2.4 billion people rely on traditional biomass fuels for cooking" and "at least 1.6 billion people do not have access to electricity"). The United Nations adopted eight millennial goals in September 2000. Now known as the *Millennium Development Goals*, the goals were intended to be a global partnership to reduce extreme poverty with a deadline of 2015. The goals range from ending poverty and hunger, to universal education, to gender equality and children's health. The goals also include environmental sustainability. Energy touches each one of the goals at some point. The goals can be found at http://www.un.org/millenniumgoals/. *See also* Jamal Saghir, *Energy and Poverty: Myths, Links, and Policy Issues* (May 2005) available at http://siteresources.worldbank.org/INTENERGY/Resources/EnergyWorkingNotes_4.pdf. This World Bank report argues that: the poor have very high prices for their energy; it negatively affects the environment; and without access to modern energy, poverty will continue.

52. The World Bank, *World Development Report 2010: Development and Climate Change* 1 (2009).

53. *Id.* at 5.

54. AMORY B. LOVINS, ET AL., WINNING THE OIL ENDGAME (2004).

55. Joskow at 46.
56. *World Development Report 2010* at 2.
57. Environmental Law Institute, *Estimating U.S. Government Subsidies to Energy Sources: 2002–2008*, 3 (2009).
58. *Id.*
59. INTERNATIONAL ENERGY AGENCY, WORLD ENERGY OUTLOOK at 89.
60. *See* MIT, THE FUTURE OF COAL: AN INTERDISCIPLINARY STUDY: OPTIONS FOR A CARBON-CONSTRAINED WORLD, chs. 6 & 8 (2007).
61. DEPARTMENT OF ENERGY, ANNUAL ENERGY REVIEW 2008 at xix.
62. *See* Council of Economic Advisors, *Economic Report of the President*, 287 (2009).
63. As a law of microeconomics, as the price of a resource such as energy declines, demand will increase. The country has increased its energy demand, but as the data indicate, we are achieving energy efficiencies as the real cost of energy declines. The behavioral characteristic that must be watched is the "rebound effect." As energy prices decrease, demand will increase, but at what rate? *Compare* PETER W. HUBER & MARK P. MILLS, THE BOTTOMLESS WELL: THE TWILIGHT OF FUEL, THE VIRTUE OF WASTE, AND WHY WE WILL NEVER RUN OUT OF ENERGY 109 (2005) ("Efficiency has come, and demand has risen apace.") *with* Kenneth A. Small & Kurt Van Dender, *Fuel Efficiency and Motor Vehicle Travel: The Declining Rebound Effect* (UC Irvine Economics Working Paper, Paper No. 05–06–03, 2007) available at http://www.economics.uci.edu/docs/2005–06/Small-03.pdf and UK Energy Research Centre, *The Rebound Effect: An Assessment of the Evidence for Economy-Wide Energy Savings From Improved Energy Efficiency* (2007) (both reports acknowledge the existence and the importance of the rebound effect but conclude that it does not negate efforts to achieve increased energy efficiency).
64. OUR COMMON FUTURE.
65. *Id.* at 8.
66. *Id.* at 9.
67. *Id.* at 169–71.
68. *Id.* at 176.
69. *Id.* at 196.
70. International Institute for Industrial Environmental Economics, Thomas B. Johansson & José Goldemberg (eds.), *Energy for Sustainable Development: A Policy Agenda* (2002).
71. THE 30-YEAR UPDATE.
72. THE 30-YEAR UPDATE at 254.
73. *Id.* at 269–71.
74. *Id.* at 259–61.
75. *See* the Energy Future Coalition's homepage at http://www.energyfuture-coalition.org/
76. *See* Bracken Hendricks, et al., *Rebuilding America: A National Policy Framework for Investment in Energy Efficiency Retrofits* (2009) available at http://www.ener-gyfuturecoalition.org/files/webfmuploads/Efficiency%20Docs/Rebuilding%20America%20White%20Paper%20Final.pdf

77. *See* 25 x '25 National Steering Committee, *25 x '25 Action Plan: Charting America's Energy Future* (2007) available at http://www.25x25.org/storage/25x25/documents/IP%20Documents/Action_Plan/actionplan_64pg_11-11-07.pdf. *See also* 25 x '25 National Steering Committee homepage at http://www.25x25.org/

78. *See* Energy Future Coalition, *The National Clean Energy Smart Grid: An Economic, Environmental, and National Security Imperative* available at http://www.energyfuturecoalition.org/files/webfmuploads/Smart%20Grid%20Docs/EFC%205-page%20Vision%20Statement%20-%20FINAL.pdf

79. *See* Energy Future Coalition, *Global Development Bonds* available at http://www.wbcsd.org/web/projects/sl/gdb.pdf

80. Energy Future Coalition, *Challenge and Opportunity: Charting a New Energy Future* (2003) available at http://www.energyfuturecoalition.org/files/webfmuploads/EFC_Report/EFCReport.pdf

81. National Commission on Energy Policy, a project of the Bipartisan Policy Center, is a non-profit organization established in 2007 by former Senate majority leaders Howard Baker, Tom Daschle, Bob Dole, and George Mitchell to serve as a forum for policy discussions. The homepage for the Bipartisan Policy Center is http://bipartisanpolicy.org/, and the homepage for the National Commission on Energy Policy is http://bipartisanpolicy.org/projects/national–commission–energy–policy

82. National Commission on Energy Policy, *Ending the Energy Stalemate: A Bipartisan Strategy to Meet America's Energy Challenges* (2004) available at http://www.bipartisanpolicy.org/sites/default/files/endi_en_stlmate.pdf (hereinafter *Ending the Energy Stalemate*).

83. *See* Natural Resources Defense Council, *A Responsible Energy Plan for America* (2005) available at http://physics.gac.edu/~huber/classes/FTS100/nrdc_report_2005.pdf

84. *Ending the Energy Stalemate* at x.

85. *See, e.g.*, CNA Military Advisory Board, *Powering America's Defense: Energy and the Risks to National Security* (2009) available at http://www.cna.org/documents/PoweringAmericasDefense.pdf

86. Energy Future Coalition, History available at http://www.energyfuturecoalition.org/about-us/history

Chapter 4

1. John P. Holdren, *Energy for Change: Introduction to the Special Issue on Energy & Climate*, 4 INNOVATIONS: TECHNOLOGY, GOVERNANCE, GLOBALIZATION 3 (Fall 2009).

2. *See, e.g.*, COMMITTEE ON AMERICA'S ENERGY FUTURE OF THE NATIONAL ACADEMY OF SCIENCES, NATIONAL ACADEMY OF ENGINEERING & NATIONAL RESEARCH COUNCIL, AMERICA'S ENERGY FUTURE: TECHNOLOGY AND TRANSFORMATION xi (2009) (hereinafter THE NATIONAL ACADEMIES REPORT).

3. *See, e.g.*, Christopher Flavin, *Low-Carbon Energy: A Roadmap* (2008) available at http://www.worldwatch.org/node/5945

4. Nicholas Stern, The Economics of Climate Change: The Stern Review 3 (2007) (hereinafter The Stern Review) available at http://www. hm-treasury.gov.uk/sternreview_index.htm; Intergovernmental Panel on Climate Change, Fourth Assessment Report: Synthesis for Policymakers 5 (2007) available at http://www.ipcc.ch/pdf/assessment-report/ar4/syr/ar4_syr_spm.pdf

5. *See, e.g.*, United States Climate Action Partnership, *A Call for Action: Consensus Principles and Recommendation from the U.S. Climate Action Partnership: A Business and NGO Partnership* (2007) available at http://www.us-cap.org/about-us/our-report-a-call-for-action/; Jason Furman, et al., *An Economic Strategy to Address Climate Change and Promote Energy Security* (October 2007) available at http://works.bepress.com/jason_bordoff/3/. Rick Duke & Dan Lashof, *The New Energy Economy: Putting America on the Path to Solving Global Warming* 5 (June 2008) available at http://www.nrdc.org/globalWarming/energy/eeconomy.pdf; *See also* Rachel Cleetus, Steven Clemmer & David Friedman, *Climate 2030: A National Blueprint for a Clean Energy Economy* 6 (May 2009) available at http://www.ucsusa.org/global_warming/solutions/big_picture_solutions/climate-2030-blueprint.html

6. *See* Frank Ackerman, et al., *The Economics of 350: The Benefits and Costs of Climate Stabilization* 13 (October 2009) available at http://www.e3network.org/papers/Economics_of_350.pdf; James Hansen, et al., *Target Atmospheric CO_2: Where Should Humanity Aim?* (2008) (350 ppm as the safe level) available at http://www.columbia.edu/~jeh1/2008/TargetCO2_20080407.pdf; Nicholas Stern, A Blueprint of a Safer Planet: How to Manage Climate Change and Create a New Era of Progress and Prosperity 39 (reduce to 400 ppm to have a 50% chance of keeping temperature rise to 2°C) (2009) (hereinafter Stern Blueprint); *see also* 350.org at http://www.350.org

7. *Compare* Frank Ackerman, et al., (350 ppm to be reached in 2200) *with* James Hansen, et al. (350 ppm to be reached in 2100).

8. *See, e.g.*, United States Climate Action Partnership; America's Energy Future: Technology and Transformation at 1; Holdren; Rick Duke & Dan Lashof.

9. Cleetus, et al., at 1 n.10 and 5–7.

10. A 2°C change in global average temperature from the pre-industrial level is equivalent to a 1.4°C change from 1990, which is another benchmark discussed in the literature. *See* Frank Ackerman, et al., at 14.

11. *See* Intergovernmental Panel on Climate Change, Climate Change 2007: Synthesis Report 30 (2007) (hereinafter IPCC Synthesis Report); National Aeronautics and Space Administration, *2009: Second Warmest Year on Record; End of Warmest Decade* (January 2010) available at http://www.nasa.gov/topics/earth/features/temp-analysis-2009.html

12. The Stern Review at 5.

13. Regardless of the high level of consensus on the human contribution to global warming, political arguments continue to attack the science. As a rebuttal to skeptics, *see* Union of Concerned Scientists Web site *Scientific Consensus on Global Warming* available at http://www.ucsusa.org/ssi/climate-change/scientific-consensus-on.html; *see also* the Web site for Skeptical Science at http://www.skepticalscience.com/

14. IPCC Synthesis Report at 45.
15. *See, e.g.*, Bill McKibben, Eaarth: Making a Life on a Tough New Planet (2010).
16. The Stern Review at 2 (2007); Stern Blueprint at 9 (2009).
17. Janet L. Sawin & William R. Moomaw, *Renewable Revolution: Low-Carbon Energy by 2030* 24 (2009) available at http://www.worldwatch.org/node/6340; International Energy Agency, *World Energy Outlook 2009 Fact Sheet: Why Is Our Current Energy Pathway Unsustainable?* (2009) available at http://www.iea.org/weo/docs/weo2009/fact_sheets_WEO_2009.pdf
18. Frank Ackerman, *Debating Climate Economics: The Stern Review vs. Its Critics* 9 (July 2007) available at http://ase.tufts.edu/gdae/pubs/rp/sterndebatereport.pdf
19. The Stern Review at 7.
20. *See* Frank Ackerman, et al., at 13.
21. Stern Blueprint at 26.
22. Frank Ackerman, et al., at 14 converting Stern Review data to carbon dioxide emissions from carbon dioxide equivalent data.
23. Frank Ackerman, et al., at 14.
24. Stern Blueprint at 26.
25. The Stern Review at 9; *see also* James Hansen, et al.
26. *See, e.g.*, Chris Wold, David Hunter & Melissa Powers, Climate Change and the Law 13–15 (2009); David W. Orr, Down to the Wire: Confronting Climate Collapse 3, 20 (2009); Donella H. Meadows, Jorgen Randers & Dennis L. Meadows, Limits to Growth: The 30-Year Update (2004).
27. *See* Frank Ackerman & Lisa Heinzerling, Priceless: On Knowing the Price of Everything and the Value of Nothing (2004); Mark Sagoff, The Economy of the Earth: Philosophy, Law and the Environment (1988); Lisa Heinzerling & Frank Ackerman, *Law and Economics in a Warming World*, 1 Harv. L. & Pol. Rev. 331 (2007).
28. *See* Douglas A. Kayser, *Discounting … On Stilts*, 74 U. Chi. L. Rev. 119 (2007).
29. Key papers in the debate over the discount rate are: The Stern Review (using a discount rate of 1.4%); William Nordhaus, *The Stern Review on the Economics of Climate Change* (November 17, 2006) (5%); Martin Weitzman, *The Stern Review of the Economics of Climate Change* (April 31, 2007) (6%); *see also* Kenneth J. Arrow, *Global Climate Change: A Challenge to Policy*, Economists's Voice (June 2007); Symposium, *Intergenerational Equity and Discounting*, 74 U. Chi. L. Rev. 1 (2007).
30. *See* Eric A. Posner & Cass R. Sunstein, *Climate Change Justice*, 96 Geo. L.J. 1565 (2008) (discussing issues of corrective and distributive justice relative to addressing climate change.)
31. Frank Ackerman, et al., at 21.
32. National Research Council, Hidden Costs of Energy: Unpriced Consequences of Energy Production and Use 21 (October 2010).
33. *See* Cleetus, et al.
34. Stern Blueprint at 94. The lower figure includes direct economic losses, and the higher figure accounts for all costs, such as environmental and health costs, other than national security costs.

35. *Id.* at 54.
36. Frank Ackerman, et al., at 21–27 (assessing the literature on the costs of climate change measures.)
37. *Id.* at 5.
38. *See* Nathaniel Keohane & Peter Goldmark, *What Will it Cost to Protect Ourselves from Global Warming? The Impacts on the U.S. Economy of a Cap-and-Trade Policy for Greenhouse Gas Emissions* 9 (2008) available at http://www.edf.org/documents/7815_climate_economy.pdf. The authors report a range of estimates for the period 2010–2030 from 0.23% to 2.15% and for the period of 2010–2050 from 0.59% to 3.59%. The wide range of estimates is dependent on assumptions and the structures of the models employed. *See, e.g.,* Daniel A. Farber, *Climate Models: A User's Guide* (November 16, 2007) available at http://papers.ssrn.com/sol3/papers.cfm?abstract_id=1030607
39. Keohane & Goldmark at 11.
40. McKinsey & Company, *Pathways to a Low-Carbon Economy: Version 2 of the Global Greenhouse Gas Abatement Cost Curve* 42 (2009).
41. *Id.* at 39.
42. *See, e.g.,* Cleetus, et al.; Andrew Simms, et al., *A Green New Deal: Joined-Up Policies to Solve the Triple Crunch of the Credit Crisis, Climate Change and High Oil Prices* (2008) available at http://www.neweconomics.org/sites/neweconomics.org/files/A_Green_New_Deal_1.pdf. As the title of the last report indicates, high job creation and economic returns are based on an assumption of continued high oil prices.
43. This study was commissioned to oppose the now defunct Lieberman-Warner climate change legislative proposal. *See* American Council for Capital Formation & National Association of Manufacturers, *Analysis of Lieberman-Warner Climate Security Act (S.2191) Using the National Energy Modeling System (NEMS/ACCF/NAM)* available at http://www.accf.org/pdf/NAM/fullstudy031208.pdf
44. *See* Frank Ackerman, et al., at 22–23; Julie A. Nelson, *How Costly is Climate Change Mitigation? A Methodological Critique of ACCF/NAM Claims* (March 18, 2008) available at http://www.e3network.org/papers/How_Costly_ACCFNAM2008.pdf; Janet Peace, *Insights From Modeling Analysis of the Lieberman-Warner Climate Security Act (S. 2191)* (May 2008) available at http://www.pewclimate.org/docUploads/L-W-Modeling.pdf
45. One review of the economic literature about the costs of climate change concludes: "There are no reasonable studies that say a 350 ppm stabilization target will destroy the economy; there are no studies that claim that it is desirable to wait before taking action on climate protection. On the contrary, there is strong, widespread endorsement for policies to promote energy conservation, development of new energy technologies, and price incentives and other economic measures that will redirect the world economy onto a low-carbon path to sustainability." Frank Ackerman, et al., at 6.
46. Shiyong Park, Winny Chen & Rudy deLeon, *Securing America's Energy Independence Through Energy Diversification* (April 2009) available at http://www.americanprogress.org/issues/2009/04/pdf/energy_security.pdf
47. John S. Duffield, Over a Barrel: The Costs of U.S. Foreign Oil Dependence 210 (2008) (based on DOE Energy Information Administration data).

48. *See* Christopher Beddor, et al., *Securing America's Future: Enhancing Our National Security by Reducing Oil Dependence and Environmental Damage* 1–2 (August 2009) available at http://www.americanprogress.org/issues/2009/08/pdf/energy_security.pdf

49. *See also* George E. Pataki & Thomas J. Vilsack, *Confronting Climate Change: A Strategy for U.S. Foreign Policy* (2008) available at http://www.cfr.org/publication/16362/

50. *See* JEFFREY D. SACHS, COMMON WEALTH: ECONOMICS FOR A CROWDED PLANET (2008); Kurt M. Campbell, et al., *The Age of Consequences: The Foreign Policy and National Security Implications of Global Climate Change* (November 2007) available at http://csis.org/files/media/csis/pubs/071105_ageofconsequences.pdf

51. *See* CNA Military Advisory Board, *Powering America's Defense: Energy and the Risks to National Security* (May 2009) available at http://www.cna.org/documents/PoweringAmericasDefense.pdf

52. *See* MICHAEL T. KLARE, BLOOD AND OIL: THE DANGERS AND CONSEQUENCES OF AMERICA'S GROWING PETROLEUM DEPENDENCY (2004).

53. DUFFIELD at ch. 5.

54. Mark A. Delucchi & James J. Murphy, *U.S. Military Expenditures to Protect the Use of Persian Gulf Oil for Motor Vehicles*, 36 Energy Policy 2252 (2008); DUFFIELD at ch. 6 and 206.

55. DUFFIELD at 205.

56. John Deutch & James R. Schlesinger, *National Security Consequences of U.S. Oil Dependency* 17 (2006). Note the twin sides of oil dependence and price inelasticity: "If the United States were to reduce its oil consumption by 10 percent (2.5 percent of world demand), the effect in current tight oil markets could be a temporary decline in global prices (about 12 percent to 25 percent) and a lowering of the anticipated rate of future increases." (At 23).

57. DUFFIELD at 43–44.

58. Toni Johnson, *Oil Market Volatility* (September 24, 2008) available at http://www.cfr.org/publication/15017/

59. *See* John D. Podesta, et al., *The Clean-Energy Investment Agenda: A Comprehensive Approach to Building the Low-Carbon Economy* (September 2009) available at http://www.americanprogress.org/issues/2009/09/clean_energy_investment.html

60. Swawin & Moomaw at 22.

61. THE NATIONAL ACADEMIES REPORT. The six general categories discussed are taken from the 2009 report of the Committee on America's Energy Future of the National Academies. The Committee is a consortium of the National Academy of Sciences, the National Academy of Engineering, and the National Research Council. The report was authored by a group of nationally prominent executives and academics from the public and private sectors. The report itself was supplemented by a series of individual studies, and this discussion is also supplemented by additional sources. These categories are standard in the energy policy literature and constitute a solid consensus energy policy. *See also* Laura Diaz Anadon, et al., *Tackling U.S. Energy Challenges and Opportunities: Preliminary Policy Recommendations for Enhancing Energy Innovation in the United States* (February 2009) (hereinafter *Tackling U.S. Energy Challenges and Opportunities*) (policy recommendations addressed to the Obama administration) available

at http://belfercenter.ksg.harvard.edu/publication/18826/tackling_us_energy_challenges_and_opportunities.html

62. *See* THE NATIONAL ACADEMIES REPORT at ch. 4.

63. McKinsey & Company, *Unlocking Energy Efficiency in the U.S. Economy* 1 (July 2009) available at http://www.mckinsey.com/clientservice/electricpow-ernaturalgas/US_energy_efficiency/ *See also* Guiseppe Bellantuono, *Law and Innovation in the Energy Sector* to be published in B. Delvaux, M. Hunt & K Talus (eds.), EU LAW AND POLICY ISSUES (2nd ed. 2009) available at http://papers.ssrn.com/sol3/papers.cfm?abstract_id=1480071

64. Robert Pollin, James Heintz & Heidi Garrett-Peltier, *The Economic Benefits of Investing in Clean Energy: How the Economic Stimulus Program and New Legislation Can Boost U.S. Economic Growth and Employment* 16 (June 2009) available at http://www.peri.umass.edu/economic_benefits/

65. J. Read Smith & William Richards, *25 x '25 Action Plan: Charting America's Energy Future* 41 (February 2007) available at http://www.25x25.org/storage/25x25/documents/IP%20Documents/Action_Plan/actionplan_64pg_11-11-07.pdf

66. *See* Diana Farrett, James Remes & Dominic Charles, *Fueling Sustainable Development: The Energy Productivity Solution* 6–7 (October 2008) available at http://www.mckinsey.com/mgi/publications/fueling_sustainable_development.asp; *see also* Jaeson Rosenfeld, et al., *Averting the Next Energy Crisis: The Demand Challenge* (March 2009) available at http://www.mckinsey.com/mgi/publications/next_energy_crisis/index.asp

67. Diana Farrell, et al., *The Case for Investing in Energy Productivity* 8 (February 2008) available at http://www.mckinsey.com/mgi/publications/Investing_Energy_Productivity/

68. Of the $24.4 billion, $7.2 billion is in direct spending, $14.4 billion in grants, $2 billion in tax incentives, and the remainder in bonds. *See* Robert Pollin, James Heintz & Heidi Garrett-Peltier, *The Economic Benefits of Investing in Clean Energy: How the Economic Stimulus Program and New Legislation can boost U.S. Economic Growth and Employment* 6 (June 2009) available at http://www.peri.umass.edu/economic_benefits/

69. Smith & Richards at 41; Pollin, et al., at 16.

70. Pollin, et al., at 16.

71. Smith & Richards at 42.

72. Smith & Richards at 41.

73. *See* Architecture 2030 Web site at http://www.architecture2030.org/home.html

74. *See* Edward Mazria & Kristina Kershner, *The 2030 Blueprint: Solving Climate Change Saves Billions* (April 7, 2008) available at http://www.architecture2030.org/pdfs/2030Blueprint.pdf

75. Cleetus, et al., at 39–42.

76. THE NATIONAL ACADEMIES REPORT at 83.

77. Center for American Progress & Energy Future Coalition, *Rebuilding America: A National Policy Framework for Investment in Energy Efficiency* 10 (August 2009) available at http://www.americanprogress.org/issues/2009/08/pdf/rebuilding_america.pdf

78. *See, e.g.*, InterAcademy Council, *Lighting the Way: Toward a Sustainable Energy Future* 29–30 (October 2007) available at http://www.interacademycouncil.net/CMS/Reports/11840.aspx

79. Regarding other human behavior barriers, *see* Amanda R. Carrico, et al., *Energy and Climate Change: Key Lessons for Implementing the Behavioral Wedge* (November 2010) available at http://papers.ssrn.com/sol3/papers. cfm?abstract_id=1612224

80. *See Tackling U.S. Challenges and Opportunities* at 7–8; American Physical Society, *Energy Future: Think Efficiency: How America Can Look Within to Achieve Energy Security and Reduce Global Warming* (September 2008) available at http://www.aps.org/energyefficiencyreport/report/aps-energyreport.pdf

81. *See* Stephen Doig, Mathais Bell & Natalie Mims, *Industrial Electric Productivity: Myths, Barriers and Solutions* (2009) available at http://www.rmi. org/rmi/Library/2009-18_IndustrialElectricProductivity

82. Doig, Bell & Mims at 5–7.

83. America's Energy Future Report, *supra* note 61 at ch. 5.

84. Smith & Richards, *supra* note 64 at 28.

85. Rachel Cleetus, Steven Clemmer & David Friedman, *Climate 2030: A National Blueprint for a Clean Energy Economy* 107–08 (May 2009) available at http://www.ucsusa.org/global_warming/solutions/big_picture_solutions/climate-2030-blueprint.html

86. *See, e.g.,* Worldwatch Institute & Center for American Progress, *The Renewable Path to Energy Security* 22–23 (September 2006) available at http://www.worldwatch.org/files/pdf/AmericanEnergy.pdf; Worldwatch Institute, *Red, White, and Green Transforming U.S. Biofuels* (2009); Worldwatch Institute, *Worldwatch Issue Brief: U.S. Biofuels: Climate Change and Policies* (July 2009) available at http://www.worldwatch.org/files/pdf/Biofuels%20Issue%20Brief.pdf

87. *See, e.g.,* Natural Resources Defense Council, *Getting Biofuels Right: Eight Steps for Reaping Real Environmental Benefits from Biofuels* (2007) available at http://www.nrdc.org/energy/files/right.pdf. The study also notes the higher environmental costs for corn ethanol versus cellulosic ethanol.

88. The National Academies Report at 246.

89. *See* Intergovernmental Panel on Climate Change, IPCC Special Report on Carbon Dioxide Capture and Storage (2005); John Deutch & Ernest Moniz, *The Future of Coal: An Interdisciplinary Study* (2007) available at http://web.mit. edu/coal/

90. *See* United States Climate Action Partnership, *A Blueprint for Legislative Action: Consensus Recommendations for U.S. Climate Protection Legislation* 18–20 (January 2009) available at http://www.us-cap.org/newsroom/blueprint-for-legislative-action/

91. *Obama Announces Steps to Boost Biofuels, Clean Coal* (February 3, 2010) available at http://www.whitehouse.gov/the-press-office/obama-announces-steps-boost-biofuels-clean-coal

92. *See* EPA, *Renewable Fuel Standard Program* at http://www.epa.gov/otaq/renewablefuels/index.htm

93. *See* Department of Agriculture, *Biomass Crop Assistance Program* available at http://www.fsa.usda.gov/FSA/webapp?area=home&subject=ener&topic=bcap

94. *See* White House, *A Comprehensive Federal Strategy on Carbon Capture and Storage* available at http://www.whitehouse.gov/the-press-office/presidential-memorandum-a-comprehensive-federal-strategy-carbon-capture-and-storage

95. THE NATIONAL ACADEMIES REPORT at ch. 6.
96. United Nations Environment Programme & New Energy Finance, Ltd, *Global Trends in Sustainable Energy Investment 2009* (June 2009) available at http://www.unep.org/pdf/Global_trends_report_2009.pdf; REN 21, *Renewables Global Status Report: 2009 Update* (2009) available at http://www.ren21.net/pdf/RE_GSR_2009_Update.pdf
97. Cleetus, et al., at 57.
98. *Id.* at 67–68; *see also* Charles F. Kutscher, *American Solar Energy Society, Tackling Climate Change in the U.S.; Potential Carbon Emissions Reductions from Energy Efficiency and Renewable Resources by 2030* (January 2007) available at http://ases.org/images/stories/file/ASES/climate_change.pdf
99. *Id.* at 59.
100. *See* American Wind Energy Association, *Year End 2009 Market Report* (January 2010) available at http://www.awea.org/publications/reports/4Q09.pdf
101. DOE Office of Energy Efficiency and Renewable Energy, *20% Wind Energy by 2030: Increasing Wind Energy's Contribution to U.S. Electricity Supply* (July 2008) available at http://www1.eere.energy.gov/windandhydro/pdfs/41869.pdf. Reaching this target would require 300,000 MW of new capacity or approximately 10x of what is currently installed and would provide an estimated 500,000 new jobs.
102. Cleetus, et al., at 69; *see also* National Renewable Energy Laboratory PowerPoint at http://www.nrel.gov/analysis/docs/cost_curves_2005.ppt-1690
103. *See* Daniel P. Krueger & Andre Begoso, *Mandating Federal Renewables: The Importance of Getting the REC Market Right*, 148 PUB. UTIL. FORT. 40 (January 2010).
104. Smith & Richards at 37.
105. *See, e.g.,* North American Reliability Corporation, *Special Report: Accommodating High Levels of Variable Generation* (April 2009) available at http://www.aeso.ca/gridoperations/17792.html
106. Amory Lovins & Lena Hansen, *Keeping the Lights on While Transforming Utilities* (2010); Lena Hansen & Jonah Levine, *Intermittent Renewables in the Next Generation Utility* (2008) both available at http://www.rmi.org/?UrlName =Library&Cat1=Energy+and+Resources&Cat2=Solar&CatType=sharepoint
107. THE NATIONAL ACADEMIES REPORT at ch. 7.
108. INTERNATIONAL ENERGY AGENCY, WORLD ENERGY OUTLOOK 80 (2009).
109. *See* DOE Energy Information Administration, *World Proved Reserves of Oil and Natural Gas, Most Recent Estimates* (March 3, 2009) available at http://www.eia.doe.gov/emeu/international/reserves.html
110. THE NATIONAL ACADEMIES REPORT at 345.
111. John W. Rowe & Ed Fortunato, *The Emerging Impact of Shale Gas Resources*, 50 INFRASTRUCTURE 1 (Fall 2010) (current estimates indicate that shale gas can provide 20 years of total U.S. energy needs).
112. IHS CERA, *Fueling North America's Energy Future: The Unconventional Natural Gas Revolution and the Carbon Agenda* (2010) available at http://www2.cera.com/docs/Executive_Summary.pdf
113. *See* JOHN DEUTCH & ERNEST J. MONIZ, THE FUTURE OF COAL: AN INTERDISCIPLINARY MIT STUDY (2007) available at http://web.mit.edu/coal/

114. *See* Massachusetts Institute of Technology, *Retrofitting Coal-Fired Power Plants for CO₂ Emissions Reductions* (March 23, 2009) available at http://web.mit.edu/mitei/docs/reports/coal-paper.pdf

115. THE NATIONAL ACADEMIES REPORT at 366–67; *see also*, DEUTCH & MONIZ at Appendix 3E.

116. *See* Calera at http://www.calera.com/index.php. This carbon sequestration company extracts carbon dioxide from power plants and transforms it into a hard material (calcium carbonate) that can be used as a building material.

117. THE NATIONAL ACADEMIES REPORT at 381–85.

118. *Id.* at ch. 8.

119. *See, e.g.*, Michael T. Burr, *Hope for Change: Can Climate-Policy Brinksmanship Create a Sustainable Nuclear Industry?* 147 PUB. UTIL. FORT. 28 (December 2009).

120. National Research Council, *Review of DOE's Nuclear Energy Research and Development Program* (2007) available at http://www.nap.edu/catalog.php?record_id=11998#toc; *see also* American Physical Society, *Nuclear Power and Proliferation Resistance: Securing Benefits, Limiting Risk* (May 2005) available at http://www.aps.org/policy/reports/popa-reports/proliferation-resistance/upload/proliferation.pdf; John Deutch & Ernest Moniz, *The Future of Nuclear Power: An Interdisciplinary Study* (2003) available at http://web.mit.edu/nuclearpower/

121. I Googled "nuclear + renaissance" and the response was more than 574,000 hits.

122. *See* Nuclear Energy Institute, *New Nuclear Plant Status* available at http://www.nei.org/resourcesandstats/documentlibrary/newplants/graphicsandcharts/newnuclearplantstatus/

123. *See, e.g.*, Paul L. Joskow & John E. Parsons, *The Economic Future of Nuclear Power* (March 2009) available at http://econ-www.mit.edu/files/3984. The authors argue that nuclear power is notably more costly than either natural gas or coal-produced electricity if there are no carbon charges on either coal or natural gas. ("Absent the imposition of explicit or implicit prices on CO₂ emissions and given the current expected costs of building and operating alternative generating technologies, it does not appear that a large nuclear "renaissance" will occur based primarily on the economic competitiveness of new nuclear power plants compared to alternative fossil-fueled base load generating technologies." at 22. *See also* Yangbo Du & John E. Parsons, *Update on the Cost of Nuclear Power*, MIT Center for Energy and Environmental Policy Research *Working Paper 09–003*; Amory B. Lovins, Imran Sheikh & Alex Markevich, *Nuclear Power: Climate Fix or Folly?* (December 2008) (nuclear power too costly) available at http://www.rmi.org/images/PDFs/Energy/E09-01_NuclPwrClimFixFolly1i09.pdf; Amory B. Lovins & Imran Sheikh, *The Nuclear Illusion* (Draft May 2008) (same) available at http://www.rmi.org/images/PDFs/Energy/E09-01_NuclPwrClimFixFolly1i09.pdf; Union of Concerned Scientists, *Nuclear Power and Global Warming* (nuclear power's continuing safety and security risks) available at http://www.ucsusa.org/assets/documents/nuclear_power/npp.pdf; Lisbeth Gronlund, David Lochbaum & Edwin

Lyman, *Nuclear Power in a Warming World: Assessing the Risks, Addressing the Challenges* (December 2007) available at http://www.ucsusa.org/assets/documents/nuclear_power/nuclear-power-in-a-warming-world.pdf

124. Cleetus, et al., at 83.

125. *See Obama Administration Announces Loan Guarantees to Construct New Nuclear Power Reactors in Georgia* (February 16, 2010) available at http://www.whitehouse.gov/the-press-office/obama-administration-announces-loan-guarantees-construct-new-nuclear-power-reactors

126. Matthew Bunn & Martin B. Malin, *Enabling a Nuclear Revival – And Managing Its Risks*, INNOVATIONS: TECHNOLOGY, GOVERNANCE, GLOBALIZATION 173, 176 (Fall 2009).

127. The United States currently operates 69 pressurized-water reactors (PWRs) and 35 boiling-water reactors (BWRs), which are collectively known as light-water reactors (LWRs).

128. Rebecca Smith, *Small Reactors Generate Big Hopes*, WALL ST.J. A1 (February 18, 2010); THE NATIONAL ACADEMIES REPORT at 456; Cleetus, et al., at 82.

129. *See* William Atkinson, *The Incredible Shrinking Reactor*, 148 PUB. UTIL. FORT. 50 (May 2010).

130. *See* Bunn & Malin at 174.

131. *See, e.g.*, Matthew L. Wald, *Loan Program May Stir Nuclear Industry*, N.Y. TIMES (December 24, 2009).

132. Blue Ribbon Commission on America's Nuclear Future, Advisory Committee Charter (March 1, 2010) available at http://brc.gov/pdfFiles/BRC_Charter.pdf

133. THE NATIONAL ACADEMIES REPORT at ch. 9.

134. *See, e.g.*, Symposium, *Greening the Grid: Building a Legal Framework for Carbon Neutrality*, 39 ENVT'L L. 927 (2009).

135. Joseph P. Tomain, *"Steel in the Ground": Greening the Grid with the iUtility*, 39 ENVT'L. L. 931 (2009).

136. *See* Michael J. Thompson, *The Conundrum of Multistate Electric Transmission Expansion: Who Will Pay?* 49 INFRASTRUCTURE 3 (Winter 2010); *Illinois Commerce Commission v. FERC*, 576 F. 3d 470 (7th Cir. 2009).

Chapter 5

1. National Petroleum Council, *Facing the Hard Truths About Energy: A Comprehensive View to 2030 of Global Oil and Natural Gas* 5 (July 18, 2007).

2. *See Obama's Remarks on Offshore Drilling*, N.Y. TIMES (April 1, 2010) available at http://www.nytimes.com/2010/04/01/science/earth/01energy-text.html

3. IHS CERA, *First Production from Newly Opened Offshore Atlantic Continental Shelf Areas could Occur as Early as Seven Years* (March 31, 2010) available at http://press.ihs.com/article_display.cfm?article_id=4225

4. The U.S. Department of Labor constructed a Web site about the disaster, which recounts events and provides government documentation at http://www.msha.gov/PerformanceCoal/PerformanceCoal.asp. *See also* Ian Urbina, *No Survivors Found After West Virginia Mine Disaster*, N.Y. TIMES (April 9, 2010) available at http://www.nytimes.com/2010/04/10/us/10westvirginia.html

5. This disaster is also subject to a federal investigation. *See* National Commission on the BP Deepwater Horizon Oil Spill and Offshore Drilling at http://www.oilspillcommission.gov/. The commission is scheduled to issue its report on January 12, 2011.

6. *See* Campbell Robertson & Leslie Kaufman, *Size of Spill is Larger Than Thought*, N.Y. Times (April 28, 2010) available at http://www.nytimes.com/2010/04/29/us/29spill.html?ref=us; Faiz Shakir, et al., *The Progress Report: Oil Spill: Gulf Kill* (May 27, 2010) available at http://pr.thinkprogress.org/2010/05/pr20100527

7. Associated Press, *2 Miners are Killed in Mine Collapse*, N.Y. Times (April 30, 2010) available at http://www.nytimes.com/2010/04/30/us/30mine.html

8. Department of Interior, Press Release, *Secretary Salazar Announces Approval of Cape Wind Energy Project on Outer Continental Shelf off Massachusetts* (April 28, 2010) available at http://www.doi.gov/news/doinews/Secretary-Salazar-Announces-Approval-of-Cape-Wind-Energy-Project-on-Outer-Continental-Shelf-off-Massachusetts.cfm

9. International Energy Agency, World Energy Outlook (2009); DOE Energy Information Administration, Annual Energy Review 2008 (July 2009) available at http://www.eia.doe.gov/aer/; DOE Energy Information Administration, Annual Energy Outlook 2010 Early Release available at http://www.eia.doe.gov/oiaf/aeo/overview.html#trends. All of the energy projections are based on the assumption that there are no changes in government laws and regulations.

10. Based on 2007 figures. *See* Energy Security Leadership Council, *A National Strategy for Energy Security: Recommendations to the Nation on Reducing U.S. Oil Dependence* 14 (September 2008) (hereinafter *National Strategy*). Chaired by General P.X. Kelly, USMC (Ret.) and Frederick W. Smith, Chairman, President and CEO of FedEx Corp., the Energy Security Leadership Council is a project of Securing America's Future Energy (*see* http://www.secureenergy.org/site/page.php?index) and was founded to address U.S. energy security. The Council is comprised of nationally prominent business and military leaders.

11. *See* Peter Jackson, *The Future of Global Oil Supply: Understanding the Building Blocks* (November 4, 2009) available at http://www.cera.com/aspx/cda/client/report/report.aspx?KID=5&CID=10720

12. Elisabeth Rosenthal, Booming China Is Buying Up World's Coal, N.Y. Times A1 (November 22, 2010) (In 2009, the U.S. exported 2,714 tons of coal to China, and in the first six months of 2010 exports increased to 2.9 million tons. Demand is so significant that new coal mines are being planned for the western U.S.)

13. Additional data on the growth of fossil fuel demand: Vice President Cheney's National Energy Policy Development Group predicted that by 2021 the demand for oil would increase 33%, the demand for natural gas would increase 50%, and the demand for electricity would increase 45%. *See* National Energy Policy Development Group, *National Energy Policy* x (May 2001).The National Petroleum Council Report projects that world demand for oil will increase between 35% and 80% and domestic demand will increase between 15% and 58%, *id.* at Chapter 1, page 6, while world demand for natural gas will increase about 50% and domestic demand will increase about 20%, at 7.The DOE

estimates that oil demand will grow about 30% between 2004 and 2030, DOE ENERGY INFORMATION ADMINISTRATION, ANNUAL ENERGY OUTLOOK 2007: WITH PROJECTIONS TO 2030 200 (February 2007) (reference case), while demand for coal will increase 44% during the same period.

14. *National Strategy* at 18.

15. Erik Shuster, *Tracking New Coal-Fired Power Plants* (January 8, 2010) available at http://www.netl.doe.gov/coal/refshelf/ncp.pdf.; Justin A. Brown, *King Coal's Uncertain Future: An Analysis of the Growing U.S. Coal Moratorium*, in Peter V. Lacoutre (ed.), RECENT DEVELOPMENTS IN PUBLIC UTILITY, COMMUNICATIONS AND TRANSPORTATION INDUSTRIES 2010 379 (2010).

16. KENNETH S. DEFFEYES, BEYOND OIL: THE VIEW FROM HUBBERT'S PEAK (2005); HUBBERT'S PEAK: THE IMPENDING WORLD OIL SHORTAGE (2002).

17. Matthew R. Simmons, *Have We "Peaked?"* (PowerPoint Presentation) (2009) available at Simmons & Company homepage at http://www.simmonsco-intl. com/research.aspx?Type=msspeeches for other speeches and presentations. *See also* MATTHEW R. SIMMONS, TWILIGHT IN THE DESERT: THE COMING SAUDI OIL SHOCK AND THE WORLD ECONOMY (2005).

18. INTERNATIONAL ENERGY AGENCY, WORLD ENERGY OUTLOOK 2009 85–86 (2009).

19. *See* Daniel Yergin, *It's Still the One*, FOREIGN POLICY (September/October 2009) available at http://www.foreignpolicy.com/articles/2009/08/17/its_ still_the_one

20. Jonathan Weisman & Jeffrey Ball, *U.S. to Toughen Drill Rules*, WALL ST. J. A1 (May 26, 2010).

21. Clifford Krauss & Elisabeth Rosenthal, *Mired in Canada's Oil Sands*, N.Y. TIMES B1 (May 19, 2010).

22. IHS CERA, *Growth in the Canadian Oil Sands: Finding the New Balance* ES-3(2009) available at http://www.cera.com/aspx/cda/client/knowledgeArea/ serviceDescription.aspx?KID=228

23. IHS CERA at ES-6.

24. See e.g., Jad Mouawad, *A Quest for Energy in the Globe's Remote Places*, N.Y. TIMES A1 (October 9, 2007).

25. DOE ENERGY INFORMATION ADMINISTRATION, ANNUAL ENERGY REVIEW 2006 8 (June 2007).

26. *National Strategy* at 22.

27. DOE, *Fact #522: Cost of Oil Dependence 2008* (June 9, 2008) available at http:// www1.eere.energy.gov/vehiclesandfuels/facts/2008_fotw522.html

28. *National Strategy* at 14.

29. *See, e.g.*, Sierra Club, *The Dirty Truth About Coal: Why Yesterday's Technology Should Not Be Part of Tomorrow's Energy Future* (June 2007) available at http:// www.sierraclub.org/coal/dirtytruth/coalreport.pdf

30. *National Strategy* at 23.

31. *See, e.g.*, Jad Mouawad, *Oil Prices Continue to Rise, With a Close Above $78*, N. Y. TIMES C3 (August 1, 2007); Matt Chambers, *Oil Soars to Exchange Record*, WALL ST. J. C9 (August 1, 2007).

32. Gregory Basheda, et al., *Why Are Electricity Prices Increasing? An Industry-Wide Perspective* (June 2006) available at http://www.eei.org/industry_issues/

electricity_policy/state_and_local_policies/rising_electricity_costs/Brattle_Report.pdf

33. John Broder, *Rule to Expand Mountaintop Coal Mining*, N.Y. Times A1 (August 23, 2007).

34. *See* Joseph P. Tomain, *The Dominant Model of United States Energy Policy*, 61 Univ. Colorado L. Rev. 355 (1990); Energy Law Group, Energy Law and Policy for the 21st Century ch. 6 (2000); John Deutch & Ernest J. Moniz (co-chairs), The Future of Coal: Options for a Carbon-Constrained World: An Interdisciplinary MIT Study (2007) (coal indispensable to our energy future).There exists the possibility of a slowing of the growth in use of fossil fuels if the commercial nuclear power industry expands. University of Chicago, The Economic Future of Nuclear Power (August 2004); John Deutch & Ernest J. Moniz, The Future of Nuclear Power: An Interdisciplinary MIT Study (2003); Edmund L. Andrews & Matthew L. Wald, *Energy Bill Aids Expansion of Atomic Power*, N.Y. Times A1 (July 31, 2007); Joseph P. Tomain, *Nuclear Futures*, 15 Duke Envt'l L. & Policy Forum 221 (2005).

35. Robert Stobaugh & Daniel Yergin (eds), Energy Future: A Report of the Energy Project of the Harvard Business School 213 (1979).

36. Andrew C. Revkin & Matthew L. Wald, *Solar Power Captures Imagination, Not Money*, N. Y. Times A1 (July 16, 2007).

37. Second quarter earnings in 2010, for example, were reported as: Exxon – $7.56 billion, up 91% over that of the first quarter; Chevron – $5.41 billion, or triple that of the first quarter; ConocoPhillips $2.5 billion, also triple that of the first quarter. Isabel Ordonez, *Strong Run From Big Oil, Not in Stocks Though* (July 30, 2010) available at http://blogs.wsj.com/marketbeat/2010/07/30/strong-run-from-big-oil-not-in-stocks-though/

38. Jacqueline Lang Weaver, *The Traditional Petroleum-Based Economy: An "Eventful" Future*, 36 Cumberland L. Rev. 505, 564 (2005–2006).

39. *See* Pew Economic Policy Group, *Subsidy Scope: Government Subsidies: Revealing the Hidden Budget* available at http://subsidyscope.com/media/pdf/Subsidyscope%20Framing%20Paper.pdf

40. The homepage for Subsidy Scope is at http://subsidyscope.com/

41. Douglas Koplow & Aaron Martin, *Fueling Global Warming: Federal Subsidies to Oil in the United States* ES-1 (1998).

42. Ronald J. Sutherland, *"Big Oil" at the Public Trough? An Examination of Petroleum Subsidies*, Policy Analysis (February 1, 2006) available at http://www.cato.org/pubs/pas/pa-390es.html

43. *See* Norman Myers & Jennifer Kent, *Perverse Subsidies: How Tax Dollars Can Undercut the Environment and the Economy* 5–9 (2001).

44. *See* Sidney A. Shapiro & Joseph P. Tomain, *Regulatory Law and Policy: Cases and Materials* 45 (3rd ed. 2003).

45. Jason Furman, et al., *An Economic Strategy to Address Climate Change and Promote Energy Security* 25 (October 2007) available at http://www.brookings.edu/papers/2007/10climatechange_furman.aspx

46. This section has drawn on Mona Hymel, *The United States Experience with Energy-Based Tax Incentives: The Evidence Supporting Tax Incentives for Renewable Energy*, 38 Loy. U. Chi. L. J. 43 (2006); Salvatore Lazzari, CRS Issue

Brief for Congress: Energy Tax Policy (April 22, 2005) available at http://kuhl. house.gov/UploadedFiles/energy%20tax.pdf; Salvatore Lazzari, CRS Issue Brief for Congress: Energy Tax Policy: An Economic Analysis (June 28, 2005) available at http://cnie.org/NLE/CRSreports/05jun/RL30406.pdf

47. David Kocieniewski, *As Oil Industry Fights a Tax, It Reaps Subsidies*, N.Y.Times (July 3, 2010) available at http://www.nytimes.com/2010/07/04/business/04bptax.html

48. Kocieniewski.

49. Hymel at 48–49.

50. Furman at 25.

51. Martin Sullivan, *Oil Driller Gain Billions from "Immoral" Tax Breaks* (June 16, 2010) available at http://www.tax.com/taxcom/taxblog.nsf/Permalink/UBEN-86GPTN?OpenDocument

52. Kocieniewski.

53. U. S. Department of the Treasury, *Press Release: Statement of Alan B. Krueger Before the Subcommittee on Energy, Natural Resources and Infrastructure* (September 10, 2009) available at http://www.ustreas.gov/press/releases/tg284.htm

54. *See* the following cases in which the government sought refunds: *Quincy Oil, Inc. v. FEA*, 468 F. Supp. 383 (D. Mass. 1979); *Plaquemines Oil Sales Corp v. FEC*, 461 F. Supp. 276 (D. La. 1978); *Naph-Sol Refining Co. v. Cities Service Oil Co.*, 495 F. Supp. 882 (D. Mich. 1980); *United States v. Exxon Corp.*, 470 F. Supp. 674 (D.D.C. 1979); *Phillips Petroleum v. DOE*, 449 F. Supp 760 (D. Del. 1978); *Standard Oil Co. v. DOE*, 596 F. 2d 1029 (TEAC 1978); *United States v. Metropolitan Petroleum Co., Inc.* 743 F. Supp. 820 (S.D. Fla. 1990).

55. Jan Laitos, Sandra Zellmer, Mary C. Wood & Daniel Cole, Natural Resources Law 991–92 (2006).

56. Federal Oil and Gas Royalty Management Act, 30 U. S. C. §§ 1701–57.

57. *See* U.S. Government Accounting Office, *Royalty Revenues: Total Revenues Have Not Increased at the Same Pace as Rising Oil and Natural Gas Prices Due to Decreasing Production Sold* (June 21, 2006) available at http://www.gao.gov/new.items/d06786r.pdf

58. Editorial, *Big Oil's Good Deal*, N.Y. Times (July 11, 2010) available at http://www.nytimes.com/2010/07/12/opinion/12mon1.html

59. *See, e.g., United States ex rel. Johnson v. Shell Oil Co.*, Dkt No. 9:96CV66 (E.D. Texas); see also a brief description of the case by one of the law firms involved in the suit at http://www.fcalawfirm.com/cases/index.html

60. Office of the Inspector General, Department of the Interior, *Minerals Management Service: False Claims Allegations* (September 19, 2007).

61. *Transmittal Letter Filed with the Report from Inspector General Earl E. Devaney to DOI Secretary Dirk Kempthorne at id.*

62. Edmund L. Andrews, *Inspector Finds Broad Failures in Oil Program*, N. Y. Times A1 (September 26, 2007).

63. DOI Office of Inspector General, *Audit Report: Mineral Management Service's Compliance Review Process* 6 (December 6, 2006).

64. Energy Information Administration, *Federal Financial Interventions and Subsidies in Energy Markets 2007* (April 2008) available at http://www.eia.doe.gov/oiaf/servicerpt/subsidy2/index.html

65. *See* Doug Kaplow, *EIA Energy Subsidy Estimates: A Review of Assumptions and Omissions* 12 (March 2010) available at http://www.earthtrack.net/documents/eia-energy-subsidy-estimates-review-assumptions-and-omissions. *See also* Jennifer Ellis, *The Effects of Fossil-Fuel Subsidy Reform: A Review of Modelling and Empirical Studies* (March 2010) (regarding the methodology for calculating subsidies) available at http://www.globalsubsidies.org/files/assets/effects_ffs.pdf

66. *Id.* at 12–13.

67. Janet L. Swain & William R. Moomaw, *Renewable Revolution: Low-Carbon Energy by 2030* 35 (2009) available at http://www.worldwatch.org/node/6340

68. Gilbert E. Metcalf, *Taxing Energy in the United States: Which Fuels Does the Tax Code Favor?* (2009) available at http://www.economicsclimatechange.com/2009/01/taxing-energy-in-united-states-which.html

69. Environmental Law Institute, *Estimating U.S. Government Subsidies to Energy Sources: 2002–2008* 3 (September 2009) available at http://www.elistore.org/reports_detail.asp?ID=11358

70. Sima J. Gandhi, *Pumping Tax Dollars to Big Oil* (April 14, 2010) available at http://www.americanprogress.org/issues/2010/04/oil_subsidies.html; Richard W. Caperton & Sima J. Gandhi, *America's Hidden Power Bill: Examining Federal Energy Tax Expenditures* (April 2010) available at http://www.americanprogress.org/issues/2010/04/pdf/energytaxexpenditures.pdf

71. *See* Stephen Breyer, Regulation and Its Reform ch. 1 (1982).

72. *See* Frank Ackerman & Elizabeth A. Stanton, *The Social Cost of Carbon* (April 1, 2010) available at http://www.e3network.org/papers/SocialCostOfCarbon_SEI_20100401.pdf

73. Mark Memmott, *BP Chief: Oil Spill is 'Relatively Tiny,'* NPR (May 14, 2010) available at http://www.npr.org/blogs/thetwo-way/2010/05/bp_chief_oil_spill_is_relative.html

74. Tom Zeller, Jr., *Federal Officials Say They Vastly Underestimated Rate of Oil Flow Into Gulf*, N.Y. Times A15 (May 28, 2010).

75. Alyson Flournoy, et al., *Regulatory Blowout: How Regulatory Failures Made the BP Disaster Possible, and How the System Can Be Fixed to Avoid a Recurrence* (October 2010) available at http://www.progressivereform.org/articles/BP_Reg_Blowout_1007.pdf

76. National Commission of the BP Deepwater Horizon Oil Spill and Offshore Drilling, *Stopping the Spill: The Five-Month Effort to Kill the Macondo Well* (November 22, 2010) available at http://www.oilspillcommission.gov/sites/default/files/documents/Containment%20Working%20Paper%2011%2022%2010.pdf

77. National Commission of the BP Deepwater Horizon Oil Spill and Offshore Drilling, *Response/Clean-Up Technology Research & Development and the BP Deepwater Horizon Water Spill* (November 22, 2010) available at http://www.oilspillcommission.gov/document/responseclean-technology-research-development-and-bp-deepwater-horizon-oil-spill

78. Ian Urbina, *Inspector General's Inquiry Faults Actions of Federal Drilling Regulators*, N.Y. Times A16 (May 25, 2010).

79. *Memorandum from Earl E. Devaney, Inspector General to Secretary Kempthorne* 3 (September 9, 2008) (including *Investigative Report: MMS Oil Marketing Group – Lakewood*) available at http://www.doioig.gov/upload/RIK%20REDACTED%20FINAL4_082008%20with%20transmittal%209_10%20date.pdf

80. *See* Center for Biological Diversity, *MMS Approved 27 Gulf Drilling Operations After BP Disaster: 26 Were Exempted From Environmental Review, Including Two to BP* (May 1, 2010) available at http://www.biologicaldiversity.org/news/press_releases/2010/post-disaster-permits-05-07-2010.html; Ian Urbina, *U.S. Said to Allow Drilling Without Needed Permits*, N.Y. TIMES A1 (May 14, 2010).

81. Ian Urbina, *Documents Show Earlier Fears About Safety of Offshore Well*, N.Y. TIMES A1 (May 31, 2010).

82. Ian Burina, *BP Chose Riskier of Two Options for Well Casing*, N.Y. TIMES A1 (May 27, 2010).

83. John M. Broder, *U.S. to Split Up Agency Policing Oil Industry*, N.Y. TIMES (May 11, 2010) available at http://www.nytimes.com/2010/05/12/us/12interior.html

84. Center for Biological Diversity.

85. *See* Ian Urbina, *Despite Obama's Moratorium, Drilling Projects Move Ahead*, N.Y. TIMES A1 (May 24, 2010).

86. Peter Baker, *Regret Mixed with Resolve*, N.Y. TIMES A1 (May 28, 2010); Clifford Krauss & John M. Broder, After Delay, *BP Resumes Effort to Plug Oil Leak*, N.Y. TIMES A1 (May 28, 2010).

87. Giovanni Russenello, *W. Va. Mine Blast: Coal Firm Had Worst Safety Record* (November 23, 2010) available at http://www.msnbc.msn.com/id/40325100/ns/business-us_business/

88. *See* Center for American Progress, Progress Report, *Profits Over Safety* (April 8. 2010) available at http://pr.thinkprogress.org/2010/04/pr20100408; *see also* MSHA list of violations 2000–2010 at http://www.msha.gov/PerformanceCoal/Violation_Summary.pdf; Steven Mufson, Jerry Markon & Ed O'Keefe, *West Virginia Mine Has Been Cited for Myriad Safety Violations*, Washington Post (April 7, 2010) available at http://www.washingtonpost.com/wp-dyn/content/article/2010/04/05/AR2010040503877.html?sid=ST2010040505519; Stephen Power, Kris Maher & Sibhan Hughes, *Mine Owner Cited on Day of Blast*, WALL ST. J. A6 (April 8, 2010).

89. *See* MICHAEL SHNAYERSON, COAL RIVER (2008).

90. Center for American Progress.

91. Matthew L. Wald, *Mine Executive Favors Outside Inquiry Into Deaths*, N.Y. TIMES (May 20, 2010) available at http://www.energytribune.com/articles.cfm/4157/Mine-Executive-Favors-Outside-Inquiry-Into-Deaths

92. *See* Adam Liptak, *Motion Ties W. Virginia Justice to Coal Executive*, WASHINGTON POST (January 15, 2008) available at http://www.nytimes.com/2008/01/15/us/15court.html?_r=3&hp&oref=slogin&oref=slogin

93. *Caperton v. A. T. Massey Coal Co., Inc.*, 129 S.Ct. 2252 (2009).

94. Michael Cooper, Gardiner Harris & Eric Lipton, *In Mine Safety, A Meek Watchdog*, N.Y. TIMES (April 10, 2010) available at http://www.nytimes.com/2010/04/11/us/11mining.html

95. Ed O'Keefe, *U.S. Mine Safety and Health Administration Faces Training and Oversight Problems*, WASHINGTON POST (April 7, 2010) available at http://www.washingtonpost.com/wp-dyn/content/article/2010/04/06/AR2010040603986.html
96. Michael Cooper, et al.
97. *Id.*
98. *Id.*
99. *Cheney v. United States District Court for D.C.*, 542 U.S. 367 (2004).
100. Michael Abramowitz & Steven Mufson, *Papers Detail Industry's Role in Cheney's Energy Report*, WASHINGTON POST (July 18, 2007) available at http://www.washingtonpost.com/wp-dyn/content/article/2007/07/17/AR2007071701987.html
101. JEFF GOODELL, BIG COAL 186–89 (2006).
102. GOODELL; ROSS GELBSPAN, THE HEAT IS ON: THE CLIMATE CRISIS, THE COVER-UP, THE PRESCRIPTION (1998); BOILING POINT: HOW POLITICIANS, BIG OIL AND COAL, JOURNALISTS AND ACTIVISTS ARE FUELING THE CLIMATE CRISIS – AND WHAT WE CAN DO TO AVERT DISASTER (2004).
103. Faiz Shakir, et al., *Environment: Climate Zombie Caucus* (November 22, 2010) available at http://pr.thinkprogress.org/2010/11/pr20101122
104. Energy Security Leadership Council, *Economic Impact of the Energy Security Leadership Council's National Strategy for Energy Security* (February 2009) available at http://www.secureenergy.org/sites/default/files/990_ESLC-Impact-Report.pdf
105. *Id.*
106. BURTON RICHTER, BEYOND SMOKE AND MIRRORS: CLIMATE CHANGE AND ENERGY IN THE 21ST CENTURY ch.3 (2010).
107. RICHTER at 99.
108. Shai Agassi, *World Without Oil: Better Place Builds a Future for Electric Vehicles*, 4 INNOVATIONS: TECHNOLOGY, GOVERNANCE, GLOBALIZATION 125, 136–37 (Fall 2009); *see also* Peter Huber, *The Million-Volt Answer to Oil* (October 2008) available at http://www.manhattan-institute.org/html/eper_03.htm
109. *See, generally*, DAVID B. SANDALOW (ED.), PLUG-IN ELECTRIC VEHICLES: WHAT ROLE FOR WASHINGTON? (2009).
110. RICHTER at 101.
111. AMORY B. LOVINS, ET AL., WINNING THE OIL ENDGAME: INNOVATION FOR PROFITS, JOBS, AND SECURITY 44–72 (2004).
112. *See* W. Ross, et al., *Analysis of Policies to Reduce Oil Consumption and Greenhouse Gas Emissions from the U.S. Transportation Sector* 2 (February 2010) available at http://belfercenter.ksg.harvard.edu/files/Policies%20to%20Reduce%20Oil%20Consumption%20and%20Greenhouse%20Gas%20Emissions%20from%20Transportation.pdf
113. National Research Council, *Transitions to Alternative Transportation Technologies – Plug-In Hybrid Electric Vehicles* 1 (2009).
114. *Id.* at 2.
115. NATIONAL ACADEMIES, LIQUID TRANSPORTATION FUELS FROM COAL AND BIOMASS: TECHNOLOGICAL STATUS, COSTS AND ENVIRONMENTAL IMPACTS (2009).

116. Shai Agassi at 135–36.
117. Ashlee Duncan, *Pulling the Plug on Greenhouse Emissions: The U.S. Power Grid Could Accommodate Plug-In Vehicles*, 3 ENVT'L & ENERGY L. & POL. J. 158 (2008); Dean Murphy, et al., *Plugging In*, 148 PUB. UTIL. FORT. 30 (June 2010).
118. PETER FOX-PENNER, SMART POWER: CLIMATE CHANGE, THE SMART GRID, AND THE FUTURE OF ELECTRIC UTILITIES 70 (2010).
119. Press Release, Better Place, *Better Place Secures $350 Million Series B Round Led by HSBC Group* (Jan. 24, 2010). *See also* Better Place homepage at http://www.betterplace.com/
120. Shai Agassi at 127–28.
121. Rocky Mountain Institute, *Smart Garage Charrette Report* 8 (2008) available at http://move.rmi.org/files/smartgarage/SmartGarageCharretteReport_2.10.pdf
122. *Id.* at 8–9.
123. Press Release, Tesla Motors, *Tesla Motors and Toyota Motor Corporation Intend to Work Jointly on EVE Development, TMC to Invest in Tesla* at http://www.teslamotors.com/media/press_room.php?id=2509
124. Ford Motor Company, *Ford Invests $450 Million More in Electric Vehicles: Next Generation Hybrid, Plug-In Hybrid and New DEV Will be Produced at Michigan Assembly Plant* (Jan. 11, 2010) available at http://media.ford.com/article_display.cfm?article_id=31805
125. *See* Nissan LEAF homepage at http://www.nissanusa.com/leaf-electric-car/index#/leaf-electric-car/specs-features/index
126. *See* Chevrolet Volt homepage at http://www.chevrolet.com/pages/open/default/future/volt.do
127. Victor B. Flatt, *Paving the Legal Path for Carbon Sequestration From Coal*, 19 DUKE ENVT'L L. & POL. FORUM 211, 212 (2009).
128. *See, e.g.*, Joe Hezir & Melanie Kenderdine, *Federal Research Management for Carbon Mitigation for Existing Coal Plants* (March 23, 2009) available at http://web.mit.edu/mitei/docs/reports/hezir-kenderdine.pdf
129. Energy Security Leadership Council & Securing America's Future Energy at 49.
130. Alexandra B. Klass, *CPR Perspective: Carbon Capture and Geological Sequestration* (July 2009) available at http://www.progressivereform.org/perspCarbonCapture.cfm
131. *See, e.g.*, MIT Energy Initiative Symposium, *Retrofitting of Coal-Fired Power Plants for CO_2 Emissions Reductions* (Marcy 23, 2009) available at http://mit.edu/MITEI/docs/reports/meeting-report.pdf
132. *Id.*
133. *See* Barbara Freese, Steve Clemmer & Alan Nogee, *Coal Power in a Warming World: A Sensible Transition to Cleaner Energy Options* (October 2008) available at http://www.ucsusa.org/assets/documents/clean_energy/Coal-power-in-a-warming-world.pdf
134. *See* Robert Bryce, *A Bad Bet on Carbon*, N.Y. TIMES (May 12, 2010) available at http://www.nytimes.com/2010/05/13/opinion/13bryce.html?pagewanted=print
135. John Deutch & Ernest J. Moniz at ix.
136. *Id.* at x.

137. Freese, et al., at 12; DOE National Energy Technology Laboratories, *IGCC Plants With and Without Carbon Capture and Sequestration* available at http://www.netl.doe.gov/energy-analyses/pubs/deskreference/B_IG_051507.pdf
138. *National Strategy* at 49.
139. *See* Duke Energy Pres Release, *Edwardsport Integrated Gasification Combined Cycle (IGCC) Station* at http://www.duke-energy.com/pdfs/igcc-fact-sheet.pdf; General Electric Press Release, *GE Achieves Cleaner Coal Energy Milestone – Advanced Technology Turbines Shift to First-of-Its-Kind Power Plant* at http://www.gepower.com/about/press/en/2010_press/043010.htm
140. *See* Freese, et al., at ch. 3.
141. *See* Freese at 16.
142. *See* DOE, *FutureGen Clean Coal Project* available at http://www.fossil.energy.gov/programs/powersystems/futuregen/
143. *See* DOE, *Clean Coal Technology and the Clean Coal Power Initiative* available at http://www.fossil.energy.gov/programs/powersystems/cleancoal/index.html. *See also* DOE National Energy Technology Laboratories, *Carbon Sequestration Technology Roadmap and Program Plan* (2007) available at http://www.netl.doe.gov/technologies/carbon_seq/refshelf/project%20portfolio/2007/2007Roadmap.pdf
144. *See* Wendy B. Jacobs, et al., *Proposed Roadmap for Overcoming Legal and Financial Obstacles to Carbon Capture and Sequestration* (March 2009) available at http://belfercenter.ksg.harvard.edu/files/2009-04_ETIP_Jacobs_et_al.pdf; Alexandra B. Klass & Elizabeth J. Wilson, *Climate Change, Carbon Sequestration, and Property Rights*, 2010 U. Ill. L. Rev. 363 (2010).

Chapter 6

1. CNA Military Advisory Board, *Powering America's Economy: Energy Innovation at the Crossroads of National Security Challenges* viii (July 2010) available at http://www.cna.org/sites/default/files/research/WEB%2007%2027%2010%20MAB%20Powering%20America%27s%20Economy.pdf
2. *See* Sidney A. Shapiro & Joseph P. Tomain, *Rethinking Reform of Electricity Markets*, 40 Wake Forest L. Rev. 497 (2005). In this article, Professor Shapiro and I argued for the continuing need for regulation because of the continuing natural monopoly characteristics of the electricity industry. *See also* Joseph P. Tomain, *Whither Natural Monopoly? The Case of Electricity*, in The End of a Natural Monopoly: Deregulation Competition in the Electric Power Industry: The Economics of Legal Relationships (Peter Grossman & Daniel H. Cole eds., 2003). This chapter adds the argument that government regulation is necessary for the transition of the electricity industry to one that promotes a low-carbon energy economy, not a traditional dirty fossil fuel economy.
3. *See also* Joseph P. Tomain, *The iUtility* in Beyond Environmental Law: Policy Proposals for a Better Environmental Future ch. 10 (David M. Driesen & Alyson C. Flournoy eds., 2010); *"Steel in the Ground": Greening the Grid with the iUtility*, 39 Envt'l L. 931 (2009); *Building the iUtility*, 146 Pub. Util. Fort. 28 (August 2008).

4. National Academy of Sciences, National Academy of Engineering & National Research Council, *Electricity from Renewable Resources* 2 (2010).

5. *See* Ronald Brownstein, *The California Experiment*, THE ATLANTIC (October 7, 2009) available at http://www.theatlantic.com/magazine/archive/2009/10/the-california-experiment/7666/; *see* Governor Arnold Schwarzenegger, Executive Order S-14-08 available at http://www.gov.ca.gov/executive-order/11072/

6. *See* MICHAEL B. GERARD (ED.), GLOBAL CLIMATE CHANGE AND U.S. LAW 7–11 (2007); *see also Connecticut v. American Electric Power*, 406 F. Supp. 265, 268 (S.D.N.Y 2005).

7. Edison Electric Institute, *Key Facts About the Electric Power Industry* v (FEB. 2007).

8. Edison Electric Institute at 4.

9. Edison Electric Institute at 5.

10. *See* WILLIAM T. GORMLEY, JR., THE POLITICS OF PUBLIC UTILITY REGULATION (1983).

11. Lester Lave, Jay Apt & Seth Blumsack, *Deregulation/Restructuring Part I: Reregulation Will Not Fix the Problems*, 20 ELECTRICITY J. 9, 10 (October 2007).

12. *See* Gregory Basheda, et al., *Why Are Electricity Prices Increasing? An Industry-Wide Perspective* (June 2006) available at http://www.thefederalregister.com/d.p/2009-05-21-E9-12029

13. LEONARD S., ANDREW S., & ROBERT C. HYMAN, AMERICAN'S ELECTRIC UTILITIES: PAST, PRESENT AND FUTURE ch. 18 (8th ed. 2005).

14. JOSEPH P. TOMAIN, NUCLEAR POWER TRANSFORMATION (1987).

15. In 1978, as part of the National Energy Act, Congress passed the Powerplant and Industrial Fuel Use Act, which prohibited the construction of power plants designed to burn oil or natural gas. In effect, the act worked to promote the construction of coal plants. In 1987, as natural gas prices fell, the provisions against natural gas plants were repealed.

16. Paul Joskow, *Deregulation* (February 28, 2009) (2009 Distinguished Lecture AEI Center for Regulatory and Market Studies) available at http://econ-www.mit.edu/files/3875

17. Public Utilities Regulatory Policies Act of 1978, Pub. L. No. 95–617, 92 Stat. 3117 (codified as amended in scattered sections of 15, 16, 26, 42, and 43 U.S.C.). For an historical review of the electricity industry, *see* Joseph P. Tomain, *networkindustries.gov*, 48 KAN. L. REV. 829 (2000); *The Past and Future of Electricity Regulation*, 32 ENV'T L. 435 (2002); Sidney A. Shapiro & Joseph P. Tomain, *Rethinking Reform of Electricity Markets*, 40 WAKE FOREST L. REV. 497 (2005).

18. Paul Joskow, *Lessons Learned From Electricity Market Liberalization*, 29 ENERGY J. 9, 10–11 (2008).

19. ANNUAL ENERGY REVIEW 2008 at 282.

20. Linda Stuntz & Susan Tomasky, *An Electric Grid for the 21st Century* 10–11 (2010).

21. Judge Richard Posner's description of a cable television grid applies with equal force to the electricity grid: "The cost of the cable grid appears to be the biggest cost of a cable television system and to be largely invariant to the number of subscribers this system has.... [O]nce the grid is in place ... the cost of adding another

subscriber probably is small, if so, the average cost of cable television would be minimized by having a single company in any given geographical area.…" *Omega Satellite Products. Co. v. City of Indianapolis*, 694 F.2d 119 (7th cir. 1982).

22. Then Judge Kenneth Starr provides a good description of the regulatory compact: "The utility business represents a compact of sorts; a monopoly on service in a particular geographical area (coupled with state-conferred rights of eminent domain or condemnation) is granted to the utility in exchange for a regime of intensive regulation, including price regulation, quite alien to the free market.… Each party to the compact gets something in the bargain. As a general rule, utility investors are provided a level of stability in earnings and value less likely to be attained in an unregulated or moderately regulated sector; in turn, ratepayers are afforded universal, non-discriminatory service and protection from monopolistic profits through political control over an economic enterprise." *Jersey Central Power & Light Co. v. FERC*, 810 F.2d 1168, 1189 (D.C. Cir. 1987).

23. James C. Bonbright, Albert L. Danielson & David R. Kamerschen, Principles of Public Utility Rates (2nd ed. 1998); Charles Phillips, Jr., The Regulation of Public Utilities: Theory and Practice (3rd ed. 1993).

24. *See* Scott Hempling, *Multi-Utility Issues at a Glance* 4–5 (March 1, 2009) available at http://www.nrri.org/pubs/multiutility/NRRI_multi_utility_issues_mar09-04.pdf

25. The traditional rate formula can be simply expressed:

$$R = O + B\,r$$

Each of these variables is significant as are the consequences on the industry as a result of the application of the formula. **R** is the *revenue requirement* that is necessary for the utility to remain in business and serve its customers. The *revenue requirement* is comprised of *operating expenses* (**O**) and a *rate base* (B) on which the utility earned a *rate of return*.

26. Harvey Averch & Leland L. Johnson, *Behavior of the Firm Under Regulatory Constraint*, 52 Am. Econ. Rev. 1052 (1962).

27. U.S.-Canada Power System Outage Task Force, *Final Report on the August 14, 2003 Blackout in the United States and Canada: Causes and Recommendations* (April 2004).

28. *See* Bruce Radford, *The Queue Quandry*, 146 Pub. Util. Fort. 28 (March 2008).

29. *See* Scott M. Gawlicki, *AMI Standards – A Work in Progress*, 146 Pub. Util. Fort. 68 (September 2008).

30. ISO/RTO Council, *2009 State of the Markets Report* (2009) available at http://www.isorto.org/atf/cf/%7B5B4E85C6-7EAC-40A0-8DC3-003829518-EBD%7D/2009%20IRC%20State%20of%20Markets%20Report.pdf; COMPETE, *RTO and ISO Markets are Essential to Meeting Our Nation's Energy and Environmental Challenges* (March 19, 2010) available at http://www.competecoalition.com/files/RTO-ISO-Benefits-White-Paper.pdf

31. Bracken Hendricks, *Wired for Progress: Building a National Clean-Energy Smart Grid* 40–48 (February 2009) available at http://www.americanprogress.org/issues/2009/02/pdf/electricity_grid.pdf

32. *See, e.g.*, Michael Burr & Bruce W. Radford & Scott M. Gawlicki, *Special Report: Selling the Smart Grid*, 146 Pub. Util. Fort. 42 (April 2008).

33. John Holdren, *The Energy Innovation Imperative: Addressing Oil Dependence, Climate Change, and Other 21st Century Energy Challenges*, 2 INNOVATIONS: TECHNOLOGY, GOVERNANCE, GLOBALIZATION 3 (Spring, 2006).

34. *See* DON TAPSCOTT & ANTHONY D. WILLIAMS, WIKINOMICS: HOW MASS COLLABORATION CHANGES EVERYTHING (2006); John Seely Brown & John Hagel III, *The Next Frontier of Innovation*, MCKINSEY QUARTERLY No. 3 (2005) available at http://www.johnseelybrown.com/pushpull.pdf

35. *See* KEMA, *Set a Course for the Future* (2009) available at http://www.kema. com/Images/KEMA%20Set%20a%20Course%20for%20the%20Future%20 Brochure%202008_tcm9-23986.pdf. ("Energy generation and storage will become more decentralized, controlled, and delivered at the micro-grid level, enabling greater efficiency in fuel conversion and delivery, as well as conformance to increased renewable regulations and emerging environmental social responsibility.")

36. *See* Alison Silverstein, *The Smart Grid and the Utility of the Future* (May 30, 2008) PowerPoint presentation available at http://www.smartgridnews.com/ artman/uploads/1/silversteinmay2008.pdf

37. John J. Marhoefer, *Intelligent Generation™: The Smart Way to Build the Smart Grid*, 23 NAT. RESOURCES & ENV'T 15, 21 (Summer 2008).

38. Gregory Basheda, et al., at 2–5.

39. COMPETE, *RTO and ISO Markets are Essential to Meeting Our Nation's Energy and Environmental Challenges* (March 19, 2010) available at http://www.com-petecoalition.com/files/RTO-ISO-Benefits-White-Paper.pdf

40. DOE Environmental Information Administration, *U.S. Electric Utility Demand-Side Management: Trends and Analysis* available at http:\\www.eia.doe.gov\ cneaf\pubs_html\feat_dsm\contents.html

41. The Department of Energy reported on efforts under EPAct 2005 to look at demand response. Their report indicated that, given the lack of uniformity in estimating and quantifying methods, it was not possible to quantify the national benefits of demand response. Nevertheless, DOE recognized that demand response continued to have promise and recommended its continuation and studied consideration. Department of Energy, *Benefits of Demand Response in Electricity Markets and Recommendations for Achieving Them: A Report to the United States Congress Pursuant to Section 1252 of The Energy Policy Act of 2005* (February 2006). *See also* Jon Wellinghoff & David L. Morenoff, *Recognizing the Importance of Demand Response: The Second Half of the Wholesale Electric Market Equation*, 28 ENERGY L. REV. 389 (2007) (federal and state regulators should increase DSM requirements); Chris McCall, et al., *Demanding More from DR*, 148 PUB. UTIL. FORT. 24 (August 2010).

42. *See, e.g.*, FERC Order No. 732, Docket No. RM09–23–000 (March 19, 2010), which expands the list of non-utility generators or qualifying facilities.

43. TOMAIN at 8.

44. TOMAIN at *id*.

45. *See* Elcon, *Utility Energy Efficiency Programs: Too Cheap to Meter?* (November 2008) available at http://www.elcon.org/Documents/112608/ UtilityEEPrograms-TooCheaptoMeter-Nov%2026,08.pdf

46. *See* NERC, *Reliability Standards for the Bulk Electric Systems in North America* available at http://www.nerc.com/page.php?cid=2|20

47. *See* The American Clean Energy and Security Act of 2009 (the Waxman-Markey bill), H.R.2454, 111th Cong., 1st Sess.
48. Elcon at 2.
49. Under standard economic theory, as prices decline consumption will increase. Energy efficiency programs are intended to reduce price; they are also intended to promote conservation. Nevertheless, as electricity prices decline there will be some increase in consumption, which is sometimes referred to as the *rebound* effect or the *takeback* effect. *Id.* at 5.
50. *See, generally,* U.S. Environmental Protection Agency, *Aligning Utility Incentives with Investment in Energy Efficiency* (November 2007).
51. Michael T. Burr, *Seeing Green: Renewables Attract Utility Investment Dollars,* 147 Pub. Util Fort. 28 (May 2009).
52. *Id.*
53. The Pew Center on Global Climate Change, *States with Renewable Portfolio Standards* available at http:\\www.pewclimate.org\what_s_being_done\in_the_states\rps.cfm (August, 2007).
54. The American Clean Energy and Security Act of 2009 §101 (also known as the Waxman-Markey Bill). A Discussion Draft can be found at http://energycommerce.house.gov/Press_111/20090331/acesa_discussiondraft.pdf
55. *See* Union of Concerned Scientists, *Renewable Electricity Standards at Work in the States* available at http://www.ucsusa.org/assets/documents/clean_energy/RES_in_the_States_Update.pdf; *see also* Union of Concerned Scientists, *Experts Agree: Renewable Electricity Standards are a Key Driver of New Renewable Energy Development* available at http:\\www.ucsusa.org\clean_energy\clean_energy_policies
56. *See, e.g.,* EPA Green Power Partnership, *Renewable Energy Certificates* (July 2008) available at http://www.epa.gov/greenpower/documents/gpp_basics-recs.pdf
57. *See* Daniel P. Krueger & Andre Begasso, *Mandating Federal Renewables: The Importance of Getting the REC Market Right,* 148 Pub. Util. Fort. 40 (January 2010).
58. Steven Ferrey, et al., *FIT in the USA,* 148 Pub. Util. Fort. 60 (June 2010).
59. Bruce W. Radford, *PURPA's Changing Climate,* 148 Pub. Util. Fort. 20 (July 2010).
60. Vermont House Bill 446 (2009) available at http://www.leg.state.vt.us/docs/2010/bills/House/H-446.pdf; California Public Utilities Code §399.2 (2009).
61. *See, generally,* Karlynn Cory, Toby Couture & Claire Kreycik, *Feed-in Tariff Policy: Design, Implementation, and RPS Policy Interactions* (March 2009) available at http://www.nrel.gov/docs/fy09osti/45549.pdf
62. *See, e.g.,* Richard P. Sedano, *Electric Product Disclosure: A Status Report* (July 2002) available at http://www.raponline.org/docs/RAP_Sedano_ElectricProductDisclosure_2002_07.pdf; Pew Center on Global Climate Change, *Greenhouse Gas Reporting and Disclosure: Key Elements of a Prospective U.S. Program* available at http://www.pewclimate.org/docUploads/policy_inbrief_ghg.pdf
63. *See, e.g.,* The Climate Registry at http://www.theclimateregistry.org/ (supports voluntary and state-mandated programs); Carbon Disclosure Project at https://www.cdproject.net/en-US/Pages/HomePage.aspx (compiling international

database that is publically available); and Global Reporting Initiative at http://www.globalreporting.org/Home (same).

64. SEC, *Commission Guidance Regarding Disclosure Related to Climate Change*, 75 Fed. Reg. 6290 (February 18, 2010) available at http://www.sec.gov/rules/interp/2010/33-9106fr.pdf

65. *See, e.g.*, Amory B. Lovins & Chris Lotspeich, *Energy Surprises for the 21st Century*, 53 J. INT'L AFFAIRS 191 (Fall 1999); Amory B. Lovins, *Designing a Sustainable Energy Future: Integrating Negawatts with Diverse Supplies at Least Cost* (November 13, 2003) available at http://www.rmi.org/images/PDFs/Energy/E03-13_DsnSusEnrgyFuture.pdf

66. *See* U.S. Environmental Protection Agency, *National Action Plan for Energy Efficiency* 2–6 to 2–7 (July 2006) available at http://www.epa.gov/cleanenergy/documents/napee/napee_report.pdf; *see also* USEPA, ALIGNING UTILITY INCENTIVES WITH INVESTMENT IN ENERGY EFFICIENCY ch. 5 (November 2007).

67. *See, e.g.*, Michael J. Beck & William Klun, *IOUs Under Pressure: Policy and Technology Changes are RE-shaping the Utility Business Model*, 147 PUB. UTIL. FORT. 37 (June 2009). ("Electric utilities are at the confluence of a once-in-a-lifetime economic, technology and regulatory forces [sic] that will, over time, re-shape utility business model. The demonstrable need for massive investment in electricity delivery infrastructure, along with the construction of new, clean production technologies, presages enormous cost pressures on rate payers – and utilities that serve them.")

68. Lori Bird, Claire Kreycik & Barry Friedman, *Green Power Marketing in the United States: A Status Report (2008 Data)* (September 2009); NREL *Highlights Leading Utility Green Power Leaders* (April 13, 2009) available at http://www.nrel.gov/news/press/2009/679.html. Recently, utilities have been able to reduce their premiums. *See* Associated Press, *Green Mountain to Reduce Green Power Premium* (April 13, 2009) available at http://www.boston.com/news/local/vermont/articles/2009/04/13/green_mountain_power_to_reduce_green_power_premium/; *see also* NCGreenPower, *Notice of Change in Premium for New Solar Photovoltaic (PV) Agreements* available at http://www.ncgreenpower.org/resources/

69. NREL *Highlights Leading Utility Green Power Programs* (April 3, 2007). Available at http:\\www.eere.energy.gov\greenpower\resources\tables\pdfs\0405_topten_pr.pdf

70. US DOE and US EPA Leadership Group, *National Action Plan for Energy Efficiency*, ES-5 (July, 2006). *See also*, Amory B. Lovins, *Energy End-Use Efficiency* (September 2005).

71. *See also* Edward H. Comer, *Transforming the Role of Energy Efficiency*, 23 NAT. RESOURCES & ENV'T. 34 (Summer 2008).

72. *See* Steven Mitnick, *Making Efficiency Cool: A New Business Plan for Capturing Big Savings*, 147 PUB. UTIL. FORT. 35 (April 2009).

73. Michael Price, *Bringing Customers on Board – Part II*, 148 PUB. UTIL FORT. 30 (August 2010).

74. *See* Ahmad Faruqui, Ryan Hledik & Sanem Sergici, *Rethinking Prices: The Changing Architecture of Demand Response in America*, 148 PUB. UTIL. FORT. 30 (January 2010).

75. *See, e.g.,* Thomas Friedman, Hot, Flat and Crowded: Why We Need a Green Revolution and How It Can Renew America 224–28 (2008).

76. Efficiency Vermont 2006: *Preliminary Executive Summary; Efficiency Vermont 2006: Preliminary Results in Saving Estimates Report* (March 30, 2007).

77. Vermont Energy Investment Corporation, *Efficiency Vermont: Annual Plan: 2009–2011* 7 (December 16, 2008) available at http://www.efficiency-vermont.com/stella/filelib/EVT%20Annual%20Plan%202009–2011.pdf

78. Robert L. Borlick, *Pricing Negawatts*, 148 Pub. Util. Fort. 14 (August 2010).

79. Revis James, et al., *The Power to Reduce CO_2*, 145 Pub. Util. Fort. 60 (October 2007).

80. *See* Stephen Maloney, *When the Price is Right*, 145 Pub. Util. Fort. 24 (October 2007); Terry Pratt, et al., *Rating the New Risks: How Trading Hazards Affect Enterprise Risk Management at Utilities*, 145 Pub. Util Fort. 28 (June 2007); *See also* Timothy P. Gardner & James C. Hendrickson, *Carbon Wargames: U.S. Utilities Gain Strategic Insights by Playing Out a Carbon-Constraint Scenario*, 145 Pub. Util. Fort. 46 (December 2007); Mike Gettings, *A Prescription for Regulatory Agreements Regarding Energy Commodity Price Risk Mitigation* (July 18, 2008) available at http://nrri.org/pubs/gas/PACE_Final_Regulatory_Paper_9-9-08.pdf

81. Paul Joskow, *Lessons Learned From Electricity Market Liberalization*, 29 Energy J. 9, 25–28 (2008).

82. Aligning Utility Incentives with Investment in Energy Efficiency, *supra* note 68 at ch. 3.

83. *See* Faruqui, Hledik & Sergici.

84. *See* Hossein Haeri, Jim Stewart & Aaron Jennings, *New York Negawatts: Balancing Risks and Opportunities in Efficiency Investments*, 148 Pub. Util. Fort. 12 (January 2010) (The authors describe a bonus system put in place by the New York Public Service Commission that estimates that six New York utilities will earn about $27 million annually over three years.).

85. *See* Severin Borenstein, *The Trouble With Electricity Markets (and Some Solutions)*, (January 2001) (discussion of real-time pricing as more efficient in light of the market manipulation in California in 2000) available at http://www.ucei.berkeley.edu/PDF/pwp081.pdf

86. The attraction of marginal cost pricing for rate setting has been long known. Then Wisconsin Public Service Commissioner, now United States Court of Appeals Judge, Richard Cuhady, made the point forcefully in 1974: "Our decision in this vintage preceding marks a new and constructive departure in the establishments of rates – one which gives adequate emphasis to the formulation of the prices themselves as distinguished from related aggregates such as revenue requirement or return.... The instant case in its later phases, however, primarily concerns the structure or design of prices and the relationship of such structure to demand, to the efficient allocation of resources, to wasteful uses of resources, to conservation, to environmental protection, to revenue erosion and also to more conventional (albeit vital) concerns such as revenue requirement." *In Re Madison Gas and Electric Co.*, 5 PUR 4th 28, 50 (1974).

87. *See National Action Plan for Energy Efficiency* at 5–5 to 5–6 (July 2006) available at http://www.epa.gov/cleanenergy/documents/napee/napee_report.pdf

88. Ahmad Faruqui & Ryan Hledik, *Transition to Dynamic Pricing*, 147 Pub. Util. Fort. 26, 27–29 (March 2009).
89. Ren Orens, et al., *Inclining the Climate: GHG Reduction via Residential Ratemaking*, 147 Pub. Util. Fort. 41 (May 2009). *See also* Ray Palmer, *Status Report of FERC National Assessment of Demand Response* (February 15, 2009) available at http://www.narucmeetings.org/Presentations/ DR%20Collaborative%20Ray%20Palmer.pdf. Section 529 (a) of the Energy Independence and Security Act of 2007 and Section 1252(e)(3) of the Energy Policy Act of 2005 require FERC to conduct a national assessment of demand response. *See* Federal Energy Regulatory Commission, *Assessment of Demand Response and Advanced Metering* (September 2009).
90. Lisa Wood, *Efficiency Close-Up*, 148 Pub. Util. Fort. 34 (July 2010).
91. *See* Wayne Shirley, Jim Lazar & Frederick Weston, *Revenue Decoupling: Standards and Criteria* (June 30 2008) available at http://www.raponline.org/Pubs/ MN-RAP_Decoupling_Rpt_6-2008.pdf; National Association of Regulatory Utility Commissioners, *Decoupling for Electric & Gas Utilities: Frequently Asked Questions* (September 2007) available at http://www.epa.gov/statelocalclimate/ documents/pdf/supp_mat_decoupling_elec_gas_utilities.pdf
92. David Magnus Boonin, *A Rate Design to Encourage Energy Efficiency and Reduce Revenue Requirements* (July 2008) available at http://nrri.org/pubs/ electricity/rate_des_energy_eff_SVF_REEF_jul08-08.pdf
93. Ahmad Faruqui, *Inclining Toward Efficiency*, 146 Pub. Util. Fort. 22 (August 2008) ("Dynamic pricing lowers peak-period demands and avoids expensive peaking capacity.... One recent study quantified at $31 billion the national savings that would accrue from just a 5-percent reduction in peak demand." At 23 citing Ahmad Faruqui, et al., Electricity J. 68 (October 2007).
94. Peter Cappers, et al., *Financial Analysis of Incentive Mechanisms to Promote Energy Efficiency: Case Study of a Prototypical Southwest Utility* (March 2009) available at http://eetd.lbl.gov/ea/emp/reports/lbnl-1598e-app.pdf
95. A key element in any decoupling involves setting revenue targets and adjusting those targets over the course of a period of time such as the year to "true up" the charges to customers. *See National Action Plan for Energy Efficiency* at 2–4.
96. *Id.* at 1–8. Regarding long-term incremental cost rates, *see* Alfred E. Kahn, The Economics of Regulation: Principles and Institutions, Volume I ch. 4 (1988); *see also* Severin Borenstein, *The Long-Run Efficiency of Real-Time Electricity Pricing* (February 2005) (Economic model of long-run real-time pricing demonstrates significant economic benefits for large users and may also reveal economic benefits for small users depending on their demand elasticity.) Available at http://repositories.cdlib.org/cgi/viewcontent. cgi?article=1036&context=ucei/csem
97. *Id.* at 9–15.
98. In order to function properly, the REEF must be continuously adjusted along a known timeframe, and the information must be readily and easily available to customers. Smart meters will be necessary to perform these functions efficiently and reliably. *See* Rick Morgan, *Rethinking Dumb Rates*, 147 Pub. Util. Fort. 34 (March 2009). The SFV rate design is a form of dynamic pricing that depends on an advanced metering infrastructure with smart meters.

99. Another rate adjustment is to eliminate the "hedge premium." *Id.* at 36 ("The 'hedge premium' reflects the costs of guaranteeing a flat rate around the clock."); Ahmad Faruqui & Ryan Hledik, *Transition to Dynamic Pricing*, 147 Pub. Util. Fort. 26, 33 (March 2009).

100. *See, e.g.*, Ahmad Faruqui & Ryan Hledik, *Transition to Dynamic Pricing*, 147 Pub. Util. Fort. 26 (March 2009).

101. *Aligning Utility Incentives With Investment in Energy Efficiency* at ch. 4.

102. *See* Rick Morgan, *Rethinking Dumb Rates*, 147 Pub. Util. Fort. 34 (March 2009).

103. Recent studies indicate pricing elasticity for residential consumers. The Brattle Group estimates that short-run residential price elasticities due to price changes can be as low as –0.01 and as high as –0.39. Whereas long-run elasticities due to equipment changes can range between –0.03 and –1.17. *See* Ahmad Faruqui, *Inclining Toward Efficiency*, 146 Pub. Util. Fort. 22, 25–26 (August 2008). A study by the Electric Power Research Institute indicates that short-run price elasticity ranges from –0.2 to –0.6 whereas long-run elasticities range from between –0.7 to –1.4 with a mean of –0.9. EPRI, *Price Elasticity of Demand for Electricity: A Primer and a Synthesis* (January 2008) ("A compelling conclusion is that a wide variety of consumers exhibit price response when provided an opportunity to do so." At 2); Sanem I. Sergici & Ahmad Faruqui, *Experimental Design Considerations in Evaluating the Smart Grid* (December 15, 2008) (finding consumer demand response from 17 dynamic pricing models) available at http://www.brattle.com/_documents/uploadlibrary/upload733.pdf. An earlier study by the RAND Corporation examined 30 years of data and estimated residential price elasticity at between –0.24 and –0.32. Mark A. Bernstein & James Griffin, *Regional Differences in the Price-Elasticity of Demand for Energy* (2005).

104. *See, generally*, The Capitol.Net, Smart Grid (2009).

105. *See, generally*, Jeff Guldner & Meghan Grabel, *Dealing with Change: The Long-Term Challenge for the Electric Industry*, 23 Nat Resources & Env't 3 (Summer 2008).

106. The Energy Independence and Security Act of 2007 §1307 defines the smart grid as having these characteristics: (1) use of digital information to improve reliability, security, and efficiency; (2) optimize grid operations securely; (3) deploy and integrate distributed and renewable resources; (4) incorporate demand response and energy efficiency; (5) deploy smart technologies; (6) integrate smart appliances and consumer devices; (7) deploy and integrate advanced electricity storage and electric vehicles; (8) provide timely information to consumers; (9) develop interoperability standards; and (10) reduce barriers to adopting smart grid technologies.

107. In a report by the Energy and Environment Program of the Aspen Institute, the authors list as benefits: (1) deferred generation decisions; (2) promote consumer choice; (3) timely information on energy costs; (4) reduce regressive cross-subsidies; (5) mitigate potential market power; (6) enhanced system reliability and price responsiveness; (7) improve predictability of demand; (8) improve distribution system reliability; (9) reduce outages; (10) reduce carbon emissions; (11) substitute demand response for generators providing

ancillary services; (12) account for transmission line losses; and (13) create a platform for innovation. Linda Stuntz & Susan Tomasky, *An Electric Grid for the 21st Century* 49–50 (2010). *See also* Ahmad Faruqui, Peter Fox-Penner & Ryan Hledik, *Smart-Grid Strategy: Quantifying Benefits*, 148 PUB. UTIL. FORT. 32 (July 2009).

108. U.S. Department of Energy, *Smart Grid System Report* iii–iv (July 2009) available at http://www.oe.energy.gov/DocumentsandMedia/SGSRMain_090707_lowres.pdf

109. Electric Power Research Institute, *Assessment of Achievable Potential from Energy Efficiency and Demand Response Programs in the U.S. (2010–2030)* (January 2009) available at http://mydocs.epri.com/docs/public/000000000001018363.pdf

110. ENERGY INFORMATION ADMINISTRATION, ANNUAL ENERGY OUTLOOK 2009 71 (March 2009) available at http://www.eia.doe.gov/oiaf/aeo/. Note that due to increase cost of adding energy efficiency and demand response, there is no dollar-for-dollar reduction in required investment. Still, the cost of investment will decline about 15%. At chs 3 & 4.

111. Marc W. Chupka, et al., *Transforming America's Power Industry* ch. 2 (November 2008) available at http://www.brattle.com/_documents/UploadLibrary/Upload725.pdf

112. North American Electric Reliability Corporation, *2008 Long-Term Reliability Assessment: 2008–2017 – The Reliability of the Bulk Power System* 15–17 (October 2008) available at http://www.nerc.com/files/LTRA_2008_v1.2.pdf

113. Marc W. Chupka, et al., *Transforming America's Power Industry: The Investment Challenge 2010–2030* vi (November 2008) available at http://www.eei.org/ourissues/finance/Documents/Transforming_Americas_Power_Industry.pdf; *see also* US DOE, *National Transmission Grid Study* (May 2002) available at http://www.pi.energy.gov/documents/TransmissionGrid.pdf

114. *See* Susan F. Tierney, *A 21st Century "Interstate Electric Highway System" – Connecting Consumers and Domestic Clean Power Supplies* (October 31, 2008) available at http://www.analysisgroup.com/uploadedFiles/Publishing/Articles/Tierney_21st_Century_Transmission.pdf; Rob Gramilch, Michael Goggin & Katherine Gensler, *Green Power Super Highways: Building a Path to America's Clean Energy Future* (February 2009) available at http://www.awea.org/GreenPowerSuperhighways.pdf

115. *See, e.g.*, Mike Heyeck, *AEP's I-765 Proposal and the Future of America's Transmission Grid* 16 (August 11, 2006) available at http://www.netl.doe.gov/moderngrid/docs/Heyeck.pdf

116. James P. Fama, *The New Grid: How to Build and Pay for It* (March 19, 2010).

117. *See, e.g.*, Charles River Associates, *CRA International Consultants Study High Voltage Transmission in the Southwest Power Pool* (November 17, 2008) available at http://www.crai.com/News/listingdetails.aspx?id=9236; Gramilch, et al., at 8 ("As a result, a 765-kV grid overlay could reduce U.S. peak load electricity losses by 10 GW or more, the equivalent output of 20 typical 500 MW coal-fired power plants, and reduce annual CO_2 emissions by 16 million tons.")

118. *See* Lynne Kiesling, *Smart Policies for a Smart Grid: Enabling a Consumer-Oriented Transactive Network* (March 12, 2009) available at http://www.

hks.harvard.edu/hepg/Papers/2009/Lynne_Kiesling_March09.pdf;   Bernie Neenan, *Smart Policies for a SmartGrid (or, the Other Way Around)*, (March 2009) available at http://www.hks.harvard.edu/hepg/Papers/2009/Bernie_Neenan_March09.pdf

119. *See* Michael T. Burr, *Beyond Intermittency*, 148 Pub. Util. Fort. 24 (May 2010).

120. *See* Energy Future Coalition, *The National Clean Energy Smart Grid: An Economic, Environmental, and National Security Imperative* available at http://www.energyfuturecoalition.org/files/webfmuploads/Transmission%20EFC%20Vision%20Statement%20-%20Exec%20Sum%20-%20FINAL.pdf; Reid Dechton, *Testimony before the Senate Committee on Energy and Natural Resources* (March 12, 2009) available at http://www.energyfuturecoalition.org/files/webfmuploads/Smart%20Grid%20Docs/Reid%20Detchon%20Testimony%20Package%203.12.pdf. The Edison Electric Institute, an association of power providers, estimates current investments in solar and wind transmission projects at $21 billion. *See* Edison Electric Institute, *Transmission Projects: Supporting Renewable Resources* (February 2009) available at http://www.eei.org/ourissues/ElectricityTransmission/Documents/TransprojRenew_web.pdf

121. *See, e.g.*, North American Reliability Corporation, *Electric Industry Concerns on the Reliability of Climate Change Initiatives* (November 2008).

122. U.S. Department of Energy, *20% Wind Energy by 2030: Increasing Wind Energy's Contribution to U.S. Electricity Supply* (July 2008) available at http://www1.eere.energy.gov/windandhydro/pdfs/41869.pdf. A recent study indicates that, currently, almost 300,000 MW of wind projects exist, which is more than sufficient capacity to satisfy the 2030 goal.

123. Rob Gramilch, Michael Goggin & Katherine Gensler, *Green Power Super Highways: Building a Path to America's Clean Energy Future* 5 (February 2009). This report also notes that the solar industry can create 440,000 jobs and $325 billion in economic development over the next eight years. at 6.

124. Edison Electric Institute, *EEI Survey of Transmission Investment: Historical and Planned Capital Expenditures – 1999–2008* (May 2005) (noting the reversal of an historic trend of under-investment) available at http://www.eei.org/ourissues/ElectricityTransmission/Documents/Trans_Survey_Web.pdf; James P. Fama, *The New Grid: How to Build and Pay for It* (March 19, 2010) (PowerPoint presentation ABA 39th Annual Conference on Environmental law) (Edison Electric Institute data on actual and planned transmission investment from $5.6 billion in 2001 to an estimated $12 billion in 2011, for a total of nearly $80 billion).

125. The Energy Independence and Security Act of 2007, P.L. 110–140 (2007) (EISA).

126. Notice issued pursuant to EISA §1304; *see* Notice of Intent to Issue Funding Opportunity Announcement No.:DE-FOA-0000036 available at http://www.asertti.org/newsletter/2009-03-24/FOA_SmartGrid.pdf

127. American Recovery and Reinvestment Act, P.L. No. 111–5 (February 17, 2009).

128. Under the act, $11 billion is specified for the grid and more than $10 billion specified for energy efficiency in transmission and reliability. *See American Recovery and Reinvestment Act: Moving America Toward a Clean Energy Future* (February 17, 2009) available at http://www.whitehouse.gov/assets/documents/Recovery_Act_Energy_2-17.pdf. *See also* EISA, §130. *See also* Notice: Proposed Rules, Smart Grid Policy, Dkt. No. E9–12029 (May 21, 2009) available at http://www.thefederalregister.com/d.p/2009-05-21-E9-12029

129. The American Clean Energy and Security Act of 2009, H.R. 2454, 111th Congress, 1st Sess. §§ 141–146 (smart grid advancement) and § 151 (transmission planning).

130. *See* Mason Willrich, *Electricity Transmission Policy for America: Enabling a Smart Grid, End-to-End* (July 2009) available at http://web.mit.edu/ipc/research/energy/pdf/EIP_09-003.pdf

131. *See* Renewable Energy Transmission Company, *The US Electric Transmission Grid: Essential Infrastructure in Need of Comprehensive Legislation* (April 2009).

132. Pursuant to the Energy Policy Act of 2005, FERC has the authority to "backstop" transmission siting at the state does not meet certain criteria. That jurisdiction, however, has been constrained by a court decision that overturned a FERC ruling that sought to override a state denial of a siting permit. The court ruled that the statute empowered FERC to act only when a state withheld a decision, not when they issued a denial. *Piedmont Environmental Council v. FERC*, 558 F. 3d 304 (4th Cir. 2009). Regarding siting authority, *see* Ashley C. Brown & Jim Rossi, *Siting Transmission Lines in a Changed Milieu: Evolving Notions of the "Public Interest" in Balancing State and Regional Considerations* (December 2010) available at http://papers.ssrn.com/sol3/papers.cfm?abstract_id=1444111

133. Stuntz & Tomasky at 11–12.

134. James M. Seibert, *Paradox of Thrift: Economic Barriers Complicate T&D Modernization*, 148 Pub. Util. Fort. 10 (July 2009).

135. *See* International Energy Agency, *Energy Market Experience: Tackling Investment Challenges in Power Generation in IEA Countries* (2007).

136. One area of concern for moving forward with an improved transmission grid involves siting authority. Historically, states have exercised significant jurisdiction over the siting of transmission and distribution lines. The Energy Policy Act of 2005 amended the Federal Power Act to give FERC siting authority. That grant of power, however, may not be as broad a grant of authority as might be necessary. In *Piedmont Environmental Council v. FERC*, 558 F. 3d 304 (4th Cir. 2009) (Dkt. No. 07–1651) the court ruled that FERC lacked authority to permit the construction of an electric transmission facility in an area designated as a national interest or when the state had denied a permit for that facility. Federal siting authority may well be necessary for the full and effective realization of the smart grid, especially regarding connections with renewable resources. *See, e.g.*, Bruce W. Radford, *Federalizing the Grid: Renewable Mandates Will Shift Power to FERC but Pose Problems for RTOs*, 147 Pub.Util. Fort. 20 (April 2009); Catherine R. Connors, et al., *Transmission Preemption: Federal Policy Trumps State Siting Authority*, 148 Pub. Util. Fort. 46 (November 2010).

As part of FERC's charge under EISA §1305, with the cooperation of other organizations and stakeholders, the commission must "coordinate development of a framework that includes protocols and model standards for information management to achieve interoperability of smart grid devices and systems." *See, e.g.*, David A. Wollman, *Status of NIST's EISA Smart Grid Efforts* (March 26, 2009) available at http://www.researchcaucus.org/docs/Smart%20Grid%20 Congressional%20briefing%20DAW%20ASME%20IEEE%2026Mar2009%20 draft.pdf

137. Energy Policy Act 2005 §1241, Pub. L. No. 109–58, 119 Stat. 594, 315 and 1283 (2005) adding §219 to the Federal Power Act, 16 U.S.C. § 2621 (2010).

138. FERC, *Smart Grid Policy Statement*, 128 FERC ¶61,060 (FERC Dkt. No. PL09-4-000) (July 16, 2009) available at http://www.nerc.com/files/Smart_Grid_7-16-09_Policy_Statement.pdf

139. CFR §35.35 (2008).

140. *See Pacific Gas and Electric Company*, 129 FERC ¶ 61,251 (FERC Dkt. No. EL09–72–000) (December 17, 2009) available at http://www.ferc.gov/whats-new/comm-meet/2009/121709/E-4.pdf. *See also* Order on Petition for Declaratory Order, FERC Dkt. No. EL08–75–000, 125 FERC ¶61, 076 (October 21, 2008) (approving 2% additional return; 100% of construction-work-in-progress into the rate base; 100% of prudently incurred costs associated with cancellation or abandonment, not as a result of utilities activities to various projects); Order Granting Incentives, And Accepting Proposed Rate Formula Modifications, Subject to Conditions, FERC Dkt. No. ER07–1415–000, 121 FERC ¶61,284 (December 21, 2007) (similar orders for various projects).

141. *See* Scott H. Strauss & Jeffrey A. Schwartz, *Transmission Incentive Overhaul*, 147 Pub. Util. Fort. 32 (February 2009) (arguing that FERC has been too generous with its incentive rates especially in regard to the return on equity add-ons.)

142. The U.S. grid is roughly divided between the East and the West, and ERCOT serves the State of Texas. *See* U.S. DOE, *Overview of the Electric Grid* available at http://sites.energetics.com/gridworks/grid.html

143. *See* Renewable Energy Transmission Company, *The US Electric Transmission Grid: Essential Infrastructure in Need of Comprehensive Legislation* (April 2009) available at http://www.renewabletrans.com/retcopaper1.pdf

144. *See, e.g.*, Stuntz & Tomasky at 4.

145. *Illinois Commerce Comm'n v. FERC*, 576 F. 3d 470 (7th Cir. 2009).

146. *See* Bruce W. Radford, *Wellinghoff's War: FERC Fights for the Green-Grid Superhighway – Even if Congress Won't*, 148 Pub. Util. Fort. 24 (January 2010).

147. *Smart Grid System Report* at iv.

148. Mason Willrich, *Electricity Transmission Policy for America: Enabling a SmartGrid, End-to-End* (July 2009) available at http://web.mit.edu/ipc/research/energy/pdf/EIP_09-003.pdf

149. *See, e.g.*, Paul L. Joskow, *Challenges for Creating a Comprehensive National Electricity Policy* (September 26, 2008) available at http://econ-www.mit.edu/files/3236.; L. Lynne Kielsing, Deregulation, Innovation and Market Liberalization: Electricity Regulation in a Continually Evolving Environment (2009); Paul L. Joskow, *Lessons Learned From Electricity Market Liberalization*, 29 Energy J. 9 (December 2008).

Chapter 7

1. COMMITTEE ON AMERICA'S ENERGY FUTURE OF THE NATIONAL ACADEMY OF SCIENCES, NATIONAL ACADEMY OF ENGINEERING & NATIONAL RESEARCH COUNCIL, AMERICA'S ENERGY FUTURE: TECHNOLOGY AND TRANSFORMATION 30 (2009) (hereinafter AMERICA'S ENERGY FUTURE REPORT).

2. *See* John Holdren, *Federal Energy Research and Development for the Challenges of the 21st Century* in LEWIS M. BRANSCOMB & JAMES H. KELLER, INVESTING IN INNOVATION: CREATING A RESEARCH AND INNOVATION POLICY THAT WORKS 299, 299–300 (1998).

3. JOSH LERNER, BOULEVARD OF BROKEN DREAMS: WHY PUBLIC EFFORTS TO BOOST ENTREPRENEURSHIP AND VENTURE CAPITAL HAVE FAILED – AND WHAT TO DO ABOUT IT ch. 3 (2009) (why the public sector is more successful with new firms than old ones).

4. *See* Stephanie Strom, *Foundations Find Benefits in Facing Up to Failures*, N.Y. TIMES (July 26, 2007) available at http://www.nytimes.com/2007/07/26/us/26foundation.html. Mark A. Smylie & Stacy A. Wenzel, *The Chicago Annenberg Challenge: Successes, Failures, and Lessons for the Future* (August 2003). (The Annenberg Challenge consisted of a $500 million gift targeted at improving student performance. The final report on the Chicago project concluded: "Our research indicates that student outcomes in Annenberg schools were much like those in demographically similar non-Annenberg schools and across the Chicago school system as a whole, indicating that among the schools it supported, the Challenge had little impact on student outcomes." At 1.); *see also* The Annenberg Challenge, *The Annenberg Challenge: Lessons and Reflections on Public School Reform* 11 (2002).

5. *See* PAUL BREST & HAL HARVEY, MONEY WELL SPENT: A STRATEGIC PLAN FOR SMART PHILANTHROPY (2008); MATTHEW BISHOP, PHILANTHROCAPITALISM: HOW THE RICH CAN SAVE THE WORLD (2008); Michael Porter & Mark R. Kramer, *Philanthropy's New Agenda: Creating Value*, HARVARD BUSINESS REV. 121 (November–December 1999).

6. *See* Virgin Atlantic Press Release at http//www.virgin-atlantic.com/en/gb/allaboutus/environment/bransonpledge.jsp

7. The concept of the theory of change has been applied in the philanthropic and foundation world to measure the impact of those organizations on the communities that they serve. *See, e.g.,* ActKnowledge, *A Theory of Change* at http://www.theoryofchange.org/pdf/tocII_final4.pdf; Organizational Research Services, *Theory of Change: A Practical Tool for Action, Results and Learning* (2004) available at http://www.issuelab.org/research/theory_of_change_a_practical_tool_for_action_results_and_learning; and, Andrea A. Anderson, *Theory of Change as a Tool for Strategic Planning: A Report on Early Experience* (October 2004) available at http://www.theoryofchange.org/pdf/tocII_final4.pdf

8. *See* Mario Morino & Bill Shore, *High-Engagement Philanthropy: A Bridge to a More Effective Social Sector* (June 2004) available at http://www.vppartners.org/report2004.html; Roger L. Martin & Sally Osberg, *Social Entrepreneurship: The Case for Definition*, STANFORD SOCIAL INNOVATION REVIEW 29 (Spring 2007).

9. Regarding the need for energy technology innovation, *see* ANTHONY GIDDENS, THE POLITICS OF CLIMATE CHANGE 11 (2009) ("Without such innovation, it is impossible to see how we can break our dependency upon oil, gas and coal, the major sources of environmental pollution. A turn to renewable sources of energy is essential, and it has to be on a very large scale."); Nicholas Stern, A BLUEPRINT FOR A SAFER PLANET: HOW TO MANAGE CLIMATE CHANGE AND CREATE A NEW ERA OF PROGRESS AND PROSPERITY 7 (2009). Regarding a definition of "energy technology," *see* Kelly Sims Gallagher, John P. Holdren & Ambuj D. Sagar, *Energy-Technology Innovation*, 31 ANNUAL REV. ENV'T & RESOURCES 193, (2006) ("*[E]nergy technology* refers to the means of locating, assessing, harvesting, transporting, processing, and transforming the primary energy forms found in nature (e.g., sunlight, biomass, crude petroleum, coal, uranium-bearing rocks) to yield either direct energy services (e.g., heat from fuelwood or coal) or secondary forms more convenient for human use (e.g., charcoal, gasoline, electricity)." The authors also include in their definition methods of distribution and conversion as well as both hardware and software.)

10. Gallagher, Holdren & Sagar at 195.

11. *See* J. Scott Holladay & Michael A. Livermore, *CLEAR & The Economy: Innovation, Equity and Job Creation* (April 2010) available at http://www.policyintegrity. org/documents/ClearandTheEconomy.pdf

12. WILLIAM J. BAUMOL, THE FREE MARKET INNOVATION MACHINE: ANALYZING THE GROWTH MIRACLE OF CAPITALISM (2002); WILLIAM J. BAUMOL, ROBERT E. LITAN & CARL J. SCHRAMM, GOOD CAPITALISM, BAD CAPITALISM, AND THE ECONOMICS OF GROWTH AND PROSPERITY (2007).

13. NATIONAL ACADEMY OF SCIENCES, RISING ABOVE THE GATHERING STORM: ENERGIZING AND EMPLOYING AMERICA FOR A BRIGHTER ECONOMIC FUTURE 1 (2007).

14. RISING ABOVE THE GATHERING STORM at 11–12.

15. World Economic Forum, *Energy Vision Update 2010: Towards a More Energy Efficient World* (2010) available at http://www.weforum.org/pdf/ip/energy/ Energy_VisionUpdate2010.pdf

16. CNA Military Advisory Board, *Powering America's Economy: Energy Innovation at the Crossroads of National Security Challenges* (July 2010) available at http:// www.google.com/search?sourceid=navclient&ie=UTF-8&rlz=1T4ADBF_enU S284US286&q=Powering+America%e2%80%99s+Economy%3a+Energy+Inn ovation+at+the+Crossroads+of+National+Security+Challenges

17. This book concentrates on domestic energy policy. Climate change, of course, is an international matter. For a study of technological innovation in the international arena, *see* Center for American Progress & Global Climate Network, *Breaking Through on Technology: Overcoming the Barriers to the Development and Wide Deployment of Low-Carbon Technology* (July 2009) available at http://www.americanprogress.org/issues/2009/07/pdf/gcn_report.pdf; RICHARD B. STEWART, BENEDICT KINGSBURY & BRUCE RUDYK, CLIMATE FINANCE: REGULATING AND FUNDING STRATEGIES FOR CLIMATE CHANGE AND GLOBAL DEVELOPMENT (2009).

18. For a general description of public choice, *see* DENNIS C. MUELLER, PUBLIC CHOICE III (2003). For an analysis and critique, *see* JERRY L. MASHAW, GREED,

Chaos, and Governance: Using Public Choice to Improve Public Law (1997); Daniel A. Farber (ed.), Research Handbook on Law and Public Choice (2010).

19. *See, e.g.*, James M. Landis, *Report on Regulatory Agencies to the President-Elect* (December 1960).

20. *See Piedmont Environmental Council v. FERC*, 558___ F. 3d 304___ (USCA 4th Cir. Dkt. No. 07–1651) (February 18, 2009) (limiting FERC transmission-siting jurisdiction) and *Illinois Commerce Comm'n v. FERC*, 576 F. 3d 470 (7th Cir. 2009) (limiting FERC jurisdiction to allocate transmission costs).

21. *See* Jason Yackee & Susan Webb Yackee, *Administrative Procedures and Bureaucratic Performance: Is Federal Rulemaking "Ossified"?* (April 12, 2009) available at http://papers.ssrn.com/sol3/papers.cfm?abstract_id=1371588. http://papers.ssrn.com/sol3/papers.cfm?abstract_id=1371588

22. Joseph P. Tomain, *Institutionalized Conflicts Between Law and Policy*, 22 Houston L. Rev. 661 (1985); Alfred C. Aman, Jr., *Institutionalizing the Energy Crisis: Some Structural and Procedural Lessons*, 65 Cornell L. Rev. 491 (1980); Clark Byse, *The Department of Energy Organization Act: Structure and Procedure*, 30 Admin. L. Rev. 193 (1978).

23. *See* Department of Energy *Organizational Chart* at http://www.energy.gov/organization/orgchart.htm

24. *See, e.g.*, Richard K. Lester, *Reforming the U.S. Energy Innovation System* (September 2008) available at http://web.mit.edu/ipc/publications/pdf/EIP%20WP%2008-001.pdf

25. *See* Daniel M. Kammen & Gregory F. Nemet, *Reversing the Incredible Shrinking Energy R&D Budget*, Issues in Science and Technology 84 (September 2005) available at http://www.climatetechnology.gov/stratplan/comments/Kammen-2.pdf

26. Josh Freed, Avi Zevin & Jesse Jenkins, *Jumpstarting a Clean Energy Revolution with a National Institutes of Energy* 4–5 (September 2009) available at http://thebreakthrough.org/blog/Jumpstarting_Clean_Energy_Sept_09.pdf

27. Gallagher, Holdren & Sagar at 222.

28. Kammen & Nemet at 8.

29. John P. Holdren, *The Energy Innovation Imperative: Addressing Oil Dependence, Climate Change, and Other 21st Century Energy Challenges*, Innovations: Technology, Governance, Globalization 3 (Spring 2006) available at http://belfercenter.ksg.harvard.edu/files/innovations_the_imperative_6_06.pdf

30. *See, e.g.*, Matthew L. Wald & Tom Zeller, Jr., *Cost of Tapping Green Power Makes Projects a Tougher Sell*, N.Y. Times A1 (November 8, 2010) (State regulators reluctant to add renewable wind resources even at the low increased cost of a .02% per month rate increase.).

31. Stuart Minor Benjamin & Arti K. Rai, *Fixing Innovation Policy: A Structural Perspective*, 77 Geo. Wash. L. Rev. 1, 13–14 (2008).

32. *See, e.g.*, Bill Gates, 2010 Annual Letter 2 (January 2010) ("Society underinvests in innovation in general.") available at http://www.gatesfoundation.org/annual-letter/2010/Pages/bill-gates-annual-letter.aspx

33. America's Energy Future Report at 77.

34. Gallagher, Holdren & Sagar at 223–27.

35. AMERICA'S ENERGY FUTURE REPORT at 77–78.
36. Venkatesh Narayanamurti, Laura D. Anadon & Ambuj D. Sagar, *Institutions for Energy Innovation: A Transformational Challenge* 2 (Fall 2009) available at http://belfercenter.ksg.harvard.edu/publication/19572/institutions_for_energy_innovation.html
37. HANS H. LANDSBERG, ET AL., ENERGY: THE NEXT TWENTY YEARS 544 (1979).
38. Benjamin & Rai at 3.
39. *See* Jim Watson, *Setting Priorities in Energy Innovation Policy: Lessons for the UK*, Discussion Paper, Energy Technology Innovation Policy Research Group, Belfer Center for Science and International Affairs (October 2008) available at http://belfercenter.ksg.harvard.edu/publication/18593/setting_priorities_in_energy_innovation_policy.html; Kevin Smith, *Climate Change and Radical Energy Innovation: The Policy Issues*, Working Papers on Innovation Studies No. 200090101, Centre for Technology, Innovation and Culture (2009) available at http://econpapers.repec.org/paper/tikinowpp/20090101.htm
40. Peter Ogden, John Podesta & John Deutch, *A New Strategy to Spur Energy Innovation* 3–4 (January 2008) available at http://www.americanprogress.org/issues/2008/01/energy_innovation.html. This paper is a reprint of the article that appeared under the same title in ISSUES IN SCIENCE AND TECHNOLOGY (Winter 2008) available at http://www.issues.org/24.2/ogden.html
41. William B. Bonvillian & Charles Weiss, *Taking Covered Wagons East: A New Innovation Theory for Energy and Other Established Technology Sectors*, 4 INNOVATIONS: TECHNOLOGY, GOVERNANCE, GLOBALIZATION 289 (Fall 2009).
42. Venkatesh Narayanamurti, Laura D. Anadon & Ambuj D. Sagar, *Institutions for Energy Innovation: A Transformational Challenge* 8–11 (Fall 2009) available at http://belfercenter.ksg.harvard.edu/publication/19572/institutions_for_energy_innovation.html
43. *Id.* at 8–9.
44. Ogden, Podesta & Deutch at 6–8.
45. For a definition of an "energy innovation system," *see* Lester at 4 (The term "refers to the complex of markets, direct government support mechanisms, indirect government incentives, banks, venture capital funds, and other financial institutions, public and private research and educational organizations and programs, government regulations, codes, and standards in which new technologies are developed, commercialized, and taken up by energy suppliers and users."); *see also* Guiseppe Bellantuono, *Law and Innovation in the Energy Sector* to be published in B. Delvaux, M. Hunt & K Talus (EDS.) EU LAW AND POLICY ISSUES (B. Delvaux, M. Hunt & K Talus ed., 2nd ed. 2009) available at http://papers.ssrn.com/sol3/papers.cfm?abstract_id=1480071; Gallagher, Holdren & Sagar at 200–06.
46. *See* Laura Diaz Anadon, et al., *U.S. Public Energy Innovation Institutions and Mechanisms: Status and Deficiencies* 1 (January 14, 2010) available at http://belfercenter.ksg.harvard.edu/publication/19877/us_public_energy_innovation_institutions_and_mechanisms.html; *see also* Laura Diaz Anadon & John P. Holdren, *Policy for Energy Technology Innovation*, in Kelly Sims Gallagher (ed.) ACTING IN TIME ON ENERGY POLICY ch. 5 (Kelly Sims Gallagher ed., 2009).

47. Venkatesh Narayanamurti, Laura D. Anadon & Ambuj D. Sagar, *Transforming Energy Innovation*, ISSUES IN SCIENCE AND TECHNOLOGY 57 (Fall 2009) (hereinafter *Transforming Energy Innovation*) available at http://www.issues.org/26.1/narayanamurti.html

48. *See* Adam B. Jaffe, Richard G. Newell & Robert N. Stavins, *Technology Policy for Energy and the Environment*, in Adam B. Jaffe, Josh Lerner & Scott Stern (eds.), INNOVATION POLICY AND THE ECONOMY ch 2. (Adam B. Jaffe, Josh Lerner & Scott Stern (ed.)) (2004).

49. *See* Simona O. Negro, Marko P. Hekkert & Ruud Smits, *Stimulating Renewable Energy Technologies by Innovation Policy*, Innovations Studies Utrecht (ISU) Paper No. 08–13 (April 2008) available at http://www.geo.uu.nl/isu/isu.html#_#08.13_(PDF); Gallagher, Holdren & Sagar at 195.

50. Venkatesh Narayanamurti, Laura D. Anadon & Ambuj D. Sagar, *Institutions for Energy Innovation: A Transformational Challenge* 2 (Fall 2009) (hereinafter *Institutions for Energy Innovation*) available at http://belfercenter.ksg.harvard.edu/publication/19572/institutions_for_energy_innovation.html

51. *Id.*

52. *Id.* at 5.

53. *Institutions for Energy Innovation*; Laura Anadon, et al., *U.S. Public Energy Innovation Institutions and Mechanisms: Status and Deficiencies* (January 14, 2010) (hereinafter *U.S. Public Energy Innovations Institutions*) available at http://belfercenter.ksg.harvard.edu/publication/19877/us_public_energy_innovation_institutions_and_mechanisms.html?breadcrumb=%2Fexperts%2F1839%2Fcharles_jones; and *Transforming Energy Innovation*.

54. The following general discussion of these five principles is taken from the papers cited at *id.* More specific references are also provided.

55. *Transforming Energy Innovation* 57–58.

56. Ogden, Podesta & Deutch at 11.

57. Ogden, Podesta & Deutch at 9.

58. Gallagher, Holdren & Sagar at 210–14.

59. *See* David Popp, Richard G. Newell & Adam B. Jaffe, *Energy, The Environment, and Technological Change* 8 (NBER Paper No. W14832) (April 2009) available at http://papers.ssrn.com/sol3/papers.cfm?abstract_id=1373342

60. *See* ActKnowledge; Anderson.

61. *See* SUZANNE SCOTCHMER, INNOVATION AND INCENTIVES ch. 8 (2004).

62. David Orr argues that climate change presents a unique problem that threatens us with what he refers to as the "long emergency." In order to respond to the "long emergency," government must develop the "capacity necessary to solve multiple problems that cross the usual lines of authority, departments, and agencies as well as those between federal, state, and local governments.... We will need to build new alliances between the public, nongovernmental organizations, local and state governments and business. Above all, government must enable creative leadership at all levels of society, it must lead first by example, not simply by fiat." David W. Orr, DOWN TO THE WIRE: CONFRONTING CLIMATE CHANGE 42 (2009).

63. *See* David B. Spence, *Can Law Manage Competitive Energy Markets?* 93 CORNELL L. REV. 765, 809–17 (2008) (discussing the necessary elements for competitive energy markets).

64. *See* Web site for the National Renewable Energy Laboratory at http://www. nrel.gov/

65. *See, e.g.,* John Podesta, et al., *The Clean-Energy Investment Agenda: A Comprehensive Approach to Building the Low-Carbon Economy* (September 2009) available at http://www.americanprogress.org/issues/2009/09/clean_ energy_investment.html

66. Donella Meadows, Jorgen Randers & Dennis Meadows, Limits to Growth: The 30-Year Update 237 (2004).

67. Ben Geman, *Energy Department Sees $2 Billion Boost Under White House Plan,* The Hill (February 1, 2010) available at http://thehill.com/blogs/e2-wire/677-e2-wire/79035-energy-dept-avoids-the-budget-ax-sees-2-billion-boost-under-white-house-plan

68. American Association for the Advancement of Science, *AAAS Report XXXIV: Research and Development FY 2010* ch. 8 (2009) available at http:// www.aaas.org/spp/rd/rdreport2010/

69. R&D for geothermal technology grows 13.6% to $50 million; building technology increases by 69.8% to $238 million; industrial technology increases 11.1% to $100 million; and the vehicle technology increases 21.9% to $333 million.

70. Ian Talley & Siobhan Hughes, *Energy: Reduced Tax Breaks Sought for the Oil Industry,* Wall Street J. (February 1, 2010) available at http://online. wsj.com/article/SB10001424052748704107204575039200881610236. html?mod=WSJ_Energy_leftHeadlines

71. Peter R. Orszag, Director, Office of Management and Budget, *Memorandum for the Heads of Executive Departments and Agencies* (August 4, 2009) available at http://www.nextgov.com/nextgov/ng_20090805_2153.php

72. *Id.* at 2

73. *Id.* at 2–3.

74. Laura Diaz Anadon, et al., *DOE FY 2011 Budget Request for Energy Research, Development, Demonstration and Deployment: Analysis and Recommendations* (April 2010) available at http://belfercenter.ksg.harvard.edu/files/Harvard_ ERD3_FY2011Analysis.pdf

75. John Seely Brown & John Hagel III, *The Next Frontier of Innovation* (2005) available at http://www.anti-bertelsmann.de/sozialtechnik/frontierofinnovation.pdf

76. Bonvillian & Weiss at 293.

77. *Id.* at 294.

78. *See* DOE Basic Energy Sciences Committee, *Basic Research Needs To Assure a Secure Energy Future* (February 2003) available at http://www.sc.doe.gov/production/bes/reports/files/SEF_rpt.pdf

79. The list of 46 projects can be found at http://www.er.doe.gov/bes/EFRC/EFRC_ awards_FY09.html

80. *See also* U.S. Department of Energy, Energy Frontier Research Centers: Tackling Our Energy Challenges in a New Era of Science available at http://www.er.doe. gov/bes/EFRC/EFRC_awards_FY09.html

81. *Energy Innovation Institutions and Mechanisms* at 3.

82. DOE *Hubs Overview* at http://www.energy.gov/hubs/overview.htm

83. American Clean Energy and Security Act of 2009, H.R.2454, 111th Cong.

84. America Competes Act, P.L. 110–69.

85. U.S. Department of Energy, *About ARPA-E* available at http://arpa-e.energy.gov/About.aspx
86. U.S. Department of Energy, *Funded Projects*, available at U.S. Department of Energy, *About ARPA-E* available at http://arpa-e.energy.gov/FundedProjects.aspx
87. Ogden, Podesta & Deutch at 10.
88. Lester at 12–24.
89. *The American Clean Energy Leadership Act of 2009*, S. 1462 111th Cong. Available at http://energy.senate.gov/public/index.cfm?FuseAction=IssueItems.Detail&IssueItem_ID=1fbce5ed-7447-42ff-9dc2-5b785a98ad80
90. Ogden, Podesta & Deutch at 10–14.
91. A similar proposal calls for the creation of the Office of Innovation Policy to be either an independent agency or housed within the White House. *See* Benjamin & Rai at 56–79.
92. Ogden, Podesta & Deutch at 11–14.
93. Josh Freed, Avi Zevin & Jesse Jenkins, *Jumpstarting a Clean Energy Revolution with a National Institutes of Energy* (September 2009) available at http://the-breakthrough.org/blog/Jumpstarting_Clean_Energy_Sept_09.pdf
94. Kammen & Nemet at 9–11.
95. James Duderstadt, et al., *Energy Discovery-Innovation Institutes: A Step Toward America's Energy Sustainability* (February 2009) available at http://www.brookings.edu/reports/2009/0209_energy_innovation_muro.aspx. *See also* Jonathan Sallet, Ed Paisley & Justin Masterman, *The Geography of Innovation: The Federal Government and the Growth of Regional Innovation Clusters* (September 2009) available at http://www.scienceprogress.org/2009/09/the-geography-of-innovation/
96. *See Symposium: Harnessing the Power of Information for the Next Generation of Environmental Law*, 86 Texas L. Rev. 1347 (2008).
97. Duderstadt et al. at 3–4.

Chapter 8

1. Remarks by the President to the Nation on the BP Oil Spill (June 15, 2010) available at http://www.whitehouse.gov/the-press-office/remarks-president-nation-bp-oil-spill
2. John Dryzek, *Ecology and Discursive Democracy: Beyond Liberal Capitalism and the Administrative State* in Martin O'Connor (ed.), Is Capitalism Sustainable? Political Economy and the Politics of Ecology 176 (1994); Christopher H. Schroeder & Rena Steinzor (eds.), A New Progressive Agenda for Public Health and the Environment: A Project of the Center for Progressive Regulation (2005).
3. Additionally, we may well be underpricing carbon in our current energy planning. Frank Ackerman & Elizabeth A. Stanton, *The Social Cost of Carbon* (April 1, 2010) available at http://mrzine.monthlyreview.org/2010/as190410.html; *see also* National Research Council, Hidden Costs of Energy: Underpriced Consequences of Energy Production and Use (2010).
4. Nicholas Stern, A Blueprint for a Safer Planet: How to Manage Climate Change and Create a New Era of Progress and Prosperity 11(2009) (hereinafter Blueprint).

5. David W. Orr, Down to the Wire: Confronting Climate Collapse 6–7 (2009).

6. Orr at 7.

7. *Barack Obama's Inaugural Address*, N.Y. Times (January 20, 2009) available at http://www.nytimes.com/2009/01/20/us/politics/20text-obama.html

8. Jon Gertner, *Capitalism to the Rescue*, N.Y. Times Magazine 54 (October 5, 2008).

9. Thomas L. Friedman, Hot, Flat and Crowded: Why We Need a Green Revolution – and How It Can Renew America (2008).

10. Clifford Krauss & Jad Mouawad, *Oil Industry Backs Protest of Emissions Bill*, N.Y. Times (August 19, 2009); Kate Sheppard, *The Chamber of Commerce vs. Climate Science* (March 1, 2010) available at http://motherjones.com/politics/2010/02/chamber-commerce-vs-climate-science; Anthony Giddens, The Politics of Climate Change 119 (2009) ("The American Petroleum Institute, an industry research organization, claimed as its main goal to make sure that 'climate change becomes a non-issue.'" [footnote omitted]). As counterpoint to junk climate science, *see* Union of Concerned Scientists, *Global Warming Science* at http://www.ucsusa.org/global_warming/science_and_impacts/science/

11. Frank Ackerman & Lisa Heinzerling, Priceless: On Knowing the Price of Everything and the Value of Nothing (2004).

12. *See* Green VC Directory at http://www.ecobusinesslinks.com/green_venture_capital.htm. *See also Greentech Media* homepage at http://www.greentechmedia.com/. Greentech Media provides research and information and programming on a full range of clean energy initiatives including solar, the smart grid, VC investments, energy efficiency, and other topics.

13. Ernst & Young, *Electric Vehicle and Energy Efficiency Developments Drive & 733 Million Venture Capital Investment in US Cleantech Market Q1 2010* available at http://www.ey.com/US/en/Newsroom/News-releases/Electric-vehicle-and-energy-efficiency-developments-drive-USD-733-million-venture-capital-investment

14. Gertner.

15. Al Gore, An Inconvenient Truth: The Planetary Emergency of Global Warming and What We Can do About It (2006); Our Choice: A Plan to Solve the Climate Crisis (2009).

16. KPCB, *Innovative Portfolio Companies* available at http://www.kpcb.com/portfolio/portfolio.php?greentech

17. *Verdiem* homepage at http://www.verdiem.com/

18. Press Release, *California State Agencies Turn to Verdiem to Achieve Energy Savings Goal Set by Governor Schwarzenegger* (June 7, 2010) available at http://www.verdiem.com/news-press/press-detail.aspx?prid=62. *See also* Verdiem, *Immediate, Measurable Results for Meeting California Government Mandates* available at http://www.verdiem.com/docs/PCPowerMgmt_CAL_Whitepaper051210.pdf

19. *Bloom Energy* homepage at http://www.bloomenergy.com/

20. *Khosla Ventures* homepage at http://www.khoslaventures.com/khosla/firm.html

21. Energy Future Coalition, Challenge and Opportunity: Charting a New Energy Future (2003) available at http://www.energyfuturecoalition.org/files/webfmuploads/EFC_Report/EFCReport.pdf

22. NATIONAL COMMISSION ON ENERGY POLICY, ENDING THE ENERGY STALEMATE: A BIPARTISAN STRATEGY TO MEET AMERICA'S ENERGY CHALLENGES (December 2004) available at http://bipartisanpolicy.org/sites/default/files/endi_en_stlmate.pdf

23. *Bipartisan Policy Center* homepage at http://bipartisanpolicy.org/

24. *Climate Change Policy Partnership* homepage at http://www.nicholas.duke.edu/ccpp/partners.html

25. *Yale School of Forestry and Environmental Studies* publications homepage at http://environment.yale.edu/pubs/

26. *Energy and Resources Group* homepage at http://erg.berkeley.edu/

27. *Harvard Electricity Policy Group* homepage at http://www.hks.harvard.edu/hepg/

28. *Carbon Mitigation Initiative* homepage at http://cmi.princeton.edu/about/

29. Stephen W. Pacala & Robert Socolow, *Stabilization Wedges: Solving the Climate Problem for the Next 50 Years with Current Technologies*, 305 SCIENCE 968 (August 13, 2004); Carbon Mitigation Initiative, *Stabilization Wedges* available at http://cmi.princeton.edu/wedges/

30. Carbon Mitigation Initiative, *Annual Report* 2009 (2009) available at http://cmi.princeton.edu/annual_reports/pdfs/ninth_year.pdf

31. *USCAP* homepage at http://www.us-cap.org/

32. United States Climate Action Partnership, *A Blueprint for Legislative Action: Consensus Recommendation for U.S. Climate Protection Legislation* (January 2009) available at http://www.us-cap.org/PHPages/wp-content/uploads/2010/05/USCAP_Blueprint.pdf

33. ACORE, *2009 Renewable Energy Finance Forum – Wall Street: Executive Summary Report* (2009) available at http://www.acore.org/files/REFF-WS_Report.pdf; *RETECH 2009: Executive Summary Report* (2009) available at http://www.acore.org/files/RETECH_Report.pdf

34. ACORE, *Reinventing Renewable Energy: Toward a Technology Strategy for Improving Security, Creating Jobs and Reducing Emissions* (July 2009) available at http://www.acore.org/files/ACORE-EPRI-FINAL.pdf

35. ACORE, *An Overview of the U.S. Renewable Energy Field in 2009* (2009) available at http://www.acore.org/files/re_overview2009.pdf

36. Available at ACORE Publications page at http://www.acore.org/publications

37. *Climate Action Network* homepage at http://www.climatenetwork.org/

38. GIDDENS at 8–16 and ch. 6.

39. Peter Lehner, *Environment, Law, and Nonprofits: How NGOs Shape Our Laws, Health, and Communities*, 26 PACE ENVTL. L. REV. 19 (2009); Chira Giorgetti, *Organizational Summary: The Role of Nongovernmental Organizations in the Climate Change Negotiations*, 9 COLO. J. INT'L ENVTL. L. & POL'Y 115 (1998).

40. *Focus the Nation* homepage at http://www.focusthenation.org/

41. *350.org* homepage at http://www.350.org/

42. James G. McGann, *The Global "Go-To Think Tanks:" The Leading Public Policy Research Organizations in the World* 10 (January 31, 2010) ("Think tanks, or public policy research institutions, have begun to prove their utility in the domestic policy sphere as information transfer mechanisms and agents of change by aggregating and creating new knowledge through collaboration with

diverse public and private actors.") available at http://www.fpri.org/research/thinktanks/mcgann.globalgotothinktanks.pdf

43. Christine MacDonald, Green, Inc.: An Environmental Insider Reveals How a Good Cause Has Gone Bad (2008).

44. Pew Center Global Climate Change, *In Brief: The Business Case for Climate Legislation* (June 2010 Update) available at http://www.pewclimate.org/docUploads/business-case-for-climate-legislation-06-2010.pdf

45. Kate Galbraith, *Companies Add Chief Sustainability Officers*, N.Y. Times (March 2, 2009) available at http://green.blogs.nytimes.com/2009/03/02/companies-add-chief-sustainability-officers/

46. Daniel C. Esty & Andrew S. Winston, Green to Gold: How Smart Companies Use Environmental Strategy to Innovate, Create Value and Build Competitive Advantage (2009).

47. Peter Senge, et al., The Necessary Revolution: How Individuals and Organizations are Working Together to Create a Sustainable World (2008).

48. *Carbon Principles* homepage at http://carbonprinciples.org/

49. *American Carbon Registry* homepage at http://www.americancarbon-registry.org/

50. *Carbon Disclosure Project* homepage at https://www.cdproject.net/en-US/Pages/HomePage.aspx

51. Securities and Exchange Commission, *Commission Guidance Regarding Disclosure Related to Climate Change* (February 2, 2010) available at http://www.sec.gov/rules/interp/2010/33-9106.pdf

52. REN21, Renewables Global Status Report: 2009 Update 8 (2009).

53. Clinton Global Initiative homepage at http://www.clintonglobalinitiative.org/

54. *Pegasus Sustainable Century* homepage at http://www.pcalp.com/news/news-detail.php?item=68

55. Cleantech Group LLP, *Investment Reports* available at http://cleantech.com/research/investmentreports.cfm

56. Navigant Consulting, *Jobs Impact of National Renewable Electricity Standard* (February 2, 2010) available at http://www.americanbiogascouncil.org/wp-content/uploads/2010/04/RESAllianceNavigantJobsStudy1.pdf

57. McKinsey & Company, *The Case for Investing in Energy Productivity* (February 2008) available at http://www.mckinsey.com/mgi/publications/Investing_Energy_Productivity/

58. *Id*. at 8.

59. *Id*. at 22–31. *See also* McKinsey & Company, *Fueling Sustainable Development: The Energy Productivity Solution* (October 2008) available at http://www.mckinsey.com/mgi/publications/fueling_sustainable_development.asp

60. McKinsey & Company, *Averting the Next Energy Crisis: The Demand Challenge* (March 2009) available at http://www.mckinsey.com/mgi/reports/pdfs/next_energy_crisis/MGI_next_energy_crisis_full_report.pdf

61. Pew Center Global Climate Change, *In Brief: Clean Energy Markets: Jobs and Opportunities* (April 2010 Update) available at http://www.pewclimate.org/docUploads/Clean_Energy_Update__Final.pdf

62. DB Climate Change Advisors, *Investing in Climate Change 2010: A Strategic Asset Allocation Perspective* (January 2010) available at http://www.banking-on-green.com/docs/InvestingInClimateChange2010.pdf

63. Paul Sonne, *Lockerbie Release Flawed*, WALL ST. J. A1 (August 6, 2010); Robin Pagnamenta, *Al-Megrahi's Release "Would Free BP" to Join the Rush for Libya's Oil*, THE SUNDAY TIMES (August 15, 2009) available at http://www.timeson-line.co.uk/tol/news/uk/article6797118.ece

64. TOM BOWER, OIL: MONEY. POLITICS, AND POWER IN THE 21ST CENTURY xiii (2009).

65. CNN, *Massive Ice Island Breaks Off Greenland* (August 9, 2010) available at http://www.cnn.com/2010/WORLD/americas/08/07/greenland.ice.island/index.html

66. Max H. Bazerman, *Barriers to Acting in Time On Energy and Strategies for Overcoming Them* in KELLY SIMS GALLAGHER (ED.), ACTING IN TIME ON ENERGY POLICY 162, 176–77 (2009).

67. GIDDENS; BLUEPRINT.

68. *See* Jim Rossi, *The Political Economy of Energy and Its Implications for Climate Change Legislation*, 84 TULANE L. REV. 379 (2009).

Chapter 9

1. ANTHONY GIDDENS, THE POLITICS OF CLIMATE CHANGE 4 (2009).

2. GEOFF NUNBERG, THE YEARS OF TALKING DANGEROUSLY (2009).

3. News Corporation, *The Language of a Clean Energy Economy* (2910) available at http://www.hks.harvard.edu/hepg/Papers/2010/Luntzpreso.pdf

4. Editorial, *Try Something Hard: Governing*, N.Y. TIMES, Opinion Section 7 (November 14, 2010).

5. John M. Broder & Marjorie Connelly, *Even on Gulf Coast, Energy and Economy Surpass Spill Concerns, Poll Finds*, N.Y. TIMES A16 (June 22, 1020).

6. Yale Project on Climate Change & George Mason University Center for Climate Change Communication, *Climate Change in the American Mind: Americans' Climate Change Beliefs, Attitudes, Policy Preferences, and Actions* (March 2009) available at http://www.climatechangecommunication.org/images/files/Climate_Change_in_the_American_Mind.pdf

7. Yale Project on Climate Change & George Mason University Center for Climate Change Communication, *Climate Change in the American Mind: Americans' Global Warming Attitudes and Beliefs in June 2010* (June 2010) available at http://www.climatechangecommunication.org/images/files/ClimateBeliefsJune2010(1).pdf

8. Jon A. Krosnick, *The Climate Majority*, N.Y. TIMES A21 (June 9, 2010).

9. Yale Project on Climate Change & George Mason University Center for Climate Change Communication, *Climate Change in the American Mind: Public Support for Climate & Energy Policies June 2010* (June 2010) available at http://www.climatechangecommunication.org/images/files/PolicySupportJune2010.pdf

10. TED NORDHAUS & MICHAEL SHELLENBERGER, BREAK THROUGH: FROM THE DEATH OF ENVIRONMENTALISM TO THE POLITICS OF POSSIBILITY 5–8 (2007).

11. Pew Global Climate Change, *Workshop Summary Report: Assessing the Benefits of Avoided Climate Change: Cost-Benefit Analysis and Beyond* (March 16–17, 2009) available at http://www.pewclimate.org/docUploads/benefits-workshop-summary.pdf

12. Max Bazerman, *Barriers to Acting in Time On Energy and Strategies for Overcoming Them*, in KELLY SIMS GALLAGHER (ED.), ACTING IN TIME ON ENERGY POLICY 174–76 (2009).

13. Pew Center Global Climate Change, *In Brief: Update on the 10–50 Solution: Progress Toward a Low-Carbon Future* (January 2010) available at http://www.pewclimate.org/docUploads/10-50-brief-update.pdf.; *see also* National Commission on Energy Policy, *Clean Energy Technology Pathways: An Assessment of the Critical Barriers to Achieving a Low-Carbon Energy Future* (March 2010) available at http://www.bipartisanpolicy.org/sites/default/files/58280_BPC_.pdf

14. *See* GIDDENS at 4 (2009); ORR at 6.

15. ORR at xiv–xv.

16. *See* RICHARD A. POSNER, A FAILURE OF CAPITALISM: THE CRISIS OF '08 AND THE DESCENT INTO DEPRESSION (2009); THE CRISIS OF CAPITALIST DEMOCRACY (2010); JOSEPH E. STIGLITZ, FREEFALL: AMERICA, FREE MARKETS, AND THE SINKING OF THE WORLD ECONOMY (2010); DAVID WESSEL, IN FED WE TRUST; BEN BERNANKE'S WAR ON THE GREAT PANIC (2009); JOHN CASSIDY, HOW MARKETS FAIL: THE LOGIC OF ECONOMIC CALAMITIES (2009); JUSTIN FOX, THE MYTH OF THE RATIONAL MARKET: A HISTORY OF RISK, REWARD, AND DELUSION ON WALL STREET (2009).

17. ROBERT B. REICH, AFTERSHOCK: THE NEXT ECONOMY AND AMERICA'S FUTURE (2010).

18. PAUL KRUGMAN, THE CONSCIENCE OF A LIBERAL ch. 1 (2007); *see also supra* note 16.

19. *See, e.g.*, NORDHAUS & SHELLENBERGER at ch. 9.

20. Pew Center Global Climate Change, *In Brief: Clean Energy Markets: Jobs and Opportunities* (February 2010) available at http://www.pewclimate.org/docUploads/Clean_Energy_Update_Final.pdf

21. Bazerman at 169–73.

22. ORR at 37–40.

23. GIDDENS at 91–96.

24. GIDDENS at 8–9 and 68–72.

Index